49.50/67C

Neutrino Physics
and Astrophysics

ETTORE MAJORANA INTERNATIONAL SCIENCE SERIES
Series Editor:
Antonino Zichichi
European Physical Society
Geneva, Switzerland

(PHYSICAL SCIENCES)

Recent volumes in the series:

Volume 5 **PROBING HADRONS WITH LEPTONS**
Edited by Giuliano Preparata and Jean-Jacques Aubert

Volume 6 **ENERGY FOR THE YEAR 2000**
Edited by Richard Wilson

Volume 7 **UNIFICATION OF THE FUNDAMENTAL PARTICLE INTERACTIONS**
Edited by Sergio Ferrara, John Ellis, and Peter van Nieuwenhuizen

Volume 8 **CURRENT ISSUES IN QUANTUM LOGIC**
Edited by Enrico G. Beltrametti and Bes C. van Fraassen

Volume 9 **ENERGY DEMAND AND EFFICIENT USE**
Edited by Fernando Amman and Richard Wilson

Volume 10 **INTERACTING BOSE — FERMI SYSTEMS IN NUCLEI**
Edited by F. Iachello

Volume 11 **THE SOLUTION OF THE INVERSE PROBLEM IN GEOPHYSICAL INTERPRETATION**
Edited by R. Cassinis

Volume 12 **NEUTRINO PHYSICS AND ASTROPHYSICS**
Edited by Ettore Fiorini

Neutrino Physics
and Astrophysics

Edited by
Ettore Fiorini
University of Milan
Milan, Italy

Plenum Press · New York and London

Library of Congress Cataloging in Publication Data

International Conference on Neutrino Physics and Astrophysics (10th: 1980: Erice, Italy) Neutrino physics and astrophysics.

(Ettore Majorana international science series. Physical sciences; v. 12)
"Proceedings of Neutrino '80, an International Conference on Neutrino Physics and Astrophysics, held June 23—28, 1980, in Erice, Sicily."
Bibliography: p.
Includes index.
1. Neutrino—Congresses. 2. Astrophysics—Congresses. 3. Neutrino astrophysics—Congresses. I. Fiorini, Ettore. II. Title. III. Title: Neutrino '80. IV. Series.
QC793.5.N42I57   1980   539.7'215   81-11999
ISBN 0-306-40746-9   AACR2

Proceedings of Neutrino '80, an International Conference on
Neutrino Physics and Astrophysics, held June 23-28, 1980,
in Erice, Sicily

© 1982 Plenum Press, New York
A Division of Plenum Publishing Corporation
233 Spring Street, New York, N.Y. 10013

All rights reserved

No part of this book may be reproduced, stored in a retrieval system, or transmitted, in any form or by any means, electronic, mechanical, photocopying, microfilming, recording, or otherwise, without written permission from the publisher

Printed in the United States of America

PREFACE

Neutrino '80 held at the Ettore Majorana Center for Scientific Culture in Erice, was the tenth of a series of International Conferences on Neutrino Physics and Astrophysics. It also marked the fiftieth anniversary of the first mention, by Wolfgang Pauli, of a neutral particle emitted in beta decay. The conference occurred at a very propitious time in neutrino physics: the possibility of a non-zero neutrino mass and of neutrino oscillations has obvious implications of great importance in neutrino astrophysics and cosmology, as well as in the grand unified theories.

In order to encourage contacts and discussions among the various experts in different branches of neutrino physics and astrophysics, the conference was based only on plenary sessions, and mainly on review talks. Short communications were accepted only if they bore new and unexpected results which could not be covered in the appropriate review. I would like to thank the participants for their understanding of this often unpopular rule.

I take this opportunity to express my gratitude to the members of the International Advisory Committee, to George Marx, Secretary of the on-going International Neutrino Committee, to the rapporteurs and session chairmen. Thanks are especially due to Antonino Zichichi, Director of the Ettore Majorana Centre for Scientific Culture, for the warm and generous hospitality extended to us, and to Alberto Gabriele and Pinola Savalli for their untiring efforts to make our stay in Erice as enjoyable as fruitful.

I want to thank in addition Enrico Bellotti, Scientific Secretary to the conference and Luciana Brogiato for the excellent job done in the organization of the conference and in the preparation of the proceedings.

Ettore Fiorini

NEUTRINO '80

International Conference on Neutrino Physics and Astrophysics held in Erice (Italy), June 23 through 28, 1980.

International Advisory Committee:

C. BALTAY	Irvington - U.S.A.	
S. BLUDMAN	Philadelphia - U.S.A.	
N. CABIBBO	Roma - Italy	
H. FAISSNER	Aachen - Germany	
E. FIORINI	Milano - Italy (Conference Chairman)	
C. JARLSKOG	Bergen - Norway	
G. MARX	Budapest - Hungary	
D. PERKINS	Oxford - United Kingdom	
H. PIETSCHMANN	Wien - Austria	
B. PONTECORVO	Moscow - U.S.S.R.	
L. RADICATI	Pisa - Italy	
F. REINES	Irvine - U.S.A.	
A. ROUSSET	Paris - France	
A. SALAM	Trieste - Italy	
J. STEINBERG	Geneva - Switzerland	
V. TELEGDI	Zürich - Switzerland	
G.T. ZATZEPIN	Moscow - U.S.S.R.	
A. ZICHICHI	Geneva - Switzerland	
E. BELLOTTI	Milano - Italy (Scientific Secretary)	

Sponsored by:
European Physical Society, Italian Physical Society, Istituto Nazionale di Fisica Nucleare, Università di Milano.

CONTENTS

50 YEARS OF NEUTRINO PHYSICS

The Early Days of Neutrino Physics 1
   R. Peierls

Fifty Years of Neutrino Physics: Early
      Experiments 11
   F. Reines

Fifty Years of Neutrino Physics: A Few Episodes 29
   B. Pontecorvo

NEUTRINO ASTROPHYSICS

Low-Energy Neutrinos in Astrophysics 61
   W. Hampel

WEAK CHARGED CURRENTS

New Particles 81
   P. Musset

Weak Charged Currents: Hadron Final States 107
   B. Saitta

Charged Current Inclusive Interactions 143
   F. Eisele

WEAK NEUTRAL CURRENTS

Weak Neutral Current-Semileptonic Interactions 165
   B. Borgia

Review on Purely Leptonic Interactions of Weak
      Neutral Currents 191
  L.W. Mo

LOW AND INTERMEDIATE ENERGY INTERACTIONS

Parity Violation in Nuclei 219
  R.G.H. Robertson

Low Energy Neutrino Interactions 241
  H.W. Sobel, F. Reines, and E. Pasierb

CONSERVATION LAWS

Lepton Conservation 259
  F. Boehm

Grand Unification of Quarks and Leptons from
      One of Preons 275
  J.C. Pati

Baryon Conservation (Experiments) 301
  M. Goldhaber

Comments on "The Electron Neutrino Mass from
      ^{3}H-^{3}He β-Decay $^{(+)}$
      14 eV $\leqslant m_{\bar{\nu}_e} \leqslant$ 46 eV (99% C.L.)" 317
  K.E. Bergkvist

WEAK INTERACTIONS IN ELECTRON AND HADRON EXPERIMENTS

Electro-Weak Physics in e^+e^- Interactions 321
  M. Chen

Beam-Dump Experiments 341
  F. Dydak

PRESENT AND FUTURE

Future Accelerators 361
  K. Winter

CONTENTS

New Neutrino Detection Technology: Application
    of Massive Water Detectors to
    Accelerator Neutrino Physics   381
  L. Sulak

Concluding Remarks   399
  D.H. Perkins

Index   417

THE EARLY DAYS OF NEUTRINO PHYSICS

Rudolf Peierls

Nuclear Physics Laboratory
Keble Road
Oxford, England

This is a birthday party, to celebrate the neutrino reaching what is, from my frame of reference, early middle age. Our celebrations are a little premature because, as the Chairman mentioned, the real birthday is only in December, but this is an accuracy which in this area of physics is surely acceptable. I have been asked to say something about the early history of the neutrino, but you have to accept that I am not a trained historian, and what I have to say will not be a scholarly analysis. My qualification for addressing you is rather that I am one of the relics of old times, someone who was around at the time of the birth of the neutrino, a time when many of you were not born yet. So I shall talk largely about things which I remember; but I cannot rely on memory entirely. You all know how it goes with one's memory, and whenever one looks back to the original papers, things are a little different from what one thought. Sometimes reading your own writings or letters of 50 years ago can give unexpected surprises. So my remarks will be about what I remember, but with occasional help or correction from the literature.

We should start by recalling what the situation in physics was like when Pauli made his famous suggestion. At that time people were convinced that nuclei consisted of protons and electrons. Why were they so sure? First of all, these were the only elementary particles known, and one was, rightly, reluctant to postulate new particles without strong evidence. But another reason was that in β-decay, electrons come out of the nucleus, and one naturally tended to assume that they must have been in there before.

As a result, one was faced with enormous difficulties. Not only was there no force known which might be strong enough to confine an electron to a region as small as the nucleus, but, from what was known of relativistic quantum theory of the electron, such as the Dirac equation, it followed that no force, however strong, could confine an electron that closely.

This situation gave rise to the idea, of which Niels Bohr was the chief exponent, that there must be a breakdown of the laws of quantum mechanics, as we knew them, for phenomena taking place in a region as small as the nucleus. Bohr pointed out in several papers the reasons why one had to suspect such a breakdown. For one thing, the radius of the electron was smaller than its Compton wavelength, a state of affairs for which we had no precedent.

There there were also the more specific difficulties, that the spin and statistics of nuclei did not agree with the hypothesis that they were made of protons and electrons. Indeed they agreed with the spin and statistics to be expected from the protons alone (half-integral spin and Fermi statistics for nuclei of odd mass number) and this seemed to give support to the idea that electrons in the nucleus had lost their individuality. But the most serious difficulty was the continuous β-ray spectrum. Experimentalists were not too worried about the problems of spin and statistics which seemed rather abstract, but the continuous β spectrum was concrete enough. Bohr felt it was not surprising that things were rather strange, because the physics of an electron inside the nucleus was such an unprecedented situation, and the fact that even the conservation of energy seemed to go by the board, fitted in with that picture.

Bohr drafted a paper along these lines and sent a copy to Pauli for his opinion[1,2]. This was in July 1929, more than a year before Pauli made his famous suggestion. In the same letter he sent another note which related to the difficulty of measuring the magnetic moment of an electron in a Stern-Gerlach type of experiment. In his reply,[3] Pauli said he liked the paper about the electron spin very much and hoped Bohr would publish this as soon as possible. (I think it was never published.) But then followed a devastating criticism of the note on energy non-conservation, urging Bohr not to publish this on any account, but to think again. He did not agree with the idea, but it also was not clearly stated.

Bohr indeed did not publish this note, but he stuck to the idea that energy conservation was violated in β decay, and mentioned it on a number of occasions, for example in his Faraday Lecture[4], given in 1930 and published in 1932.

One may conjecture that this correspondence with Bohr intensified Pauli's thinking about the β-ray problem. He was in any case concerned about this problem, like almost everybody else. So he wrote his famour letter of December 1930, in which he pointed out that the contradictions over spin and statistics, as well as over the continuous β spectrum could be resolved by the "desperate remedy" of postulating a new particle. This was a private letter, addressed to Hans Geiger and Lise Meitner, who were attending a conference in Tübingen. Pauli could not come to this conference, as he said, because of a ball in Zurich at which he was indispensable.

He asked for people's comments on the suggestion that a second particle was emitted together with the electron. He knew this particle had to be neutral, because all charges were accounted for, and because it would have been detected if it carried a charge. He realised it had to be light, and I believe he implied that the particle must exist inside the nucleus before its emission.

Pauli still did not dare publish this suggestion, but he repeated it at an American Physical Society meeting in Pasadena the following year, where he went into some more detail. I do not propose to go into all the variations undergone by Pauli's viewpoint on these various occasions, because it would take too long and also because these are very well covered in the excellent review by Laurie Brown in Physics Today[5], in which Pauli's original letter and much other information is reproduced.

At the Pasadena metting Pauli first suggested that, while the new particle had no electric charge - it was still called the "neutron" - it might have a magnetic moment. It had, of course to have spin $\frac{1}{2}$, or at least half-integral spin, to resolve the difficulty about conservation of angular momentum in β decay, and then it seemed natural that it might have a magnetic moment. Pauli had previously shown that the Dirac equation allowed one to include a term representing a magnetic moment, independently of the charge (now usually called the "Pauli moment term") and no doubt this thought was prompted by the fact that, if the new particle was bound inside the nucleus, there had to be some force to hold it there. In fact, for some time the particle was referred to as the magnetic neutron.

The new particle had to be somewhat elusive, because it had never shown up in cloud chamber photographs of β decay. Moreover, there was the experiment of Ellis and Wooster[6], who inserted a β ray source in a block of metal, which would absorb the electrons and any γ rays, and measured the temperature rise of the metal. This was found to agree with the total electron energy, i.e. with the mean energy of the spectrum for each decaying nucleus. This 1927 experiment was repeated with greater accuracy by Meitner and Orthmann in 1930[7]. As the energy carried away by the new particle did not show up in this experiment, it had to have greater penetrating power than the γ rays, which, as we now know, is a rather modest estimate.

Meanwhile the study of β decay continued. In 1933, Ellis and Mott published an important paper suggesting that the maximum energy of the β spectrum represented the energy difference between the initial and the final nucleus. They deduced this from a branching reaction, in which a β decay either precedes or follows α emission. The energy of the α particle plus the maximum electron energy adds up to exactly the same total for the two branches. They conclude that some energy is being lost in the process, and remark that this is consistent with Pauli's idea of a second particle being emitted.

Incidentally, Ellis and Mott point out, in the same paper, that the different shapes of the β spectra could be due to their being composite. If there is an appreciable probability of the final nucleus remaining in an excited state after β decay, the electrons emitted in that event would belong to a spectrum with a lower end point. The observed spectrum would then be a superposition of a number of similar spectra with different maximum energies. They obtained some qualitative agreement, but the data available at the time were not accurate or extensive enough to carry this analysis any further.

Meanwhile the neutron had been discovered - that is the particle we call neutron today - in 1932. We tend to think now that, once the existence of the neutron was known, it was immediately evident that nuclei consist of neutrons and protons only, but if one looks back at what was being said and thought at the time, it was far from obvious. People had the greatest difficulty in seeing the simplicity of this new point of view. For example, in his first paper on the subject, Heisenberg[9] still takes it for granted that the neutron consists of a proton and an electron, only they are so tightly bound together that in discussing nuclear dynamics we may in practice treat the neutron as an elementary particle. He drew many conclusions from the fact that the neutron had such a structure. They turned out to be wrong conclusions.

One finds in the literature around 1932 remarks to the effect that it was possible to assume the neutron was an elementary particle, but then one had to assume that the heavier nuclei contained electrons in addition to the neutrons and protons, because electrons were seen to come out. The earliest mention of the possibility that there were no electrons in the nucleus which I could find was in a paper by Iwanenko[10] in 1932. He said perhaps the electron was created in the emission process just as a photon is created when an atom makes a radiative transition. This is of course the way we look at it today.

With all this it was a long time before Pauli's idea was taken seriously to the extent of drawing quantitative conclusions from it. The first to attempt this was probably Francis Perrin. He published a paper[11] based on Pauli's hypothesis in which he argued that there should be an approximate symmetry between the electron and the new particle, which he already calls the neutrino, and that they should tend to have equal and opposite momentum. He concludes that if the neutrino is lighter than the electron, which he regards as plausible because of the absence of electromagnetic self-energy, the mean kinetic energy of the electron will be less than half the maximum. The rather crude model gives approximate agreement for Ra E for zero neutrino mass.

Meanwhile there were still many attempts to elaborate a theory based on energy non-conservation. Perhaps the most elaborate, and the last, attempt of this kind was by Guido Beck and Kurt Sitte[12]. In their model β decay involves the creation of an electron-positron pair, of which the positron then disappears without trace, except that its charge returns to the nucleus, but its energy disappears. This model gives a prediction for the shape of the spectrum, given by the electron-positron phase space, which is qualitatively like the electron-neutrino phase space factor occurring in Fermi's theory for allowed transitions. It is characteristic of the situation at that time that Beck, usually a critical physicist of sound judgment, got carried away into defending his theory against Fermi's, claiming that it not only agreed better with the experiments, but was also inherently a more satisfactory theory[13,14].

But in fact Fermi's paper transformed the situation. Fermi had heard Pauli's suggestion at the Rome Conference in 1931 and liked it. At the same conference, Bohr was strongly opposed to the idea, and preferred to believe in energy non-conservation. The matter was again discussed at the Solvay Congress in 1933, by which time the neutron had been discovered, and Fermi's proposal to name Pauli's particle the neutrino had found general acceptance. In his remarks to that Congress[15], Pauli withdrew the suggestion that the neutrino should have a magnetic moment. He also pointed out that the sharp end point of the spectrum, which by this time had been established, favoured his view.

Fermi's belief in the neutrino hypothesis led him to formulate a quantitative theory, which he published briefly in Ricerca Scientifica late in 1933, and in Zeits.f.Physik early in 1934[16]. The contents of his paper are well known and I need not summarise it here. It did not take long before this approach was taken very seriously. More careful experiments were done trying to detect neutrinos, e.g. that of Chadwick and Lee[17], who concluded that the neutrino must have a small mass and no magnetic moment.

On the theoretical side, many people were getting into the act, and they included Hans Bethe and myself, who wrote two letters to Nature. One of these mentioned that according to the Fermi theory there should also be K capture, and that this very much restricted the possible energy difference between adjacent stable isobars, so that one could understand the very rare occurrence of pairs of stable isobars with a difference of one in their atomic numbers. We also considered what would be the capture cross section for neutrinos by inverse β decay, which would be its only interaction with matter if it had no magnetic moment, and there were no other unknown forces acting on it. If a neturino can be emitted in β decay the inverse process must evidently also be possible. In practice it is more likely to be the inverse of K capture. The cross section can easily be estimated from detailed balancing. We found what seemed to use the fantastically small value of $10^{-44} cm^2$, and we concluded that this meant one obviously would never be able to see a neutrino. We did not of course allow for the existence of nuclear reactors producing neutrinos in vast quantities, or for the ingenuity of experimentalists. It seemed to us therefore that the most direct proof of the existence of the neutrino might have to be the observation of the recoil in β decay. If the momentum balance showed that the missing momentum correlated with the missing energy like the energy and momentum of a particle, then this would be strong circumstantial evidence for the neutrino.

This thought evidently had also occurred to Leipunski, and he set up a recoil experiment.[18] He did not succeed in reaching an accuracy which would have been a real test, though the results were consistent with the neutrino hypothesis. More conclusive experiments were done by Crane and Halpern in 1938[19].

One development of some interest was that, as soon as the neutrino hypothesis was taken seriously, people tried to give it a wider role in physics, and to link it with other phenomena. For example, many people assumed that virtual β decay must be responsible for the nuclear force. Looking back today it seems a rather misguided idea, because of the weakness of the Fermi coupling constant, which can hardly be expected to account for the strong nuclear forces. Indeed the forces calculated from such a model were extremely weak at any reasonable distance, but since they were very singular at the origin one could always get as strong an interaction as one liked; but this was not a very appealing picture.

Another way of making theoretical use of the neutrino was the so-called neutrino theory of light. Assuming the neutrino mass exactly zero, this theory presented the photon as consisting of two neutrinos. They could not very well be bound together, because if two particles of mass zero are coupled to make up an object of zero mass, their binding energy is also zero, and the system cannot have any stability. Instead, the theory envisaged the neutrinos as free, but emitted simultaneously. However, if the resultant mass is to be zero, the two neutrinos have to be emitted with mathematical accuracy in the same direction. By the uncertainty principle this cannot be achieved without a source of infinite extent. This caused no trouble in the many papers on this theory, since they all considered for simplicity only problems in one dimension, where the collinearity is assured. It may be instructive to remember such attempts, although looking back after all these years some of them look somewhat ridiculous. One may wonder whether some of today's attempts to get the whole of physics out of the pieces we already know about, may not in fifty years' time look equally ridiculous.

Apart from these speculations, much work was going on to look at individual β spectra in the light of Fermi's theory. Not all the data seemed to fit, and this led Konopinski and Uhlenbeck[20] to modify the theory by replacing the neutrino wave function by its derivative. This amounts to an extra factor of neutrino momentum in the matrix element, and therefore shifts the distribution towards higher neutrino, i.e. lower electron energies. Since most observed spectra contained too many slow electrons by comparison with Fermi's prediction, this modification improved the agreement. One minor reason why this change was popular was that it helped a little with the β-ray theory of the nuclear forces, which I have mentioned. Since the virtual states contributing to that force can contain quite high energies, and therefore also high neutrino momenta, the extra factor helped (but not nearly enough).

It took a long time to realise that this modification was not really necessary, and that the original Fermi interaction was adequate. The point is that many observed spectra are really composite, i.e. they include transitions which leave the nucleus in an excited state, when there is less energy available for the actual β decay. This was pointed out in a letter by Bethe, Fred Hoyle and myself[21]. (This letter does not represent joint work, as we were on opposite sides of the Atlantic, but discussion between Hoyle and myself led to the same conclusions that Bethe had reached, and we pooled our ideas.) The point had already been made in the paper by Ellis and Mott[8] quoted earlier, but it evidently was forgotten again.

Another refinement of the theory was the remark by Gamow and Teller[22] that, while the Fermi interaction implied that the spin of electron and neutrino coupled to zero, there were alternative forms of the interaction in which the spins were parallel, and these would give different selection rules. As information about the spins of nuclear states accumulated, it was found that there were allowed transitions which obeyed the Fermi selection rule as well as some which required the Gamow-Teller rule, and that both types of transition occurred with about the same strength. This showed that in the interaction both Fermi and Gamow-Teller terms had to be present in comparable strength, and, as we know today, this is indeed the case for the "V-A" coupling.

It was evident from Fermi's theory that the shape of the upper end of the spectrum was sensitive to the value of the neutrino mass, which was conjectured, but not proved, to be zero. As the precision with which the end of the spectrum could be observed became greater, the upper bound for the neutrino mass decreased, but the question of the mass is still topical today, and it will be discussed at this Conference.

One paper of the late 1930s deserves special mention: Tomonaga and Tamaki[23] drew attention to the extremely rapid rise with energy of the cross section for inverse β decay. While in the region of a few MeV the cross section is extremely small, as has been mentioned, they point out that neutrinos of 10^{12}eV (1 TeV) should be able to produce showers. So in a sense this 1937 paper foreshadowed the present role of high-energy neutrinos as practical projectiles.

About that time, Neddermeyer and Anderson[24] had shown that the difficulties in interpreting penetrating tracks in the cosmic radiation indicated the existence of a new particle, which we now call the muon, though it was initially believed to be the particle predicted by Yukawa, the pion, or π meson. For a few years a variety of names were in use for this particle, but finally the name meson prevailed.

Quite soon after its discovery it was suspected of being unstable. This was in part due to prejudice, because Yukawa had suggested that his particle was also an intermediary in β decay. But there was also accumulating evidence that the attenuation of the penetrating component of cosmic rays, which consisted of these particles, depended not only on the amount of matter traversed, but also on the distance. The arguments and the evidence were reviewed by Blackett[25] who already quotes a figure of 2.10^{-6} sec for the lifetime. Further work, for example that of Cocconi[26], gave clear evidence for the decay and more accurate estimates of the half-life. Rossi[27] showed evidence of electrons resulting from the decay.

This was the state of knowledge reached in 1940, when the War interrupted or at least curtailed, work in this field. I shall not review the post-war developments in equal detail, because they are much better known, and indeed many members of this audience were by that time not only alive, but in touch with physics.

It will be recalled that the "meson" did not behave as the particle postulated by Yukawa should do. It did not seem to interact strongly with nuclei. These impressions were confirmed by Conversi, Pancini and Piccioni[28] in a brilliant experiment. Amongst the papers prompted by this result, I might mention that of Pontecorvo[29], who suggests that the cosmic-ray meson might have spin $\frac{1}{2}$ and might decay into an electron and a photon, or an electron and two neutrinos. You will recall that Marshak and Bethe[30] formulated the "two-meson hypothesis", by which a strongly-interacting particle was produced by the primary cosmic rays, and then decayed into the longer-lived but weakly interacting particle which was observed. As is well known this conjecture was proved to be true immediately by the work of Lattes, Powell and Occhialini[31], who discovered the two-stage decay of the particles they labelled π and μ, names we still use today.

We recall how soon the nature and decay modes of these particles were identified, finding more employment for neutrinos. Since the muon decay was into three particles, the spectrum of the resulting electrons was continuous and therefore an interesting object of study. Michel[32] showed that this distribution depends on a single parameter. The similarity of the muon decay with ordinary β decay invited comparison, and the similarity of the coupling constants fitting these two processes led to the postulate of universality.

We remember the other major steps in the growth of the neutrino to maturity: the detection of neutrino-induced reactions, of which we shall hear more in the next talk; the discovery of parity violation, first suspected in the K decay not involving neutrinos, and leading to our picture of a neutrino with completely (or approximately) fixed helicity; the use of accelerator-made neutrinos as projectiles, leading to the differentiation between the muon and electron neutrinos.

I should not carry my enumeration of major steps beyond this point, otherwise I would risk becoming involved with current neutrino physics, which every member of this Conference knows better than I do.

REFERENCES

1. Niels Bohr, letter to Pauli, 1 July 1929, with draft about β-decay, Niels Bohr Archive, Copenhagen.
2. Niels Bohr, note Beta Ray Spectra and Energy Conservation, Niels Bohr Archive.
3. W. Pauli, letter to Niels Bohr, 17 July 1929, Niels Bohr Archive.
4. Niels Bohr, Faraday Lecture, J. Chem. Soc. (1932): 349.
5. Laurie Brown, Physics Today, Sept. 1978: p.23.
6. C.D. Ellis and W.A. Wooster, Proc. Roy. Soc. A117: 109 (1927).
7. Lise Meitner and W. Orthmann, Z. Physik 60: 143 (1930).
8. C.D. Ellis and N.F. Mott, Proc. Roy. Soc. A147: 502 (1934).
9. W. Heisenberg, Z. Physik, 77: 1, (1932).
10. D. Iwanenko, Comptes Rendus Acad. Sci. Paris 195: 439 (1932).
11. F. Perrin, ibid. 197: 1625 (1930).
12. G. Beck and K. Sitte, Z. Physik 86: 105 (1933).
13. G. Beck and K. Sitte, ibid. 89: 259 (1934).
14. E. Fermi, ibid. 89: 522 (1934).
15. W. Pauli, Rapports du VII Conseil Solvay, Gauthier-Villars, Paris (1934), p.324 (Translation in ref. 5).
16. E. Fermi, Ricerca Scientiifica, 2, No. 2 (1933); Z.Physik 88: 161 (1934).

17. J. Chadwick and D.E. Lea, Proc. Camb. Phil. Soc. 30: 59 (1934).
18. A.I. Leipunski, ibid. 32: 301 (1936).
19. H.R. Crane and J. Halpern, Phys. Rev. 53: 789; 56: 239 (1938).
20. E.J. Konopinski and G.E. Uhlenbeck, ibid. 60: 308 (1941).
21. H. Bethe and R. Peierls, Nature, 133: 532; 689 (1934).
22. G. Gamow and E. Teller, Phys. Rev. 49: 895 (1936).
23. S. Tomonaga and H. Tamaki, Sci. Papers, Inst. Phys. Chem. Research, 733: 288, (1937).
24. S.H. Neddermeyer and C.D. Anderson, Phys. Rev. 51: 881 (1937).
25. P.M.S. Blackett, Nature 142: 692 (1938).
26. G. Cocconi, Ricerca Scientifica 11: 58 (1940).
27. B. Rossi, Phys. Rev. 57: 469 (1940).
28. M. Conversi, E. Pancini and O. Piccioni, ibid. 63: 232 (1945).
29. B. Pontecorvo, ibid. 72: 246 (1947).
30. R.E. Marshak and H.A. Bethe, ibid. 72: 506 (1947).
31. C.M.G. Lattes, G.P.S. Occhialini and C.F. Powell, Nature, 160: 453 (1947).
32. L. Michel, Proc. Phys. Soc. A63: 514 (1950).

50 YEARS OF NEUTRINO PHYSICS

EARLY EXPERIMENTS

Frederick Reines

Department of Physics
University of California
Irvine, California 92717

It is now 50 years since Pauli (1) formulated the neutrino hypothesis and 47 years since Fermi (2) cast it in its essentially modern form. In 1953, Clyde L. Cowan and I and our colleagues at Los Alamos made the first, tentative observation of the free neutrino at the fission reactor at Hanford, Washington, through the inverse beta process

$$\bar{\nu}_e + p \rightarrow n + e^+ \tag{1}$$

where $\bar{\nu}_e$, p, n, and e^+ are an electron antineutrino, proton, neutron, and positron, respectively. Our choice of this reaction was felicitous because of its simplicity, distinctive products, and the scintillation properties of some liquid hydrocarbons. Three years later, in 1956, we completed the job in a definitive way at the Savannah River Plant and experimental neutrino physics was launched (3).

In the summer of 1951 I decided that the detection of the elusive neutrino was a goal worth striving for. At the time, the neutrino hypothesis was already firmly fixed in the lexicon of physics. Physicists generally believed that the neutrino had been demonstrated indirectly and that, in fact, it was not directly observable. It was argued that the neutrino hypothesis explained the apparent lack of energy and momentum conservation in beta decay (for instance, the shapes of decay spectra and nuclear recoil in K-electron capture) and hence that the neutrino existed. In fact, had any measurements made on the beta decay process been found to be inconsistent with the neutrino hypothesis, then it could have been argued that the neutrino did not exist; the converse is untrue.

However attractive the neutrino was as an explanation for beta decay, the proof of its existence had to be derived from an observation at a location other than that at which the decay process occurred - the neutrino had to be observed in its free state to invert beta decay or otherwise interact with matter at a remote point.

In effect, observation of a free neutrino would provide incontrovertible proof of the validity of the energy and momentum conservation laws in nuclear beta decay. We tend to regard these principles as universally applicable, yet in the days when the neutrino hypothesis was put forward, agreement on this point was hardly universal. As we have heard, Niels Bohr (4) pointed out in 1930 that no evidence "either empirical or theoretical" existed that supported the conservation of energy in this case. He was, in fact, willing to entertain the possibility that energy conservation must be abandoned in the nuclear realm.

Fermi in his Nuclear Physics Course Notes (University of Chicago, 1950) stated succinctly: "Perhaps the most conclusive proof for the existence of the neutrino, and the most remote of attainment, would be to observe β decay with the recoil of nucleus and momentum of electron known so as to give the direction of the neutrino, and then on the path of the neutrino to detect almost simultaneously an inverse β reaction whose energy relations agree with the energy of the neutrino emitted in the first reaction."

The miniscule interactions cross-section ($\sim 10^{-44} cm^2$) dictates a very intense source in which it is not known how to perform the detailed measurements described. Accordingly it is fortunate that a logically satisfactory proof of the neutrino's existence requires less than the "complete" experiment of Fermi. As noted above, the essence of a "sufficient" proof is the observation of inverse beta decay at a location remote from the source of beta emitters.

We also recognized that the existence of an elastic scattering reaction $\bar{\nu}_e + e^- \to \bar{\nu}_e + e^-$ (5), not called for by the Pauli-Fermi theory nor disallowed either, could be checked by a differential shielding test, assuming that penetrating particles other than neutrinos are not emitted from the fission fragments.

SOURCE OF NEUTRINOS

A few rough estimates indicated to me that a nuclear explosion, which might yield a pulse of particles intense enough to override the background, was the most promising source of neutrinos and that a suitably shielded detector with a mass of about 1 ton might do the job. Actually all we could promise with good confidence at the outset was to detect a cross-section $\sim 10^{-39} cm^2$, i.e. four orders of magnitude greater than expected! In any event, the experiment would

EARLY EXPERIMENTS

be vastly more sensitive than any previously imagined. However, I had no idea how such an incredibly large detector could be made and thought it might be helpful to talk with Fermi, who was spending the summer at Los Alamos. As it turned out, although he agreed with the suggestion of an explosion, he also had no idea how to build the detector, and that almost ended the matter.

Some months later, while discussing with Cowan various problems on which it would be interesting to work, I mentioned my thoughts on the neutrino. He immediately felt that there must be a way to make such a detector. Our partnership began at that point, and our ideas flowed together in a mutually reinforcing manner that often made it difficult to decide who thought of what. I recall one instance that illustrates the depth of our collaboration.

We gave a talk to the Physics Division at Los Alamos in which we described our ideas for a large liquid scintillator that we had constructed (6) for use in the vicinity of a nuclear explosion. We mentioned the delayed coincidence between the positron and neutron pulses as a label for the reaction; it had not yet occurred to us that the label could be used to reduce the background. J.M.B. Kellogg asked whether it might not be possible to use a fission reactor instead of a bomb. We argued that it would not be - and besides, Fermi and Bethe had agreed with us that a bomb was the most promising source. That night I telephoned Cowan and we told each other how the delayed coincidence could be used to reduce the background, which would make the reactor an attractive source. We immediately altered our plans and the next morning met with Kellogg to cancel our bomb preparations and arrange to develop and build a detector suitable for the Hanford reactor. We learned later that others attending the talk had considered Kellogg's question and concluded that the bomb was better suited than the reactor. You can well imagine how embarrassing it would have been had the roles been reversed.

A letter to Fermi telling him of our reactor proposal (7) elicited the response shown in Fig. 1.

THE HANFORD EXPERIMENT

Viewed from the perspective of today's computer-controlled kiloton detectors, sodium iodide crystal palaces, giant accelerators and 50-man groups, our efforts to detect the neutrino appear quite modest. In the early 1950's, however, our work was thought to be large-scale. The idea of using 90 photomultiplier tubes and detectors large enough to enclose a human was considered to be most unusual. We faced a host of unanswered questions. Was the scintillator sufficiently transparent to transmit its light for the necessary few meters? How reflective was the paint? How could one

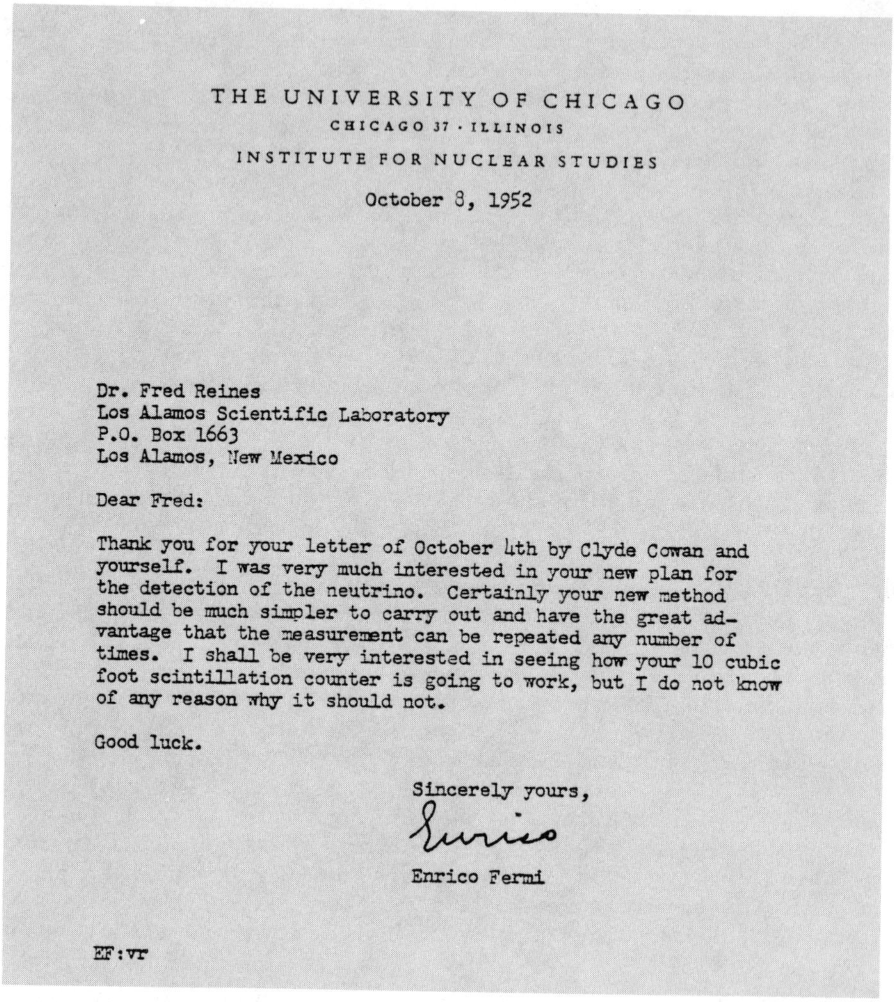

> THE UNIVERSITY OF CHICAGO
> CHICAGO 37 · ILLINOIS
> INSTITUTE FOR NUCLEAR STUDIES
>
> October 8, 1952
>
> Dr. Fred Reines
> Los Alamos Scientific Laboratory
> P.O. Box 1663
> Los Alamos, New Mexico
>
> Dear Fred:
>
> Thank you for your letter of October 4th by Clyde Cowan and yourself. I was very much interested in your new plan for the detection of the neutrino. Certainly your new method should be much simpler to carry out and have the great advantage that the measurement can be repeated any number of times. I shall be very interested in seeing how your 10 cubic foot scintillation counter is going to work, but I do not know of any reason why it should not.
>
> Good luck.
>
> Sincerely yours,
>
> Enrico Fermi
>
> EF:vr

Fig. 1. Letter from Fermi on hearing about our plan to use the Hanford reactor to attempt to observe the neutrino.

add a neutron capturer without poisoning the scintillator? Would the tube noise and afterpulses from such a vast number of photomultiplier tubes mask the signal? And besides, were we not monopolizing the market on photomultiplier tubes? (As it turned out we were not; headlight dimmers on Cadillacs consumed far more of them than we did.) In the search for answers to these questions we received strong support from various scientists at Los Alamos, and Cowan made good use of his undergraduate training as a chemist and his considerable abilities with electronics.

EARLY EXPERIMENTS

It soon became clear that this new detector designed solely for neutrinos had unusual properties with regard to other particles as well - for instance, neutron and gamma-ray detection efficiencies near 100 percent. We recognized that detectors of this type could be used to study such diverse quantities as neutron multiplicities in fission, muon capture, muon decay lifetimes, and the natural radioactivity of humans. We measured radioactivities of some humans, pointed out other uses, and continued with our neutrino search. (These applications have since been made by other workers.)

Our entourage arrived at Hanford in the spring of 1953 (Figs. 2 and 3).

After a few months of operation, during which we made several restackings of hundreds of tons of lead bricks and specially fabricated boron-paraffin boxes, we concluded that we had done all we could in the face of an enormous reactor-independent background. We turned off the equipment and took the train back to Los Alamos, knowing that we had done our best but not knowing that we had actually measured a hint of a signal (Table 1).

Back home we puzzled over the origin of the reactor-independent signal. Was it due to natural neutrinos? Was the world made predominantly of neutrinos? Could it be due to fast neutrons from the nuclear capture of cosmic-ray muons? The easiest way to find out was to put the detector underground. We did so and showed that the background was from cosmic rays.

Fig. 2. First large (0.3 m^3) liquid scintillation detector in shield. The liquid was viewed by 90 2-inch photomultiplier tubes. Before the development of this detector a 0.02 m^3 volume was considered large.

Table 1. Listing of data from the Hanford experiment.

Run	Pile status	Length of run (sec)	Net delayed pair time	Accidental background rate
			Counts per minute*	
1	On	4000	2.56	0.84
2	On	2000	2.46	3.54
3	On	4000	2.58	3.11
4	Off	3000	2.20	0.45
5	Off	2000	2.02	0.15
6	Off	1000	2.19	0.13

*Delayed coincidence rates: reactor on (10,000 seconds), 2.55 ± 0.15 count/min.; reactor off (6,000 seconds), 2.14 ± 0.13 count/min. Reactor-associated delayed coincidence rates, 0.41 ± 0.20 count/min.

Fig. 3. Shield configuration. The note on the blackboard indicates that we were within a factor of 75 of the required sensitivity. The members of the group for the Hanford phase of the search are listed on the "Project Poltergeist" sign.

EARLY EXPERIMENTS

THE SAVANNAH RIVER EXPERIMENT

Encouraged by the tentative result at Hanford (8), we designed and constructed a detector that would employ the detailed characteristics of the anti-neutrino-proton ($\bar{\nu}_e + p$) reaction and so discriminate more selectively against reactor-independent and reactor-associated backgrounds. The detector was completed in 1955 and, at the suggestion of J.A. Wheeler, was taken to a newly completed heavy water moderated reactor at the Savannah River Plant in South Carolina. There a definitive observation of $\bar{\nu}_e + p$ was made in 1956 (9).

The Savannah River reactor was admirably suited (10) for neutrino studies by virtue of its great power (~ 700 megawatts at that time) and relatively small physical size, and the availability of a well-shielded location 11 meters from the reactor center and some 12 meters belowground in a massive building. The high $\bar{\nu}_e$ flux, 1.2×10^{13} per square centimeter per second, and the reduced cosmic-ray background were essential to the success of the experiment, which even under those favorable conditions involved a running time of 100 days over a period of approximately 1 year.

Fig. 4a is a schematic of the detection experiment. An antineutrino from fission products in the reactor is incident on a water target containing cadmium chloride. According to the $\bar{\nu}_e + p$ reaction, a positron and a neutron are produced. The positron slows down and is annihilated with an electron, producing two 0.5-MeV gamma rays, which penetrate the water target and are detected in coincidence by two large scintillation detectors on opposite sides of the target (Fig. 4b). The neutron is slowed down by the water and captured by the cadmium, producing multiple gamma rays, which are also observed in coincidence by the two scintillation detectors. The antineutrino signature is therefore a delayed coincidence between the prompt pulses produced by e^+ annihilation and those produced microseconds later by the neutron capture in cadmium. A characteristic oscilloscope record is shown in Fig. 5. The experiment was composed of a series of measurements in which the delayed coincidences were studied in detail to show that (i) the reactor-associated signal rate was consistent with theoretical expectations, (ii) the first pulse of the delayed coincidence signal was due to positron annihilation, (iii) the second pulse of the delayed coincidence signal was due to neutron capture, (iv) the signal was a function of the number of target protons, and (v) radiation from the reactor other than antineutrinos could not be the cause of the signal.

The detection system required for these measurements is shown by the cutaway drawing of the detector assemblage (Fig. 6). The small interaction cross section ($\sim 10^{-43} \text{cm}^2$ per proton) and the detailed nature of the questions posed were primarily responsible

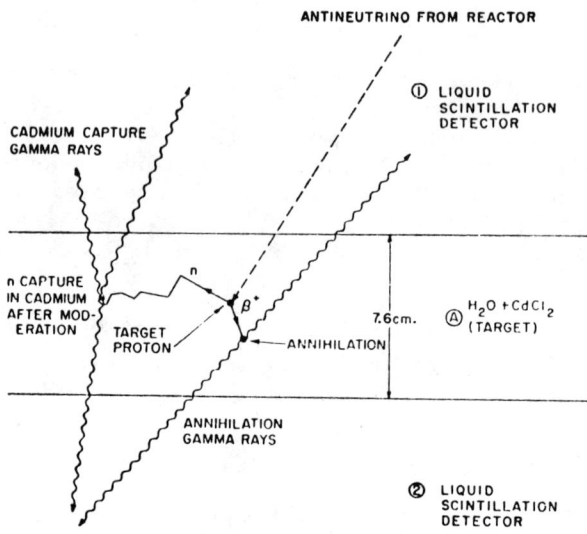

Fig. 4a. Schematic of neutrino experiment.

for the size of the detector, which exclusive of lead shielding weighed about 10 tons.

Translating our ideas into hardware was a formidable task. We calculated that we would need some 8000 liters of scintillator and realized that we would have to find a solvent less dangerous than toluene, which we had used in the smaller detectors at Hanford. A triethylbenzene-based scintillator proved to be suitable, and we set up a small pilot plant to prepare it. Construction of the detectors was complicated by the requirement that they support a 58-cm. depth of scintillator over an area of a few square meters and yet be thin enough to transit the positron annihilation radiation. We solved this problem by using a slab that consisted of two thin metal sheets mounted on opposite sides of corrugated cardboard.

It is difficult in these days of sophisticated commercially available solid-state electronics to appreciate the magnitude of the effort required to build the electronics involved in our experiment. Pulse height and time delay analyzers, linear amplifiers, coincidence circuits, gates, and so on were designed and built at Los Alamos. Indeed, the resources of a large laboratory were essential for the task.

EARLY EXPERIMENTS

Fig. 4b. Inside end view of detector tank showing 55, 5" photomultiplier tubes.

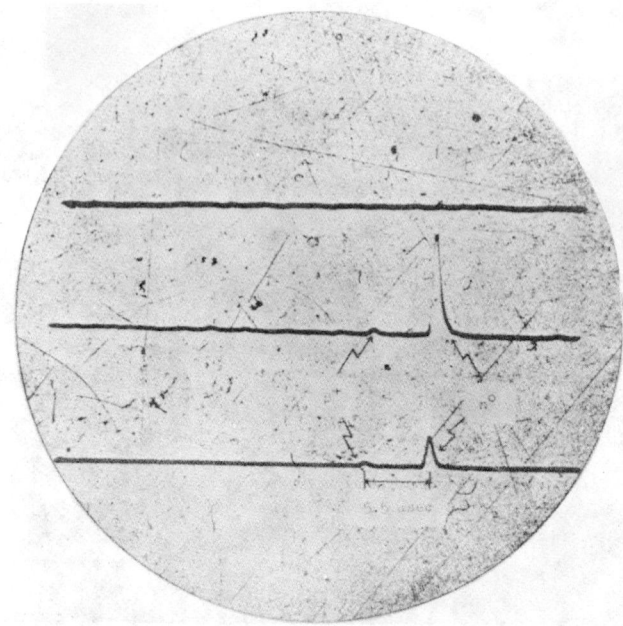

Fig. 5. A characteristic record. Each of the three oscilloscope traces corresponds to a detector tank. The event recorded occurred in the bottom triad. First seen in coincidence are the positron annihilation gamma-ray pulses in each tank followed in 5.5 μsec by the larger "neutron" pulses. A second oscilloscope with higher amplification was operated in parallel to enable measurement of the positron pulses. In Figs. 5 and 6 the positron is denoted by β^+ and the neutron by n^0.

Fig. 7 shows the electronics, which, along with the remainder of the detector and special handling equipment, was built or modified, assembled, and tested at Los Alamos before shipment to Savannah River. Today's electronics would accomplish the same task more expeditiously with less than one-tenth of the space and of the cost.

On the 1500-mile trip to the Savannah River Plant the equipment was transported in three large trucks - one of them, the electronics van, oversized. We were so concerned that the cold weather would cause precipitation of the scintillator solute that we wrapped electrical heaters around our specially constructed insulated storage tanks. Then, to ensure safe passage, we drove at low speed in a convoy preceded by one of us, who stopped and checked each bridge and tunnel. I remember the first turn-on of the detector at Savannah River - no signals were seen. It was a

EARLY EXPERIMENTS

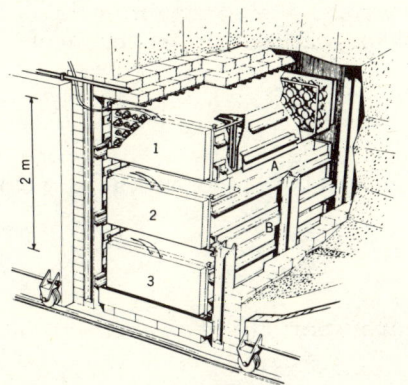

Fig. 6. Sketch of detectors inside their lead shield. The tanks marked 1, 2, and 3 contained 1400 liters of triethyl-benzene (TEB) liquid scintillator solution, which was viewed in each tank by 110 5-inch photomultiplier tubes. The TEB was made to scintillate by the addition of p-terphenyl (3 grams per liter) and POPOP [1,4-bis-2-(5-phenyloxazolyl) benzene] wavelength shifter (0.2 g per liter). The tubes were immersed in pure nonscintillating TEB to make light collection more uniform. Tanks A and B were polystyrene and contained 200 liters of water, which provided the target protons and contained as much as 40 kilograms of dissolved $CdCl_2$ to capture the product neutrons.

most peculiar feeling: maybe there were no neutrinos, or maybe they existed but were unstable and did not reach our detector. We continued tuning the detector and the signal appeared - but what a heady, if unwarranted, flight of fancy.

OBSERVATION OF THE NEUTRINO

Signal Rate

A reactor-associated correlated signal rate of 3.0 ± 0.2 events per hour was observed. This represented a ratio of signal to total accidental background of 4:1, a ratio of signal to correlated (as in neutron capture) reactor-independent background of 5:1, and a ratio of signal to reactor-associated accidental background greater than 25:1. Using positron and neutron sources and knowing the reactor flux, we determined that this signal was

within a factor of about 2 of the expected value. The measured
cross section for fission antineutrinos on protons was

$$\bar{\sigma}_{exp} = (12^{+7}_{-4}) \times 10^{-44} \text{cm}^2$$

compared to the theoretically expected value (11)

$$\bar{\sigma}_{th} = (5 \pm 1) \times 10^{-44} \text{cm}^2$$

In retrospect it appears that the $\bar{\nu}_e$ spectrum and hence $\bar{\sigma}_{th}$ was
not this well known.

First and Second Pulses

We determined that the first pulse of the delayed coincidence
pair was due to a positron by varying the thickness of a lead
sheet interposed between the water target and one liquid scintillation detector, so reducing the positron detection efficiency.
After we corrected for the small associated drop in the neutron
detection efficiency, we observed that the signal diminished as
it should if the first pulse were due to positron annihilation
radiation. Table 2 shows the expected and observed signal rates
as a function of lead thickness. A further check was provided by
the spectrum of first pulses, which showed better agreement

Fig. 7. Inside view of electronics van showing equipment
required to select and record neutrino signals.

Table 2. Lead absorption test of signal.

Lead thickness (cm)	Signal	
	Predicted	Observed
0	1.00	1.00
0.16	0.40	0.50 ± 0.13
0.48	0.12	0.32 ± 0.14
0.95	0.02	0.03 ± 0.06

with that from a positron test source than with the spectrum of the background.

That the second pulse was due to a neutron was clearly demonstrated by a series of experiments in which the cadmium concentration was varied. The most striking measurements were those made with and without cadmium in the water target. As expected for neutrons, removal of the cadmium totally removed the correlated count rate, giving a rate above accidental counts of 0.2 ± 0.07 hour^{-1}. In addition, the distribution of time intervals between the first and second pulses was found to be that for neutron capture, and the spectrum of the second pulses was consistent with that expected for neutron-capture gamma rays. Having proved that the second pulse was due to a neutron, we considered it necessary to show that the first pulse was not also caused by a neutron. We showed that such a false pulse sequence was unlikely by experiments with fast neutrons, which caused primarily an increase in the accidental rather than the correlated rate. As pointed out above, the reactor-associated rise in the accidental signal was less than 1/25 of the correlated signal, ruling out neutrons from the reactor as the cause of the delayed coincidence pair.

Signal as a Function of Target Protons

In this experiment, the number of target protons was reduced without drastically changing the detection efficiency of the system for events produced by antineutrinos. This was accomplished by mixing light and heavy water and so replacing about half of the protons by deuterons. The ratio of the measured rate for the diluted target to that for 100 percent H_2O was 0.4 ± 0.1, compared to an expected ratio of 0.5. This comparison with expectation showed the dependence of the signal on the presence of protons in the target. It was further noted that although the antineutrino signal changed significantly with dilution of the target by deuterons, the detection efficiency for background events was only slightly altered by this dilution, supporting the conclusion that the signal was, in fact, due to antineutrinos.

Absorption Test

The only known particles, other than antineutrinos, that are produced by the fission process can be heavily discriminated against by means of a gamma-ray and neutron shield. Accordingly, it was possible to test the signal for neutron and gamma-ray contamination by the addition of bulk shielding between the reactor and the detector assembly. It was shown by calculations and separate measurements with neutron and gamma-ray sources that the shield reduced gamma rays and neutrons by at least an order of magnitude. The signal, on the other hand, remained constant; that is, it was 1.74 ± 0.12 hour^{-1} with and 1.69 ± 0.17 hour^{-1} without the shield.

TELEGRAM TO PAULI

After convincing ourselves by this redundant series of tests that we were observing the neutrino, we decided to let Pauli know how correct he was and sent him the telegram shown in Fig. 8. Pauli was at the Eidgenossische Technische Hochschule (not Zurich University, to which the telegram was addressed), but the message was forwarded to him at CERN, where he interrupted the meeting he was attending to read the telegram to the conferees and then made some impromptu remarks regarding the discovery. We learned later that Pauli and some friends consumed a case of champagne in celebration.

PARITY

Soon after detection of the antineutrino, evidence was found by Wu et al., Garwin et al., and Friedmann and Telegdi (12) for parity nonconservation in beta decay. This was explained by Lee and Yang (13), Landau (14), and Salam (15) as being associated with the two-component character of the neutrino, which predicts the factor of 2 increase in the $\bar{\nu}_e + p$ cross section over that of the parity-conserving, four-component neutrino. It is interesting to speculate on the credibility of this explanation for the violation of parity had it been put forward before proof of the neutrino's existence. It is also interesting to reflect on the fact that evidence for the parity factor would have been obtained in due course (16) independently of the θ-τ puzzle that led Lee and Yang (17) to question the conservation of parity in the weak interaction.

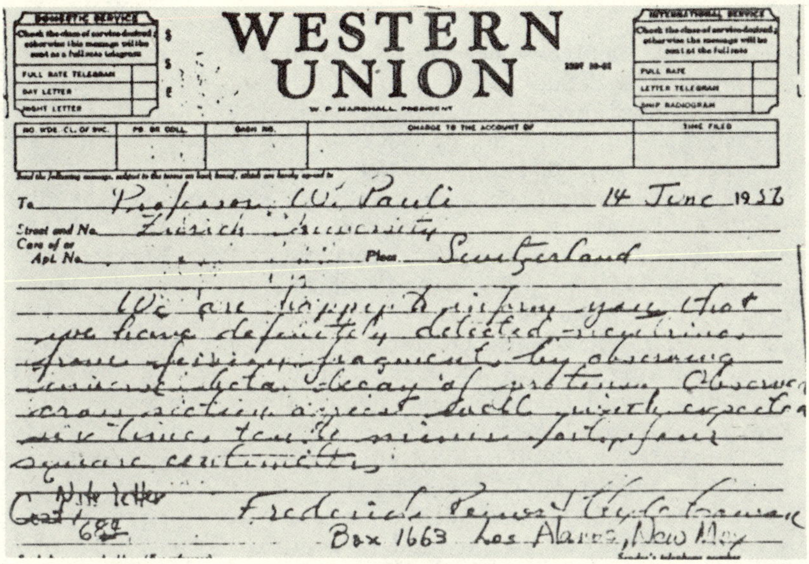
WESTERN UNION

W. P. MARSHALL, PRESIDENT

DOMESTIC SERVICE

FULL RATE TELEGRAM

DAY LETTER

NIGHT LETTER

INTERNATIONAL SERVICE

FULL RATE

LETTER TELEGRAM

SHIP RADIOGRAM

CHARGE TO THE ACCOUNT OF

TIME FILED

To Professor W. Pauli 14 June 1956

Street and No. Zurich University

Care of or Apt. No. Place Switzerland

We are happy to inform you that we have definitely detected neutrinos from fission fragments by observing inverse beta decay of protons. Observed cross section agrees well with expected six times ten to minus forty-four square centimeters.

Frederick Reines Clyde Cowan

Box 1663 Los Alamos, New Mex

Fig. 8. Telegram to Pauli informing him of our results. The text read: "We are happy to inform you that we have definitely detected neutrinos from fission fragments by observing inverse beta decay of protons. Observed cross section agrees well with expected six times ten to minus forty-four square centimeters."

ACCELERATORS

It occurred to us that the free neutrino could be used to probe the weak interaction in energy ranges outside those of ordinary beta decay if sufficiently potent accelerators could be constructed. Also, we puzzled, why should the neutral particle (or "neutretto" as it was first called) produced in the decay of a pion to a muon ($\pi \rightarrow \mu$) be the same as the neutrino of nuclear beta decay. We suggested investigating this at the Brookhaven accelerator with a suitable detector, but were unsuccessful in persuading the authorities at Los Alamos to let us continue our search.

The idea of using an accelerator was later conceived independently by Pontecorvo (18) and Schwartz (19). It was demonstrated at Brookhaven by Schwartz and his co-workers (20) that $\nu_\pi \neq \nu_e$. Groups at CERN (21), using a meson-focusing magnet, had a neutrino beam of 100 times higher intensity. They detected not only the muonic neutrino, but also electron production by the small admixture in the neutrino beam stemming from beta decays of K mesons, thus verifying the existence of ν_e some eight years after the initial observation of $\bar{\nu}_e$.

REFERENCES

1. W. Pauli, Jr., address to the Group on Radioactivity, Tübingen, Germany, 4 December 1930 (unpublished); Rapports du Septieme conseil physique Solvay, Bruxelles, 1933 (Gautier-Villars, Paris, 1934), chap. 1, p. 324.
2. E. Fermi, Z. Phys. 88, 161 (1934).
3. A popular account of these early days was written by Cowan in 1964 ["Anatomy of an Experiment: An Account of the Discovery of the Neutrino", Smithson. Inst. Annu. Rep. 4626 (1964), p. 409]. The status of the neutrino in 1936 was reviewed by H.A. Bethe and R.F. Bacher [Rev. Mod. Phys. 8, 82 (1936)]. Attempts to detect the neutrino up to 1948 were summarized by H.R. Crane [ibid., 20, 278 (1948)]. B. Pontecorvo [Inverse Beta Decay (Division of Atomic Energy, National Research Council of Canada, Chalk River; declassified and issued by the Atomic Energy Commission in 1949)] and L.W. Alvarez [Univ. Calif. Radiat. Lab. Rep. UCRL-328 (1949)] suggested a radiochemical method using a fission reactor based on the reaction $\nu + {}^{37}Cl \rightarrow {}^{37}Ar + e^-$. They did not pursue the method. Alvarez was dissuaded by his estimates of the background to be anticipated from cosmic rays; these estimates later proved to be correct for the reactors then available. As we know now, the neutrino produced in fission is $\bar{\nu}_e$ and the neutrino required for the ^{37}Cl reaction is ν_e, so the reactor result would have been negative even though the neutrino exists.
4. N. Bohr, J. Chem. Soc., p. 349 (1932).
5. Cowan and I looked for this reaction in 1956 but only obtained upper limits. The reaction was finally detected in 1976: F. Reines, H.S. Gurr, H.W. Sobel, Phys. Rev. Lett. 37, 315 (1976).
6. The technique of scintillation counting followed the discovery by W. Crookes and by J. Elster and H. Geitel [Phys. Z. 4, 439 (1903)] of the scintillation properties of zinc sulfide exposed to alpha particles [described by E. Rutherford, J. Chadwick, and C.D. Ellis, Radiations from Radioactive Substances (Cambridge Univ. Press, Cambridge, England, 1930)]. It received great impetus from the development of the photomultiplier tube and the crucial observation [H. Kallmann, Phys. Rev. 78, 62 (1950); M. Agena, M. Chiozotto, R. Querzoli, Atti Acad. Naz. Lincei Cl. Sci. Fis. Mat. Nat. Rend. 6, 626 (1949); Phys. Rev. 79, 720 (1950); G.T. Reynolds, F.B. Harrison, G. Salvini, ibid. 78, 488 (1950)] that liquids could be made to scintillate with high efficiency when the scintillating compound was at low concentration. We recognized that with a sufficiently transparent scintillator and enough photocathode area, one should, in principle, be able to make a detector of almost

arbitrarily great size – just what was needed for neutrino detection. Our first large detector, nicknamed El Monstro, was a 1-m^3 bipyramidal brass tank containing toluene and viewed on the top and bottom by four 2-inch photomultiplier tubes. Our subsequent detectors employed many more photomultipliers to increase light collection and so obtain the desired energy resolution.

7. F. Reines and C.L. Cowan, Jr., Phys. Rev. 90, 492 (1953); C.L. Cowan, Jr., F. Reines, F.B. Harrison, E.C. Anderson, F.N. Hayes, ibid., p. 493.
8. F. Reines and C.L. Cowan, Jr., ibid., 92, 830 (1953).
9. C.L. Cowan, Jr., F. Reines, F.B. Harrison, H.W. Kruse, A.D. McGuire, Science 124, 103 (1956); Phys. Rev. 117, 159 (1960).
10. At that time R. Davis, Jr., reassessed the ^{37}Cl approach and decided to make an effort to observe the reactor neutrino by using that reaction. We called to his attention the existence of other well-shielded, powerful SRP reactors, and he placed 4000 liters of CCl_4 near one of them. He obtained a negative result (R. Davis, Jr., paper presented at the American Physical Society Meeting, Washington, D.C., 1956) which, taken together with our observation, proved that although the $\bar{\nu}_e$ existed it was incapable of inverting ^{37}Ar decay. This suggested that the neutrino emitted by neutron-rich fission fragments (e^- decay), $\bar{\nu}_e$, was different from the ν_e emitted in e^+ decay, which at that time was one of the two possibilities to be checked.
11. This prediction incorporated the then held belief that parity is conserved in the weak interaction. In view of the large experimental errors and the poorly known $\bar{\nu}_e$ spectrum, we considered this crude agreement consistent with the $\bar{\nu}_e$ origin of the signal and continued our program to make this comparison more precise. (Our initial analysis grossly overestimated the detection efficiency with the result that the measured cross section was at first thought to be in good agreement with prediction.) As commented on later in this account, the effect of parity nonconservation is to increase the predicted cross section by a factor of 2. In the two-component theory the electron neutrino has only two states – one, $\bar{\nu}_e$, with its spin angular momentum parallel, and one, ν_e, with its spin angular momentum antiparallel to its linear momentum. The old four-component theory allowed each neutrino to have two spin states.
12. C.S. Wu, E. Ambler, R.W. Hayward, D.D. Hoppes, R.F. Hudson, Phys. Rev. 105, 1413 (1957); R.L. Garwin, L.M. Lederman, M.W. Weinrich, ibid., p. 1415; J.I. Friedmann and V.L. Telegdi, ibid. 106, 387 (1957).
13. T.D. Lee and C.N. Yang, Phys. Rev. 104, 1671 (1956), (parity non-conservation in β decay).
14. L. Landau, Nucl. Phys. 3, 127 (1957).

15. A. Salam, Nuovo Cimento 5, 299 (1957).
16. In the fall of 1956, following the observation experiments, we measured the cross section with equipment built for that purpose in 1954-5, but we did not publish the result until an improved measurement had been made of the $\bar{\nu}_e$ spectrum from fission (1957), which made possible a more precise comparison with theory [F. Reines and C.L. Cowan, Jr., in Second United Nations International Conference on the Peaceful Uses of Atomic Energy (A/Conf. 15/P/1026, United Nations, New York, 1958); R.E. Carter, F. Reines, J.J. Wagner, M.E. Wyman, in ibid.; Phys. Rev. 113, 273 (1959); ibid., p. 280].

$$\frac{\sigma_{expt}}{\sigma_{2\ comp}} = 1.1 \pm 0.3$$

17. T.D. Lee and C.N. Yang, ibid., p. 254.
18. B. Pontecorvo, Electron and Muon Neutrinos (Reprint P-376, Joint Institute for Nuclear Research, Dubna, Soviet Union, 1959).
19. M. Schwartz, Phys. Rev. Lett. 4, 306 (1960).
20. G. Danby, J.M. Gaillard, K. Goulianos, L.M. Lederman, N. Mistry, M. Schwartz, J. Steinberger, ibid., 9, 36 (1962).
21. M.M. Block, H. Burmeister, D.C. Cundy, B. Eiben, E. Franzinetti, J. Keren, R. Mollerud, G. Myatt, M. Nikolic, A. Orkin Lecourtois, M. Paty, D.H. Perkins, C.A. Ramm, K. Schultze, H. Sletten, K. Soop, R. Stump, W. Venus, H. Yoshiki, Phys. Lett. 12, 281 (1964); J.K. Bienlein, A. Bohm, G. Von Dardel, H. Faissner, F. Ferrero, J.M. Gaillard, H.J. Gerber, B. Holm, V. Kaftanov, F. Krienen, M. Reinharz, R.A. Salmeron, P.G. Seiler, A. Staude, J. Stein, H.J. Steiner, ibid., 13, 80 (1964).

FIFTY YEARS OF NEUTRINO PHYSICS: A FEW EPISODES

B. Pontecorvo

Laboratory of Nuclear Problems

Joint Institute for Nuclear Research, Dubna

INTRODUCTION

On the occasion of the 70th birthday of Edoardo Amaldi, about two years ago, I was invited to give a review talk on Neutrino Physics at the International Assembly of physicists, the major part of which was certainly not composed of neutrino physicists. Then the talk was much simpler than today, since you are all professional "neutrinists". Notice that I have only 30 minutes at my disposition (instead of 2 hours at the Amaldi Conference). I must avoid the danger of being trivial by telling you the a,b,c of your work. A way out of this difficulty, maybe, is to give a few recollections of such developments in neutrino physics which either are curious and at the same time very important (Pauli, Fermi) or about which I happen to be well informed for various reasons. Thus my talk will be entirely subjective (at a variance with the one I gave at the Amaldi celebration) and will be mainly dedicated to the young generation of neutrino investigators, who are well informed about today and yesterday developments, but not so well about old ones. I shall not talk about today problems, of course, since you are all here to discuss them during almost a week. By the way, most of you are used to thinking in terms of $10^5 - 10^6$ neutrino events and forgot, if you knew it, that 16 years after the Pauli neutrino hypothesis (1930) neutrinos were still considered as undetectable particles, and, as you heard today, they were first revealed in the free state only 25 years after they had been invented.

Neutrino physics is almost a synonym of weak interactions physics, but there is a difference. I took such a difference into account, but not always.

In order to decrease the subjective character of my talk, I shall present a Table of events in neutrino physics. Of course this

Table also is not objective. In the Table events are mentioned which either had a deciding meaning or initiated a large quantity of investigations. Of course, it was impossible to list all of them, even if their significance is greater than that of the investigation which initiated them. Two words more on the Table. I prepared it, at beginning, by memory, that is not consulting any literature. When eventually it became necessary to precise all the thing, I lost lots of time, but 95% of the original events remained and very few were added. I beg your pardon for deformations and omissions: the Table reflects the way by which neutrino physics has been influencing me. The Table is divided into four parts, with a rather loose periodization. First - from the discovery of radioactivity to the neutrino hypothesis, the Fermi beta decay theory and the detection of antineutrinos in the free state. Second - from the observation of weak processes other than the beta decay to the discovery of parity non-conservation in weak processes, the V-A universal theory and the observation of PC violation. Third - from the birth of high energy neutrino physics and the discovery of two neutrinos to the discovery of neutral currents, of the tau leptons, the weak decays of charmed particles and the theory of electro-weak interactions. Fourth - neutrino in astrophysics, astronomy and cosmology.

For some reasons I started to prepare the literature for the fourth part. Later on I reduced drastically all the literature. Thus the literature has an accidental character. The reason is that I did not wish to prepare a sort of contents of the Proceedings of various International Conferences. One may find the necessary information just in such Proceedings.

An inspection of the Table indicates an amazingly fast growth of neutrino physics, which became a definitely quantitative science, healthy and powerful, and yet with lots of room for qualitative surprises. In the Table I have underlined the year relating to some curious and/or so well-known event.

Table I

From the discovery of radioactivity to the neutrino hypothesis, the Fermi beta decay theory and the detection of (anti)neutrinos in the free state.

Year	Event	Authors and/or ref.
1	2	3
1896	Discovery of radioactivity	Becquerel
1899	Discovery of beta rays	Rutherford

A FEW EPISODES

1	2	3
1908	Counter (Geiger and proportional) capable of detecting single charged particles.	Geiger, Rutherford, Müller
1912	Cloud chamber	Wilson
1914	The continuous beta spectrum	Chadwick
1925	Nuclear photoemulsions	Misovsky
1927	Measurement of the heat released by beta rays	Ellis, Wooster
1927	Quantum theory of radiation	Dirac
1928	Relativistic equation of spin 1/2 particles	Dirac
<u>1929</u>	Two component theory of massless fermions	Weil
1930	The neutrino is invented	Pauli[1,2]
1932	The discovery of the positron	Anderson
1932	The discovery of the neutron	Chadwick
1932-1933	The nucleus is made up of nucleons	Ivanenko, Heisenberg, Majorana
1933	Theory of beta decay	Fermi
1934	Artificial radioactivity	Curie, Joliot
1934	Positron emission in beta decay	Curie, Joliot
1934	First discussion of the inverse beta decay	Bethe, Peierls
1935	Meson theory of nuclear forces	Yukawa
1935	Nucleus recoil in beta decay	Leipunsky[8]
1935	First mention of the double beta decay	Geppert-Maier[4]
<u>1936</u>	Far-reaching consequences of the fact that the Fermi constant is not dimensionless	Heisenberg
1936	Kurie plot	Kurie, Richards, Paxton
1936	Gamow-Teller selection rules	Gamow, Teller
1937	Neutrino Majorana	Majorana
1937	Nuclear orbital electron capture	Alvarez

1	2	3
1938	Discovery of the muon	Anderson, Neddermeyer
1939	Diffusion chamber	Langsdorf
1942	First nuclear reactor	Fermi et al.
1944	The principle of phase stability. Few years later the era is beginning of experiments performed on new types of powerful accelerators	Veksler, McMillan
1945	Christal counters and semiconductors detectors	Van Heerdin; McKay; McKenzie, Bronlay
1946	Proposal to detect low energy neutrinos with radiochemical methods	Pontecorvo
1947	The scintillation counter	Kallman
<u>1949</u>	Upper limit of the ν_e mass from ^{3}H decay	(5)
1950	Cerenkov counter	Jelley
1952	Bubble chamber	Glaser
<u>1953</u>	Conception of lepton charge	Marx; Zeldovich; Konopinski; Mahmoud
1953-1956	First observation of free (anti)-neutrinos from a reactor	Reines, Cowan
1956	The reaction $\nu_e + {}^{37}Cl \to {}^{37}A + e^-$ is not observed ($\nu_e \neq \bar{\nu}_e$)	Davis

II

From the observation of weak processes other than the beta decay to the discovery of parity non-conservation in weak processes, the V-A universal theory and the observation of PC violation.

1	2	3
1941	Direct proof of the muon radioactivity and direct measurement of its mean life (cosmic ray experiment)	Rasetti

A FEW EPISODES

1	2	3
1947	The muon is not a hadron (cosmic rays experiment)	Conversi; Pancini; Piccioni
1947	Discovery of the pion and of the π–μ decay (cosmic ray experiment)	Lattes, Occhialini, Powell
1947-1949	Deep analogy of various four fermion interactions and the conception of weak processes	ref.(6)
1947	Discovery of strange particles in cosmic rays	Rochester, Butler; Leprince-Ringuet
1948	Absence of the process $\mu \to e\gamma$ (Cosmic ray experiment)	Hincks, Pontecorvo[7]; Sard, Althaus
1948	Observation of artificial pions; after this discovery very accurate measurements of the pion and muon masses, of their mean lives and of the energy of their charged decay products have been performed and are being performed. Similarly quantitative investigations of the strange particle properties have been performed	Gardner, Lattes
1948-1949	Discovery of the neutron radioactivity	Snell, Miller; Robson
1949	In the muon decay 3 particles are emitted, the charged one being an electron: $\mu \to e + \nu + \nu'$ (cosmic ray experiments)	Hincks, Pontecorvo; Steinberger; Jdanov; Anderson et al.
1950	The Michel parameter	Michel
1950	Strong focusing in accelerators	Christophilos et al.
1952	"The disturbing possibility remains that C and P are both only approximate and CP is the only exact symmetry law"	Wick, Wightman, Wigner[8]
1953-1954	Hadron isotopic multiplets. Strangeness	Gell-Mann, Nikishima
1953	The dual properties of neutral kaons	Gell-Mann, Pais
1954	The Yand-Mills fields	Yang, Mills

1	2	3
1954	Teorema CPT	Luders; Pauli
1955	First observation of antiprotons	Chamberlain, Segré
1955	Conservation of the vector weak current	Gerstein Zeldovich
1955-1956	The θ-τ paradox (parity non-conservation in the decay of strange particles)	Whitehead et al.; Barkas et al.; Dalitz et al.; Harris et al.; Fitch et al.
1956	Discovery of the long-live neutral kaon	Landè et al.
1956	Is parity conserved in weak interactions?	Lee, Yang
1956-1957	PC invariance	Landau; Lee, Yang
1957	P and C are violated in the ^{60}Co decay	Wu et al.
1957	P and C are violated in the π-μ and μ-e decays	Garwin, Lederman, Weinrich
1957	First mention of the unification of weak and electromagnetic interactions	Schwinger
1957	Longitudinal neutrino	Landau; Salam; Lee, Yang; Sakurai
1957	Observation of the longitudinal polarization of beta particles	Frauenfelder et al.; Alichanov et al.; Nikitin et al.
1957	The V-A universal weak interaction	Gell-Mann, Feynman; Marshak, Sudershan
1957	Electron-neutrino angular correlation in beta decay (^{35}A, ^{6}He) finally found in agreement with the V-A theory	Herrmansfelt et al.
<u>1957</u>	Neutrino oscillations	Pontecorvo

1	2	3
1958	The $\pi \to e\nu$ process finally observed with a probability in agreement with the V-A theory	Fazzini et al.; Schwartz, Steinberger et al.
1958	Neutral symmetrical currents	Bludman
1958	Ionization calorimeter	Grigorov, Murzin et al.
1958	Unitary symmetry and weak interaction	Kobzarev, Okun
1958-1963	Theory of Cabibbo	Gell-Mann; Levy; Cabibbo
1958	The role of strong interactions in weak processes	Goldberger, Treiman
1959	"Kiev symmetry", that is "prequark" lepton-hadron symmetry	Gamba, Marshak, Okubo
1962	Possibility of exciting nuclei by neutrinos (neutral currents)	Eramjian, Gershtein, Nguen-Van-Hieu
1962	Observation and investigation of the reaction $\mu^- + {}^3He \to {}^3H + \nu_\mu$	Falomkin et al.
1962	Observation and investigation of the reaction $\mu^- + p \to n + \nu_\mu$ in Hydrogen	Hildebrand
1962-1963	Observation of the decay $\pi^+ \to \pi^0 + e^+ + \nu_e$ with a probability in agreement with CVC expectations	Dunaytzev et al. Depommier et al.
1963	In an experiment suggested by Gell-Mann CVC is confirmed in ${}^{12}N$ and ${}^{12}B$ decays	Lee, Mo, Wu
1964	PC violation ($K_L^0 \to 2\pi$)	Christenson et al.
1964	Superweak interactions?	Wolfenstein
1967	Charge asymmetry in the lepton decays of K_L^0	Dorfan et al., Bennet et al.

III

From the birth of high energy neutrino physics and the discovery of two neutrino types to the discovery of neutral currents, of the tau leptons, the weak decays of charmed particles and the theory of electro-weak interactions

1	2	3
1959-1960	High energy neutrinos: a practical proposal which is opening a new field in weak interaction physics	Pontecorvo, Ryndin; Schwartz; Markov
1959	Spark chamber	Fukuni, Miyamoto
<u>1959</u>-1974	Theoretical discussion of parity non-conservation in atoms and in electron-nucleon interaction	Zeldovich, Bouchiat
1961	Theory of electro-weak interactions	Glashow
1962	$\nu_e \neq \nu_\mu$ (spark chamber experiment)	Brookhaven, Danby et al.
1963	Magnetic "horn"	Van der Meer
1963	Combination of photoemulsions with other techniques for localizing the interaction position	Dvoretsky et al.
<u>1963</u>	Localization of neutrino interactions in emulsions with the help of spark chambers	Burhop et al.
1963-1964	Streamer chamber	Chikovani et al., Dolgashein et al.
1963-1964	First neutrino experiments in which a bubble chamber is used	CERN, Block et al.
1964-1967	Weak nuclear forces	Abov et al., Lobashov et al.
1964	The fractional charge quarks (u,d,s)	Gell-Mann; Zweig
1964	The mechanism by which vector mesons acquire finite masses through spontaneous symmetry breaking	Higgs
1964	$\nu_\mu \neq \bar{\nu}_\mu$	CERN, Bernardini et al.
19<u>62</u>-1964	Theoretical introduction of charm	Maki, Nakagawa et al.; Bjorken, Glashow; Vladimirsky, Okun

A FEW EPISODES

1	2	3
1964	Every quark has three colours	Greenberg
1965	Integral charge triplet quarks	Bogoliubov, Struminsky, Tavkhelidze; Nan, Nambu
1965	Because of inelastic channels the total ν-N cross section probably will increase with energy in spite of the nucleon form factor, which limits the grow of the "elastic" ν-N cross section	Markov
1967-1972	Unified gauge model of electroweak interactions	Salam, Weinberg
1967	Quantization of massless Yang-Mills fields	Fadeev, Popov; De Witt
1967	Neutrino oscillations: concrete proposals of experiments, theory and estimates of upper limits	Pontecorvo, Gribov, Bilenky
1968	Proportional and drift chambers	Charpak et al.
1969	Scaling	Bjorken
1969	The parton model	Feynman
1971	Quantization of massive Yang-Mills fields	G't Hooft
1971	Bubble chamber Gargamelle (second generation of neutrino experiments)	CERN
1971	The idea of using a target-calorimeter in neutrino experiments (second generation of electronics neutrino experiments)	Rubbia et al.
1972	What neutrino can tell us about partons	Feynman
1972	GIM mechanism: the fourth quark is necessary to make neutral currents symmetrical	Glashow, Illiopulos, Maiani
1972-1980	Total ν_μ and $\bar{\nu}_\mu$ cross sections on nucleons are increasing linearly with energy	CERN, Gargamelle and later other facilities
1972-1980	The quark-parton model is confirmed by measurements of ν_μ and $\bar{\nu}_\mu$ charged current events	CERN, Gargamelle, and later other facilities

1	2	3
1973-1980	Observation of neutral currents in the process $\bar{\nu}_\mu + e^- \to \bar{\nu}_\mu + e^-$	CERN, Gargamelle and later other facilities
1973	Observation of neutral currents in muonless events $\nu_\mu + N \to \nu_\mu + \ldots$	CERN, Gargamelle Fermilab, HPWF; and later other facilities
1973-1974	Nucleon decay?	Pati, Salam; Giorgi, Glashow
1974	J/ψ particle	Ting et al.; Richter et al.
1975	The intermediate boson mass is < 17 GeV	Batavia, CITF
1975	Detailed proposal to detect "direct" neutrinos to study the production of charmed particles by nucleons	Pontecorvo
1975	The first charmed baryon is produced by neutrinos in the Brookhaven hydrogen bubble chamber	Cassoli et al.
1975	Pairs $\mu^+\mu^-$ produced in ν_μ and $\bar{\nu}_\mu$ events demonstrate the production of charmed particles by neutrinos	Fermilab, HPWF
1975	First observation of tau lepton	SPEAR, Pearl et al.
1976	The ν_e mass is <35 eV	ITEP, Tret'yatov et al.[5]
1976	Processes $\nu_\mu + z \to \mu^- + e^+ + \ldots$ and $\bar{\nu}_\mu + z \to \mu^+ + e^- + \ldots$ demonstrate charmed particle production (H-Ne Fermilab large bubble chamber and CERN, Gargamelle)	Fermilab, Berkeley-CERN Haway-Wisconsin; Aachen-Bruxelles-CERN-Ecole Poly-technique-Milano-Orsay-London; Fermilab, ITEP-Michigan Serpukhov
1976	Observation of $\bar{\nu}_e$ e scattering (reactor experiment)	Reines, Gurr, Sobel
1976	Observation of elastic $\nu_\mu p$ and $\bar{\nu}_\mu p$ scattering and of parity violation in the weak hadron neutral current	Brookhaven, Harvard-Pennsylvania Wisconsin; Columbia-Illinois-Rockfeller

A FEW EPISODES

1	2	3
<u>1977</u>	Practical applications of detecting ν_e (measurements of power, Pu accumulation) in reactor plants	Mikaelyan et al.
1977	Discovery of the upsilon meson, probably a bound state $(b\bar{b})$ of bottom quarks of charge $|1/3|$	Fermilab, Lederman et al. Columbia-Fermilab-Stony Brook
1977	Soon after the 400 GeV proton beam was available, a third generation of refined and good statistics high energy neutrino experiments starts at CERN	CDHS, BEBC and later CHARM
1977-1980	"Beam Dump" experiments	Serpukhov, IHEP-ITEP; CERN, Aachen-Bonn-CERN-London-Oxford-Saclay (BEBC) CERN, Gargamelle; CERN, CDHS; CERN CHARM
1978	Parity non-conservation in atoms in agreement with the Weinberg-Salam model	Barkov, Zolotarev
1978	Polarized electron scattering on deuterium conformes the Weinberg-Salam model and yields a value of $\sin^2 \theta_W$ in agreement with the results of the best neutrino experiments CDHS and CHARM	SLAC Prescott et al.
1978	The ν_μ mass is <0.57 MeV	SIN; Frosch et al.
1978	Some τ and ν_τ important properties are established: $m_\tau = 1782^{+2}_{-7}$; $m_{\nu_\tau} \leq 250$ MeV; V-A variant of τ decay	SPEAR, Kirkby et al.; Feldman et al.
1979	The polarization of muons produced by the interaction of neutrinos is found to agree with V-A expectations	CERN, CHARM
1979	Observation and investigation of the reactions $\bar{\nu}_e + d \to \bar{\nu}_e + n + p$; $\bar{\nu}_e + d \to e^+ + n + n$ reactor experiment)	Irvine group, Pasierb et al.

1	2	3
1979	When meson factory π^+ are stopped in matter, $\pi^+\to\mu^+\nu_\mu$; $\mu^+\to e^++\nu_e+\bar\nu_\mu$) the reaction $\nu_e+d\to e^-+p+p$ is observed, but not the reaction $\bar\nu_e+p\to n+e^+$, that is the decay $\mu^+\to e^++\bar\nu_e+\nu_\mu$ is forbidden (no multiplicative lepton number)	Los Alamos, Burman et al.
1979	The mean life of charmed particles produced by neutrino interactions in nuclear emulsions or bubble chambers is measured and found to agree with theoretical expectations (few times 10^{-13} sec)	CERN, Collab. Wa 17; Fermilab, Berkeley-Batavia-Haway-Seattle-Wisconsin; Brookhaven, Brookhaven-Columbia

Much relevant work, which I was not able to quote, has been done and is being done at various Institutes. Below some data about neutrino beams and neutrino detectors are summarized for the benefit of "non-professional" neutrino physicists.

High energy neutrino beam facilities

Accelerator	Proton energy (GeV)	Decay length (m)	Muon filter (m)	Neutrino energy (GeV)
ANL	12.4	30	13(Fe)	0.3-6
CERN	27	70	22 Fe)	1-12
BNL	29	57	30(Fe)	1-15
IHEP	70	140	62(Fe)	2-30
FNAL	300-400	340	1000 (Earth+Fe)	10-200
CERN SPS	400	300	400(Fe)	10-200

Large bubble chambers

Bubble chamber	Filling	Useful volume (m^3)	Weight (tons)
Gargamelle, CERN	CF_3Br	5	7-9
12', ANL (USA)	H_2, D_2	16	1-2
7', BNL (USA)	H_2, D_2	6	0.4
15', FNAL (USA)	H_2	20	1.3
	H_2+Ne 20%	20	7
	H_2+Ne 64%	20	22
SKAT, IHEP (USSR)	CF_3Br	4.5	7
BEBS, CERN	H_2, D_2, Ne	20-25	

Electronic detectors of neutrinos

Location	Collaboration	Useful target weight (tons)
CERN	Aachen-Padova (AP)	20
	CERN-Dortmund-Heidelberg-Saclay (CDHS)	900
	CERN-Hamburg-Amsterdam-Rome-Moscow (CHARM)	100
Brookhaven National Lab.	Harvard-Pennsylvania-Wisconsin (H.P.W.)	30
	Columbia-Illinois-Rockfeller (CIR)	8
IHEP, Serpukhov	ITEP-IHEP (S.S.)	30
FNAL	Harvard-Pennsylvania-Wisconsin-Fermilab (HPWF)	20
	California Inst. Technology, Fermilab (CITF)	100

(The most advanced detectors are CDHS and CHARM).

IV
Neutrino in astrophysics, astronomy and cosmology

1	2	3
1939	Emission of neutrinos in thermonuclear reactions in the Sun and other stars.	Bethe[9]
1941	Supernovaes and "Urka" processes	Gamow, Schonberg[10]
1946	Radiochemical methods for detecting neutrinos, for example, the Cl-A method used in Solar neutrino astronomy	Pontecorvo[11]
1946	The big-bang theory	Gamow[12]
<u>1958</u>	B^8 as a source of relatively high energy solar neutrinos	Fowler[13]
<u>1959</u>	Neutrino emission from hot stars due to the universal Fermi interaction (the $\nu_e + e \to \nu_e + e$ process)	Pontecorvo[14]
1960	The importance for elementary particle physics and astrophysics of performing experiments at great depths underground and under water	Markov[15] Greisen[16]
<u>1961</u>	Phenomenological considerations on the possible existence of a "neutrino sea"	Pontecorvo, Smorodinsky[17]
<u>1961</u>	Upper limits imposed by cosmological considerations on the amount of invisible energy	Zeldovich, Smorodinsky[18]
1962	Possible emission of pairs $\nu\bar{\nu}$ due to hypothetical neutral currents	Pontecorvo[19]
1963	A large detector of (atmospherical) cosmic ray neutrinos located at depth 8700 m.w.e. in a South Africa mine (8 years of measurements, ∼100 events)	Case Institute of Technology and University of California, Irvine, Reines et al.[20]
<u>1964</u>	Neutrino stars?	Markov

A FEW EPISODES

1965	Telescopes and magnetic spectrometers located at a depth 7500 m.w.e. in the Kolar Gold Fields in Southern India, aimed to detect atmospheric cosmic neutrinos (6 years measurements, ~20 events)	India-Japan Collaboration, Krisnashvami et al.; Osborne et al.[21]
1965-1966	Neutrino processes and pair formation in massive stars and supernovaes	Fowler, Hoyle[22] Colgate, White[23]
1965	Emission of detectable neutrinos ($E \gtrsim 10$ MeV) in the collapse of cooled stars, i.e., in the process of neutronization: $e^- + {}^Z A \to \nu_e + {}^{Z-1} A$	Zeldovich[24]
1965	Proposal of an experiment aimed to detect neutrino from collapsing stars	Domogatsky[25] Zatsepin
1965-1967	Discovery of the relict electromagnetic radiation, confirming the bing-bang theory and requiring the presence of a similar relict neutrino sea, with important implications for cosmological nucleosynthesis	Penzias, Wilson[26] Dicke et al.[27] Zeldovich, Novikov[28] Weinberg[29]
<u>1966</u>	Upper limit on the ν_μ mass imposed by cosmological considerations	Gershtein Zeldovich[30]
<u>1967</u>-1968	The necessity of clearing up the question about lepton charge conservation and the number of neutrino types (neutrino oscillations) for the future of solar neutrino astronomy	(31,32,33)
1972	Expectations for the ${}^{37}Cl-{}^{37}A$ solar experiment based on solar standard models	Bahcall[34]
1975-1977	Cosmic sources of ultra-high energy neutrinos	Beresinsky[35] Zatsepin
1977	A quantitative theory of Supernovaes, where neutrino heating ignites thermonuclear processes in carbon	Gershtein et al.[36]
1977	Scintillation telescope of the Institute of Nuclear Research placed at 800 m.w.e. in the	Chudakov et al.[37]

	Baksan Valley, having a total mass of 300 tons (3150 moduli)	
1977	Acoustic wave detector of ultra-high energy neutrinos	Dolgoshein et al.(38) Sulak et al.(39)
1977	The importance of neutrinos emitted during the collapse of stars for the nucleosynthesis, especially for explaining the abundance of proton-rich nuclei	Domogatsky et al.(40)
1978	Definite detection of Solar neutrinos by the Cl-A method in an experiment which lasted more than 10 years	Davis et al.(41)
1978	Čerenkov H_2O detector (∼500 tons) of star collapse neutrinos places underground in S. Dakota, Ohio and under the Mont Blanc	Landé et al.(42)
1978	INR scintillation detector of star collapse neutrinos (100 tons) placed in a salt mine at Artyomovsk (600 m.w.e.)	Zatsepin et al.(43)
1980	Scintillation detectors of star collapse neutrinos (60 moduli each 2 m^3) located under the Mont Blanc	Collaboration INR-Torino
1980	Deep Under Seas Muon and Neutrino Detector (one cubic kilometer optical-acoustic H_2O detector)	See for ex. Learned(44)

PAULI

It is difficult to find a case where the word "intuition" characterizes a human achievement better than in the case of the neutrino invention by Pauli.

First, 50 years ago there were known only two "elementary" particles, the electron and the proton, and the very idea that for the understanding of things the existence of a new particle becomes imperative was in itself a revolutionary conception. What a difference from the present day situation, when at the slightest provocation lots a people are ready to invent any number of particles!

Second, the invented particle, the neutrino, should have quite exotic properties, especially an enormous penetrating power. True, Pauli at the beginning did not recognize fully such unescapable implications of his idea and modestly conceded that the neutrino may have a penetrating power about equal or ten times larger than a γ quantum. Incidentally, a dimensional thermodynamical argument, showing that neutrinos of energy ∼ 1 MeV or wave length λ must have an astronomically large mean free path, let's say equal to a thickness of water milliard of times greater than the Earth-Sun distance, was first given by Bethe and Peierls[45] who considered the two inverse processes (I am using modern notations): $z \to (z+1)+e^-+\bar{\nu}_e$ (this is a beta process taking place with a characteristic time T) and the inverse reaction $\bar{\nu}_e+(z+1) \to z+e^+$, characterized at the mentioned neutrino energy by a cross section σ

$$\sigma \lesssim \lambda^2 \frac{1}{T} \frac{\lambda}{c}.$$

The argument, which today is self-evident (almost all good arguments look obvious "a posteriori") made a deep impression upon me. I did not forget it many years later, when I suggested how free neutrino experiments might be performed with the help of reactors[11].

Third, the neutrino, because of its fantastic penetration, appeared first as a particle which, as it were, cannot be revealed in the free state, and on the existence of which you can judge only on the basis of the laws of energy and moment conservations, by detecting the nuclear recoils in the β decay, that is with the help of a method which today is quite currently used in searches for neutral particles - the so-called "missing mass" method. Experiments of this type were suggested by Pauli and the first of these was performed in Cambridge by Leipunski. Here I would like to underline that 50 years ago there was known only one process involving the neutrino, the β decay of heavy nuclei, which is a 3 particle process. Extremely important experiments of Ellis and others showed that the average energy (measured in a calorimeter) of the beta rays is equal to the average energy of the β spectrum, measured in a magnetic spectrometer. This clue, together with the notion that there is a maximum energy of β rays was certainly not missed by Pauli. All the other processes in which, as we know now, neutrino takes part, were not known at the time. Among these several two particle decays from charged particles stopping in a track detector ($\pi^+ \to \mu^+ + \nu_\mu$; $\mu^- + {}^3He \to {}^3H + \nu_\mu$ ...) leave behind beautiful signatures, since the emitted charged particle has always the same momentum, of course equal to that of the invisible neutrino. Examples of these processes are well known to everybody present here. If in the time previous to the Pauli invention such a two particle events had been discovered, there would not have been the need of Pauli genious to invent the neutrino. However, I would like to mention here that, at the time, Bohr thought that the continuous β spectrum might arise from energy non-conservation in individual processes, so that, strictly speaking, in order to solve the dilemma neutrino versus energy non-conservation, one may not be allowed in principle to make use of conservation laws.

Some more words on the Pauli invention, about which he wrote himself a few tens of years after his famous proposal, which, incidentally, was never published in a scientific periodical. Maybe not all of you know that the first idea on the existence of the neutrino appeared in a letter[1] to a group of specialists in radioactivity, who were to meet in Tübingen, the letter starting with these words: "Dear radioactive ladies and gentlemen". At this meeting Pauli was not present because he was expecting much more from a ball which he whished to attend in Zürich, the night of December 6, 1930. But in that letter there were not only jokes. There are two ideas that only a man of great intuition could have. These ideas I will formulate in the today and the Pauli terminology.

1) In the nuclei there must exist electrically neutral particles, neutrons (Pauli also called them neutrons) having spin 1/2.

2) In the β decay together with the electron there must be emitted a neutral particle, the neutrino (Pauli called it neutron), so that the total energy of the electron, neutrino and recoil nucleus is discrete, as it should be.

Thus Pauli "invented" two particles at the same time and both were very necessary (keep in mind, among other things[+] the so-called nitrogen catastrophe, that is the proof given in the classical spectroscopic investigations of Rasetti, that nuclei ^{14}N obey the Bose statistics, so that they can hardly consist of protons and electrons only). Pauli for a time thought he had invented only one particle, because mistakenly he identified them. Soon, however, he understood his error, namely, in the first official publication [2] about the neutrino (so it was called by Fermi) at the 1933 Solvay Congress. The subsequent colossal step was done by Fermi.

FERMI

Fermi got acquainted with Pauli hypothesis in 1931 at an International Conference of Nuclear Physics, where the β decay problem was discussed. There Bohr talked in favour of energy non-conservation. Fermi was quite impressed by the Pauli particle, which he started to call "neutrino". At the 1933 Solvay Conference[2] for the first time in a discussion, which appeared in the press, Pauli told his idea. Fermi evidently was already thinking deeply about the problem: his famous paper "A Tentative Theory of β Decay"[46] appeared only 2 months after the end of the Solvay Congress. This is a quantitative theory, which had a great influence on the development of physics. Without any doubt the idea on the existence of the neutrino would have remained a vague notion without Fermi's

---

[+] Details on the theoretical thinking (Rutherford, Pauli and especially Majorana) about the neutron <u>before</u> its experimental discovery by Chadwick are most interesting, but I have not the possibility to discuss them here.

contribution. This theory amazingly resisted almost without change until now and underwent only relatively small, although quite important and numerous additions. I feel quite confident that, had been Fermi alive, he would have made himself at least most of the additions, under the pressure of new experimental facts, about which I will talk later.

I would like now to say some curious facts about the appearing of the theory, facts which I have seen with my eyes, since in that period I was working in Rome.

1) The Journal "Nature" refused the paper of Fermi, because it appeared too abstract to be of interest for the readers. I am sure the editor has regretted such episode for all his life.

2) The second curious thing has to do with the difficulties Fermi encountered. Such difficulties were not mathematical, but physical. The necessary mathematics, the secondary quantization, he learned quickly, but the most serious difficulty was to recognize the fact that the electron and the neutrino are <u>created</u> when a neutron transforms into a proton. Of course, this is a thing that every student knows today: elementary particle interactions are explained by the exchange of elementary particles. This is quantum field theory and is an unescapable consequence of the quantum theory and of the theory of relativity. Particles are created and destroyed. This was the difficult point for Fermi. Pauli, in spite of its pioneer work in quantum electrodynamics, did not formulate clearly this point, and if you read the famous Fermi article on β decay, you see how he worked making an analogy with the Dirac quantum theory of radiation (photons <u>are</u> created and destroyed!) and how by analogy he selected the V variant of the β decay.

I still remember his words: when the excited Na atom emits the 5890 Å line, the photon is not sitting in the atom (it is created), similarly the electron and the neutrino are created when a neutron is changing into a proton.

At a variance with an interaction at a distance $e\bar{\psi}_p \gamma_\mu A_\mu$, as in the case of the electromagnetic interaction (through the exchange of a photon) Fermi assumed that the two currents, the heavy particle (n,p) and the light particle (e,ν) currents have a contact interaction

$$k\bar{\psi}_p \gamma_\mu \psi_n \bar{\psi}_e \gamma_\mu \psi_\nu$$

where k is a constant of the order of 10^{-49} erg cm^3 (today we all know that $k = G/\sqrt{2}$, where $G = 10^{-5}/M_p^2$ is the Fermi constant, $h=c=1$), $\bar{\psi}_p, \psi_n$ are the creation operator of the proton and the destruction operator of the neutron, etc. Fermi assumed that weak currents, as we call them now, are four-vectors, as in electrodynamics. At the beginning, Fermi felt that the nucleon weak current $\bar{\psi}_p \gamma_\mu \psi_n$ is the analogous of the electromagnetic current $\bar{\psi}_p \gamma_\mu \psi_p$ and that the lepton weak current $\bar{\psi}_e \gamma_\mu \psi_\nu$ is the analogous of the electromagnetic field.

However, in his formulation the nucleon and lepton currents, as a matter of fact, are on identical foot. Thus Fermi created its perfect building starting from a few experimental results on the beta decay of heavy nuclei, especially RaE and from an analogy with Dirac theory of radiation.

POST FERMI

I would like to underline here that our knowledge since that time has increased tremendously; however (almost) all the new things fit wonderfully into the Fermi picture. The new things you may find in the Table, especially in parts II and III.

I shall briefly summarize some of these new things:

1) Neutrinos are not only emitted in beta decay processes. There are numerous processes in which the neutrinos take part: decay of non-strange particles, decays of strange particles, decays of charmed particles, inverse of these processes induced by high energy neutrino beams, decays of charged leptons (μ and τ), deep inelastic scattering of neutrinos by nucleons, elastic scattering of neutrinos by electrons and nucleons.

2) Even a small part of these facts was sufficient to suggest that the Fermi interaction describing the beta decay process is a special case of four-fermion interactions having about the same strength. This is how there arose the conception of weak interactions.

3) There exist at least 3 types of leptons e, μ, τ and "their" neutrinos ν_e, ν_μ, ν_τ two of which have been observed through their interactions in the free state with the help of reactors $(\bar{\nu}_e)$[47], the Sun (ν_e)[41], and accelerators $(\nu_\mu, \bar{\nu}_\mu)$[48].

4) In weak processes neither parity P nor charge conjugation C are conserved although the laws of nature are (almost) invariant with respect to the combined inversion PC, which changes simultaneously the signs of coordinates and charges. Non conservation of parity implies longitudinal polarization of particles and thus there arose the theory of two component neutrino of Landau, Lee and Yang, Salam and Sakurai, which is an old theory of Weil, made plausible by parity non-conservation. A good model of the neutrino according to this theory is a screw. Actually it was shown experimentally by Goldhaber that neutrinos are left-handed. Anti-neutrinos are right-handed. Thus we have two states only and not four, as for an actual screw: screw left-handed, screw right-handed, antiscrew left-handed, antiscrew right-handed. Now the importance of the longitudinal neutrino is that such neutrino gives us the prototype of the behaviour of all other (not massless) fermions, under weak interaction. A simple mnemonic rule is that, under weak interaction, all fermions are left-handed, all anti-fermions are right-handed. This has been incorporated in the famous universal weak interaction V-A theory of Feynman and Gell-Mann, Marshak and Sudershan. As we saw, in analogy with electrodynamics, the weak

interaction involves vector operators working on the wave-functions on particles. But there are two amplitudes – V, the original Fermi one, which has the spatial transformation properties of a polar vector (that is, it changes sign under inversion of the space coordinates), while the other, A, has those of an axial vector (it does not change sign). Namely the coexistence of V and A means non--conservation of parity. Thus the Fermi weak current, which was originally a vector one, in fact became the sum of a vector and axial vector (the last one being constructed with the help of the matrix $\gamma_\mu\gamma_5$, where $\gamma_5=i\gamma_0\gamma_1\gamma_2\gamma_3$). Now I would like to come back to Fermi and to think for a moment: what might have happened if, in 1954, the fate had granted to him few years more? I believe that probably he would have invented the two-component neutrino, but I am not certain about it. What I am certain about is that Fermi, after either he or Landau, Salam, Lee and Yang had discovered the two-component nature of the neutrino, would have created the V-A theory. Not only he had started in 1933 the all business, but in the middle fifties he, a great theoretician and experimentator, better than anybody else would have recognized that some experiments, those experiments which made difficult the formulation of the Universal theory, were wrong.

5) Hadrons are mixed, that is in the weak interaction there take part coherent mixtures of hadrons. Using quark notations, the hadron charged current is $\bar{u}(d\cos\theta + s\sin\theta)+\bar{c}(-d\sin\theta + +s\sin\theta)+...$ where θ is the Cabibbo angle $\sim 15°$, $\bar{u}$ is the creation operator of the u quark, d is the destruction operator of the d quark, etc. Thus the weak interaction Lagrangian is $L_w = \frac{G}{\sqrt{2}} J_w J_w^+$,

$$J_w = \bar{e}\nu_e + \bar{\mu}\nu_\mu + \bar{\tau}\nu_\tau + ... + \bar{u}(d\cos\theta + s\sin\theta) + \bar{c}(-d\sin\theta + s\cos\theta) + ...$$

$$J_w^+ = \bar{\nu}_e e + \bar{\nu}_\mu \mu + ... + (\bar{d}\cos\theta + \bar{s}\sin\theta)u + (-\bar{d}\sin\theta + \bar{s}\cos\theta)c + ...$$

and every member is a sum of V and A of the type $\bar{e}\gamma_\mu(1+\gamma_5)\nu_e$, etc. Once more this Lagrangian is a generalization of the Fermi one with essential, but very natural, additions, which take into account post-Fermi experimental data. It accounts wonderfully for all the data concerning <u>charged currents</u>, of which the β-decay à la Fermi is the first example.

Incidentally, it is quite likely that not only hadrons but also leptons are mixed (with important implications relating to neutrino oscillations). As I already promised, however, I am not going to talk about which will be discussed in detail at our Conference.

6) Now I should mention the most important discovery of <u>neutral currents</u>, made at CERN and confirmed at the Fermi Lab. I must say that neutral currents had been discussed as possible and even likely processes a long time before a real theory – the Glashow--Salam-Weinberg theory of electro-weak interactions – was proposed. As you know, largely as a stimulus of such a theory neutral currents

were discovered experimentally. I am not going into that now, first, because it is much more interesting for you to read the Nobel talks of Glashow, Salam and Weinberg and, second, because work on neutral currents will keep you very busy all this week. Incidentally, together with the discovery of neutral currents, I think that the most important work in neutrino physics is the marvellous investigation of neutral currents of the CDHS group. But I must say that phenomenologically neutral currents of the symmetrical type ($\bar{e}e$, $\bar{\nu}\nu$, $\bar{p}p$...) had been seriously discussed by many people. I have even discussed [19] in 1962 some astrophysical consequences of such currents. The consideration of symmetrical neutral currents was very natural. Because of the overwhelming competition of electromagnetism nobody could prove that they are not present. But how about the absence of asymmetrical neutral currents $\bar{\mu}e$, $\bar{s}u$...? I simply thought that they are ugly and the symmetrical ones beautiful. GIM had not yet been invented. I would like to conclude with two remarks:

a) The very first experiment[49] on high energy neutrino physics was designed to detect neutral currents, true, at a level 10^4 times smaller that the expected charged currents. This was all we could do with the low intensity accelerator available to us, our hope being on an anomalous ν_μ-N interaction[50].

b) In the course of many years neutral currents at the proper level have been looked for (and not found!) for example at CERN, of course before the Glashow, Salam, Weinberg strategy became popular.

THE ^{37}Cl-^{37}A METHOD

I would like now to give a subjective account of a few pages in the development of neutrino physics, in which in some ways I was involved. In 1946 neutrinos were generally considered undetectable particles. Many respectable physicists were of the opinion that the very question about detecting free neutrinos was nonsense (not only because of temporary difficulties), just as nonsense is the question as to whether the pression in a vessel is or is not, say, less than 10^{-50} atmospheres. I remembered well the Bethe-Peierls argument[45] and it occured to me at the time that the appearance of powerful nuclear reactors made free neutrino detecting a perfectly decent occupation. I was living in Canada then and was well acquainted with reactor physics. The NRX Canadian reactor, in the design of which I was taking part, was not working yet, but it was clear to me that under the very compact shield, where the cosmic ray soft component was considerably weakened, one might dispose of a neutrino flux $\sim 10^{12}$cm^{-2} sec^{-1}. At the time, scintillators, which were so successfully used many years later by Cowan and Reines to detect free reactor antineutrinos, has not yet been invented. Well, it occured to me that the problem could be solved by radiochemical methods, that is, by concentrating chemically the isotope resulting from the inverse beta process from a very large mass of matter irradiated by neutrinos. A careful inspection of the famous Seaborg

A FEW EPISODES

table of artificial radioisotopes indicated a few possible target candidates, by far the best of which was a chlorine compound, the reaction at issue being:

$$\text{neutrino} + {}^{37}\text{Cl} \rightarrow {}^{37}\text{A} + e^-, \tag{1}$$

where ^{37}A decays by K-capture with the emission of 2.8 keV X-rays. I wrote here "neutrino" and not $\bar{\nu}_e$, because at the time the question as to whether $\nu_e = \bar{\nu}_e$ was not clear, but to this point I shall come back later. Now there are lots of practical reasons why ^{37}Cl is so good and I shall not list them here. One of them, however, was not known to me "a priori" and was discovered by chance. In order to experiment on the future neutrino detector at Chalk River we were preparing in a conventional way ^{37}A, and putting it inside a detector which according to our intentions was supposed to be and in fact was, a Geiger-Müller counter. Well, once, looking at an oscilloscope connected to the counter, we saw plenty of pulses from ^{37}A about equal in amplitude at voltages on the counter much lower than the Geiger threshold, and discovered (independently of Curran et al. in Glashow) the high gas gain (up to 10^6) proportional regime. Now this was very important, of course, from the point of view of detecting neutrinos, since it permits to decrease the effective background of the counter. At the time there was a sort of dogma about proportional counters, i.e., that they cannot work at multiplication factors larger than ~ 100, which is true of course, if you have a large input ionization (alpha particles, etc.), but is absurd if you have an input ionization of a few ion pairs.

I discussed the ^{37}Cl-^{37}A method with Fermi in Chicago (1947?) and later at the Basel-Como conference in 1949 (including solar possibilities). Fermi was not at all enthusiast about neutrino applications of the method, but liked very much our proportional counters, with the help of which together with Hanna we first observed L-capture (in ^{37}A, 10 ion pairs)[51] and measured the ^{3}H spectrum going quite down at the time with the upper limit of the neutrino mass[5]. In reprospect I understand very well Fermi's reaction. As I think that Segré said, Don Quixote was not the hero of Fermi. He could not have sympathy for an experiment which, true, grace to the heroic efforts of R. Davis, terminate very brilliantly, but many many years after its conception[41,52].

Now I am coming back to the question as to whether reactor antineutrinos may induce the reaction (1). Well, passing through Zürich sometimes between 1947 and 1948 I had lunch with Preiswerk and Pauli. I told Pauli about my plans with the ^{37}Cl-^{37}A method; he liked very much the general idea and remarked that it was not clear whether "reactor neutrinos" should definitely be effective in producing the reaction (1), but he thought that they probably would. Since that time the question became very clear to me and until 1950 I continued to think about it and to test low background proportional counters in that connection and in connection with solar problems. For example I remember that Camerini, who at the time was working in Bristol and was a great specialist in cosmic ray stars, helped me to calculate the cosmic ray background in various Cl-A experiments which I was planning to do. Anyway the effective background of my counters was sufficiently low to detect solar neutrinos[11] through ^{37}A decay. Since 1950 I stopped experimenting on the problem, as there was no site deep underground enough in the USSR for a solar experiment (as you know, at the Elbrus neutrino observatory such a site will be available). However I kept thinking about counters and when I had the privilege to meet R. Davis at the first Neutrino Conference in Moscow (1968), I told him that measuring the form of the counter pulse, in addition to the amplitude, should result in a considerable decrease of the effective background in its solar experiment. As I found out later from him at the ν'72 Conference in Hungary it works really that way. But now I am going back about 15 years.

NEUTRINO OSCILLATIONS

You all know R. Davis[52] has shown in 1955 that reactor (anti)neutrinos cannot effectively produce ^{37}A from ^{37}Cl, that is $\nu_e \neq \bar{\nu}_e$. Now at the time I was told in a wrong way about such experiment. A delegation came to Moscow and someone (I do not remember who) told me that R. Davis got a positive signal in his experiment. Such result at the time seemed to me fantastic (and rightly so). Wrong rumors sometimes are useful. I tried to find a way out and invented[53] neutrino oscillations of the type $\nu_e \rightleftarrows \bar{\nu}_e$. This was all wrong, and not only because the fact which needed explanation was not there, but also, as I found out later, because of different spiralities of ν_e and $\bar{\nu}_e$. Neverveless, this wrong thinking was very useful to me ten years later, when the question about possible neutrino oscillations was investigated in a modern way from a theoretical point of view and with the aim to consider many possible experiments (reactor, accelerator, cosmic, solar neutrinos)[31-33]. Among other things, the number of neutrino types considered was ≥ 2. Now neutrino oscillations and the question about neutrino masses are very much "a la mode" from a theoretical as well as an experimental points of view, as one can see by an inspection of the programs of our Conference and of the "Rochester" Conference of this year. Thus you will discuss here oscillations and neutrino masses at length and at the proper time. As Iago, I am saying: "Demand me nothing: what you know, you know: from this time forth I never will speak a word". However I wish to acknowledge the great benefit of the collaboration with I. Kobsarev, L. Okun and especially with V. Gribov and S. Bilenky.

HIGH ENERGY NEUTRINO PHYSICS

My story here is again very personal. I am going to tell you how I came to propose experiments with high energy neutrinos from meson factories and from very high energy accelerators. At the Laboratory of Nuclear Problems of the JINR in 1958 a proton relativistic cyclotron was being designed with a beam energy 800 MeV and a beam current ∿ 500 μA. This accelerator eventually was not built

A FEW EPISODES

because of financial difficulties. In retrospect I think that not going ahead at the time with an accelerator having parameter similar to those of today meson factories was an error. Anyway at the beginning of 1959 I started to think about the experimental research program for such an accelerator. It occured to me that a healthy and relatively cheap neutrino program could be accomplished by dumping the proton beam in a large Fe block, fulfilling at the same time the function of neutrino source and shield. I would say that the ideology of the LAMPF accelerator neutrino experiments which have been initiated recently is very similar to that of various experiments planned 20 years before$^{(54,55)}$ for an accelerator which was not built. About one of them, which was intended to clear up the question as to whether $\nu_e \neq \bar{\nu}_\mu$ I would like to say a few words.

I have to come back a long way (1947-1950). Several groups, among which J. Steinberger, E. Hincks and I, and others were investigating the (cosmic) muon decay. The result of the investigations was that the decaying muon emits 3 particles: one electron (this we found by measuring the electron bremstrahlung) and two neutral particles, which were called by various people in different ways: two neutrinos, neutrino and neutretto, ν and ν', etc. I am saying this to make clear that for people working on muons in the old times, the question about different types of neutrinos has always been present. True, later on many theoreticians forgot all about it, and some of them "invented" again the two neutrinos (for example M. Markov), but for people like Bernardini, Steinberger, Hincks and me ... the two neutrino question was never forgotten. Not trivial, "a priori", was the question about how to perform the experiment, a thing which I was able to formulate clearly enough$^{(54)}$ (the use of neutrino beams).

In 1959 another problem was of great importance; is the four--fermion interaction a contact interaction or is it due to the exchange of an intermediated boson? This question is still valid today, but now we have the Glashow, Salam, Weinberg theory, which predicts masses of intermediated mesons at about 100 GeV, whereas in 1959 the intermediated boson (without serious reasons) was supposed to have a mass of a few GeV. Obviously the intermediated boson could not be produced at meson factories and at the 1959 Kiev international conference Ryndin and I proposed to look for the boson making use of neutrino beams from very high energy accelerators$^{(56)}$. The theoretical idea in the proposal was that in the cross section for the production by neutrinos of the intermediate boson at sufficiently high energies there will appear G instead of G^2. As you know, the question about intermediate bosons, which was very hot at high energy accelerators until about 1972, is not going to be solved anymore in neutrino experiments (as it seems). The question about two types of neutrinos has been solved at Brookhaven in a beautiful experiment$^{(48)}$.

AN ALTERNATIVE SCENARIO OF THE NEUTRINO PHYSICS DEVELOPMENT

Now I would like to present a scenario of the weak interaction physics development which did not take place but might actually have taken place. Since most of you are high energy neutrino physicists, such a scenario, will get on your nerves. By the way, I shall act as the devil's advocate, since I am very much for high energy neutrino physics myself.

I know a great scientist, P. Kapitza, who thinks now that if an experiment is very expensive and/or cumbersome, it should not be done: with time the problem at issue will be solved in a simpler way. Well, suppose that in the early sixties the community of physicists had decided that neutrino experiments at very high energy accelerators are too expensive and cumbersome. The community, as it were, was then of the opinion that neutrino physics in a relatively cheap way should be done at meson factories, which would be built anyway and at nuclear reactors. Which would have been the results? Let us follow the lines indicated by the actual great achievements of high energy neutrino physics.

1) $\nu_e \neq \nu_\mu$. The result would have been obtained at meson factories, true, at least 10 years later. Incidentally, electron-positron collisions gave us the ν_τ.

2) The nucleon structure. Here the situation without high energy neutrino experiments would have been catastrophic. However do not forget about the information (true, different) from deep inelastic scattering of electrons and muons, which is not bad at all.

3) Neutral currents. They have been discussed by many people from a phenomenological point of view[57] before the theory of electro-weak interactions, which clearly would have been created anyway, that is without high energy neutrino experiments. The parity non-conservation in atoms, predicted by Zeldovich[58] (for example the optical activity of Bi vapour) has been observed in Novosibirsk[59], in agreement with the theoretical expectation of Glashow-Salam-Weinberg. It is an experiment difficult and refined, but relatively cheap. The beautiful SLAC experiment[60] on the scattering of polarized electrons by nucleons gives a very accurate value of $\sin^2\theta_w$. At reactors and meson factories neutrino experiments on neutral currents have been performed and are being planned. Of course the very accurate work at CERN on neutral currents and the values of $\sin^2\theta_w$ from CDHS (and now CHARM) would be lacking, and this would be quite serious.

4) Without high energy neutrinos the production of strange and charmed particles by neutrinos could not have been investigated. However we got most of our knowledge on the subject in hadron beam investigations and in electron-positron collisions.

5) The mean life of charmed particles. This is an important and recent result of high energy neutrino investigations[61].

However, I am convinced that with the help of other methods analogous results may be obtained.

Thus I shall summarize: what actually happened in high energy neutrino physics, is very expensive, but is much more informative than the consequence of my hypothetical scenario. But one should neither underestimate the importance of high energy neutrino physics nor overestimate it. This is not a note of pessimism but an appeal to avoid routine. Now high energy neutrino physics is a very healthy field of physics, in which quantitative measurements dominate. However in neutrino physics, one feels, there is still plenty of room for new qualitative results. Some new ideas concerning neutrino beams (tagged?), detectors, and problem formulations are needed. I think that neutrino oscillations and beam dump experiments are promising, but they are no more real news. Concerning beam dump neutrino experiments I would like to draw your attention to an amusing thing.

BEAM DUMP EXPERIMENTS

In the beam dump experiments which have been performed at Serpukhov and at CERN (by four groups), the production of charmed particles in proton nuclear collisions has been observed and investigated for the first time, the method consisting in the detection of prompt neutrinos from the lepton decay of charmed particles. Now this is a miracle, which you may appreciate by making analogies of the following type: when the beta decay of nuclei was first observed, you may imagine that Rutherford, instead of detecting electrons, observed neutrinos from a 10^9 curie source! or you may imagine that Lattes, Occhialini and Powell first observed the pion decay by designing and building a modern neutrino high energy facility (with a proton accelerator, the decay tunnel, the Fe shield and a multi--ton neutrino detector), instead of observing simply, as they did, the pion-decay muons.

CONCLUSIONS

What happened in neutrino physics the last years is a miracle. Everything, that is the Glashow-Salam-Weinberg theory of electro--weak interactions, looks perfectly O.K.. It is too good. The appetite comes with eating and this means Grand Unification. But I do not believe that elementary particle physics will soon die of abundance of understanding and/or lack of problems to be solved. Let us not discuss now about unexpected things, since anyway about such things one does not talk seriously in a lecture entitled "Fifty Years of Neutrino Physics". But there are already more or less important things. One of them, finite neutrino mass (together with the instability of the proton) is in the head and in the mouth of everybody. Its implications - neutrino oscillations - are extremely informative (masses of neutrinos, number of them, and mixing angles), if something can be done, as it seems, in controllable experiments of various types (reactor, accelerator, cosmic, solar).

It is not excluded$^{(+)}$ that the ν_e mass may be measured directly from the ^{3}H beta spectrum, although I am not sure that this can be done, just because of the fantastic, I would say acrobatic, difficulty of the experiment, which incidentally, is a relatively cheap one[5].

Be as it may, finite neutrino masses not only would confirm modern theoretical thinking and give us very necessary parameters but would originate a revolution in cosmology, astrophysics and neutrino astronomy.

It is curious today most popular types of search experiments – the proton radioactivity and neutrino oscillations – are not, in the main, high energy experiments.

I finished my talk. It is time to start working.

I am very grateful to Prof. E. Bellotti for discussions and help in the preparation of this paper.

REFERENCES

(1) Pauli: letter (4 December 1930) to a meeting of physicists, among which Geiger and Meitner, taking place in Tübingen. The letter was conserved by Meitner and its content was discussed since 1930. The letter was widely published only after many years. (See, for example: Brown, Phys. Today, September, 1978).
(2) Pauli, "Septieme Conseil de Physique Solvay 1933", Gauthier--Villars, Paris (1934).
(3) For accurate measurements in the reaction $^{37}A+e^- \to ^{37}Cl+\nu$ (K-capture) see: Rodeback, Allen, Phys. Rev., 86:466 (1952). A striking picture of apparent non-conservation of momentum in the beta decay of He into Li and an electron by Csikay and Szaley appeared in Proc. of the Intern. Conf., Paris, Publ. Dunad (1959).
(4) See for reviews of the subject: Primakov and Rosen, Report on Progress in Physics, 22:121 (1959); Fiorini, Proc. of the ν'77 Conf., Ed.Nauka, Moscow (1978). Neutrino-less double beta decay has not been observed and upper limits $10^{-3} - 10^{-4}$ for the lepton non-conserving relative amplitudes have been set for the double decays $^{48}Ca \to ^{48}Ti$ (Bardin et al.), $^{76}Ge \to ^{76}Se$ (Fiorini et al.), $^{82}Se \to ^{82}Kr$ (Cleveland et al.), $^{130}Te \to ^{130}Xe$ (Zdelenko et al.).

$^{(+)}$After this paper was written, a preprint was published in which a finite mass (between 14 and 46 eV within 90% confidence limits) was claimed for the neutrino emitted in the beta decay of ^{3}H: Lubinov, Nozik, Tretjakov, Kozik, ITEP 62, 1980. The experiment is beautiful. However its difficulty and the importance of the result make imperative the continuation by the Moscow group and by other groups of similar investigations.

(5) The first measurements of the ^{3}H beta spectrum were performed by means of proportional counters in 1949: Hanna, Pontecorvo, Phys. Rev., 75:983 (1949); Curran et al., Phil. Mag., 40:53 (1949). Then substantial improvements were obtained by using as a ^{3}H source a few molecule thick layer and a magnetic spectrometer, by Bergkvist, CERN Report 69-7:91 (1969); Nucl. Phys. B39:317 (1972). The last and best result is $m_{\nu_e} \leq 35$ eV (90% C.l.) (Tret'jakov et al., Proc. of the Int. Conf. on High Energy Phys., Tbilisi, Vol.II:118 (1976)). Less accurate measurements have been performed by implanting ^{3}H in a Si(Li) detector (Simpson, Proc. ν'79 Conf., vol.2:208).

(6) Pontecorvo, Phys. Rev., 72:246 (1947); Puppi, Nuovo Cimento 5:505, 1948; Klein, Nature, 161:897 (1948); Lee et al., Phys. Rev., 75:905 (1949); Tiomno, Wheeler, Rev. Mod. Phys., 21:144 (1949).

(7) These experiments gave upper limits as bad as $R = \frac{\mu \to e\gamma}{\mu \to \text{all}} \leq 10\%$. Values of R were greatly improved in a number of experiments, the bsst of which, performed with a gamma hodoscope and a magnetic spectrometer, gives a limit as good as $R \leq 2 \cdot 10^{-10}$ at the 90% c.l. (Anderson, Hofstadter et al., Proc. of the Tokyo International Conference, (1978)). For other similar processes the following limits are obtained

$R' = \frac{\mu \to 3e}{\mu \to \text{all}} \leq 2 \cdot 10^{-9}$ (Korenchenko et al., Proc. of the ν'77 Conference) and $R'' = \frac{\mu^- + ^{32}S \to e^- + ^{32}S}{\mu^- + ^{32}S \to \nu_\mu + \ldots} \leq 1.5 \cdot 10^{-10}$ (Badertscher et al., Phys. Rev. Lett. 39:1383 (1977).

(8) Wick, Wightman and Wigner, Phys. Rev. 88:101 (1952).
(9) Bethe, Phys. Rev. 55:434 (1939).
(10) Gamow and Schonberg, Phys. Rev. 59:539 (1941).
(11) Pontecorvo, National Res. Council Canada, Rep. PD 205 (1946); Helv. Phys. Acta, Suppl. 3:97 (1950).
(12) Gamow, Phys. Rev. 70:505 (1948).
(13) Fowler, Appl. J. 127:551 (1958).
(14) Pontecorvo, JETP 36:1615 (1959).
(15) Markov, Proc. of Int. Conf. on High Energy Phys., Rochester p.578 (1960).
(16) Greisen, Proc. Int. Conf. on Instrument High Energy Phys., Interscience Publ. p.209 (1960).
(17) Pontecorvo and Smorodinsky, JETP 41:239 (1961).
(18) Zeldovich and Smorodinsky, JETP 41:907 (1961).
(19) Pontecorvo, Phys. Lett. 1:287 (1963).
(20) Reines et al., Proc. of the ν'72 Conf. 2:199.
(21) Krishnasvami et al., Proc. Roy. Soc. A323:489 (1971); Osborne et al., Proc. of the ν'72 Conf. 2:223
(22) Fowler and Hoyle, The University of Chicago Press, Chicago-London (1965).
(23) Colgate and White, Astrophys. J. 143:626 (1966).
(24) Zeldovich, Letters JETP 1:40 (1965).

(25) Domagatsky and Zatsepin, Proc. 9th Int. Conf. on Cosmic Rays 2:1030 (1965).
(26) Penzias and Wilson, Appl. J. 142:419 (1965).
(27) Dicke et al., Appl. J. 142:414 (1965).
(28) Zeldovich and Novikov, Lett. JETP 6:772 (1967).
(29) Weinberg, Gravitation and cosmology, John Wiley Inc., New York (1972).
(30) Gershtein and Zeldovich, Lett. to JETP 4:174 (1966).
For cosmological bounds on the mass of neutral leptons see:
Lee and Weinberg, Phys. Rev. Lett. 39:165 (1977);
Dolgav, Visotsky and Zeldovich, Lett. JETP 26:200 (1977);
For cosmological bounds on the number of massless neutrino see:
Schvartsman, Lett. JETP 9:315 (1969); Steigman, Schram adn Gunn, Phys. Lett. 66B:202 (1977). See also for other limits on neutrino properties from astrophysics considerations: Bernstein, Riderman and Feinberg, Phys. Rev., 132 1227 (1963); Berg, Marciano and Ruderman, Phys. Rev. D17:1395 (1978).
(31) Pontecorvo, JETP 53:1717 (1967); 13:281 (1971).
(32) Gribov and Pontecorvo, Phys. Lett. 28B:493 (1969).
(33) Bilenky and Pontecorvo, Phys. Lett. 61B:248 (1976);
Lett. Nuovo Cimento 17:569 (1976); Comments on nuclear and particle phys. 7:149 (1977); Phys. Rep. 41:226 (1978).
(34) Bahcall, Proc. of the ν'72 Conference 1:29.
(35) Berezinsky and Zatsepin, Sov. Phys. Usp. 20:361 (1977).
(36) Gershtein et al. Proc. of the ν'77 Conf. 1:106.
See also Sutherland et al. Phys. Rev. D13:2700 (1976);
Pethick, Proc. of the ν'79 Conf. 1:78 Bergen, Norway.
(37) Chudakov et al. Proc. of the ν'77 Conf. 1:155.
(38) Dolgoshein et al. Proc. of the ν'77 Conf. 2:341.
(39) Sulak et al. Proc. of the ν'77 Conf. 2:350.
(40) Domogatsky et al. Proc. of the ν'77 Conf. 1:115.
(41) Davis et al. Proc. of the ν'78 Conf. p.53. The measured rate can be compared with the theoretical rate, which is about twice larger. At this time the computed expected rate is given by Bahcall in: 40 years of stellar energy, Bethe Symposium, Stony Brook, N.Y. (1979).
(42) Lande et al. Proc. of the ν'78 Conf. p.887.
(43) Zatsepin, Proc. of the ν'78 Conf. p.881.
(44) See for example: Learned, Proc. of the ν'78 Conf. p.895.
(45) Bethe and Peierls, Nature 133:532 (1934).
(46) Fermi, Nuovo Cimento 1:11 (1934).
(47) Reines and Cowan, Phys. Rev. 98:492 (1953); 113:273 (1959).
(48) Danby et al. Proc. High Energy Conf. at CERN p.809 (1962).
(49) Vasilevsky et al. Phys. Lett. 1:345 (1962).
(50) Kobzarev and Okun, JETP 41:1205 (1961).
(51) Kirkwood, Hanna and Pontecorvo, Phys. Rev. 75:982 (1949).
(52) Davis, Phys. Rev. 97:766 (1955).
(53) Pontecorvo, JETP 33:549 (1957).
(54) Pontecorvo, JETP 37:1751 (1959).
(55) Nguyen Van Hieu and Pontecorvo, JETP Lett. 7:137 (1968).

(56) Pontecorvo and Ryndin, Proc. Kiev Int. High Energy Phys. Conf. p.233 (1959).
(57) Bludman, Nuovo Cimento 9:433 (1958).
(58) Zeldovich, JETP 36:964 (1959); Bouchiat and Bouchiat, Phys. Lett. 48B:111 (1974).
(59) Barkov and Zolotarev, Proc. of the ν'78 Conf. p.423.
(60) Prescott et al. Phys. Lett. 77B:347 (1978).
(61) See for example: Conversi, Int. Conf. of the Europ. Phys. Soc., Geneva (1979).

LOW-ENERGY NEUTRINOS IN ASTROPHYSICS

W. Hampel

Max-Planck-Institut für Kernphysik
Heidelberg, Fed. Rep. Germany

ABSTRACT

In this review we shall concentrate on the two main aspects of low-energy (∼MeV) neutrino astrophysics. The larger part summarizes the current status of solar neutrino research, whereas in the last section the prospects for the detection of collapsing star neutrinos are discussed.

INTRODUCTION

We know that neutrinos play an important role in astrophysics. Because of their extremely low interaction with matter, they are able to escape out of the interior of stars and to reach us through large amounts of otherwise non-transparent matter without any alteration. Their detection therefore provides us with information not accessible to conventional experimental techniques. At present, we are confident in being able to observe the low energy neutrinos emitted during two stages in the life of a star: during hydrogen burning on the main sequence (here, of course, the neutrino radiation at the earth is dominated by the main sequence star nearest to us, the sun) and during the gravitational collapse of massive stars after they have exhausted their nuclear fuel.

In this review the present status of low energy (∼MeV) neutrino astrophysics will be discussed. We shall first report the latest results of the Brookhaven Solar Neutrino Experiment, discuss new standard solar model calculations, compare then the measured value to the theory and present possible explanations for the observed discrepancy. We shall then focus on the recent progress in the preparation of proposed new solar neutrino detectors. Finally, the

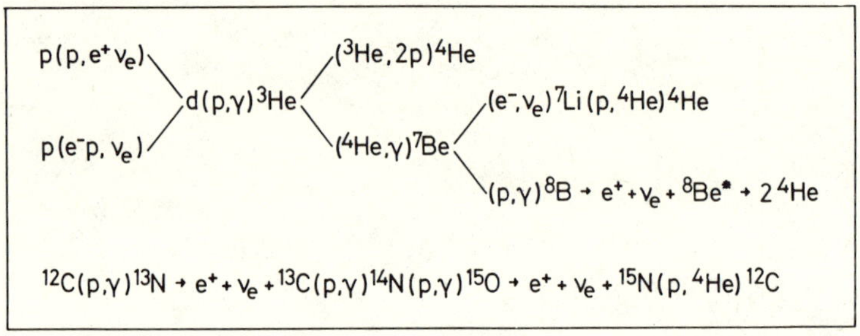

```
p(p,e⁺νₑ)                    (³He,2p)⁴He
          ⟩d(p,γ)³He⟨                   (e⁻,νₑ)⁷Li(p,⁴He)⁴He
p(e⁻p,νₑ)                    (⁴He,γ)⁷Be⟨
                                        (p,γ)⁸B → e⁺+νₑ+⁸Be* → 2 ⁴He

¹²C(p,γ)¹³N → e⁺+νₑ+¹³C(p,γ)¹⁴N(p,γ)¹⁵O → e⁺+νₑ+¹⁵N(p,⁴He)¹²C
```

Fig. 1. Reactions of the pp and CN chains.

prospects of observing the neutrino radiation from collapsing stars with these and other detectors will be discussed.

RECENT RESULTS OF THE ^{37}Cl EXPERIMENT

The sun and all main sequence stars generate their energy by a series of nuclear reactions which result in the conversion of 4 protons into a ^{4}He nucleus. In the sun this conversion proceeds mainly via the proton-proton chain, the CN cycle contributing only a few percent (Fig.1). It is well known that the observation of the neutrinos generated in these reactions provides a unique possibility to obtain direct information about the interior of the sun.

There has been one and so far only one experimental attempt to detect these solar neutrinos, the famous chlorine experiment performed now for about 15 years by Ray Davis and his colleagues from the Brookhaven National Laboratory[1]. It is based upon the neutrino capture reaction $^{37}Cl(\nu_e,e^-)^{37}Ar$. A tank with 610 tons of C_2Cl_4 located at 4400 m.w.e. depth in the Homestake Gold Mine in Lead, South Dakota, USA, is exposed to solar neutrinos. At the end of an exposure (30-100 days) the radioactive $^{37}Ar(T_{1/2}$ 35d) is removed from the tank by helium purge and its decay is finally measured in a gas proportional counter. In Fig. 2 the results of 35 single runs, from run No 18 to run No 56, obtained during the last 9 years, are plotted as a function of time. There is some scatter in the data which has led to the suggestion that the capture rate may be varying in time, but the data are fully consistent with a constant production rate, the scatter being due to statistical fluctuations only. The average ^{37}Ar production rate is 0.50+0.06 atoms per day. If the cosmic ray muon contribution, 0.08+0.03 atoms per day, is subtracted and the result is then converted into the (target-size-independent) production rate unit SNU (1 SNU=10^{-36} neutrino captures per second and target atom), which is done on the right hand side of the plot, then the remaining rate amounts to 2.2+0.4 SNU. This

LOW-ENERGY NEUTRINOS IN ASTROPHYSICS

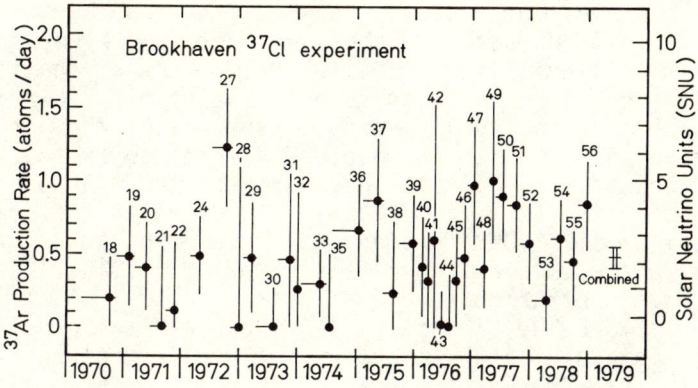

Fig. 2. ^{37}Ar production rate for 35 individual runs

is the so-called "rate above known backgrounds" which may (or may not) be attributed to solar neutrinos. The yearly averages of the ^{37}Ar production rate from 1971 to 1978 are given in Fig. 3. Again, there is no indication of a production rate varying in time.

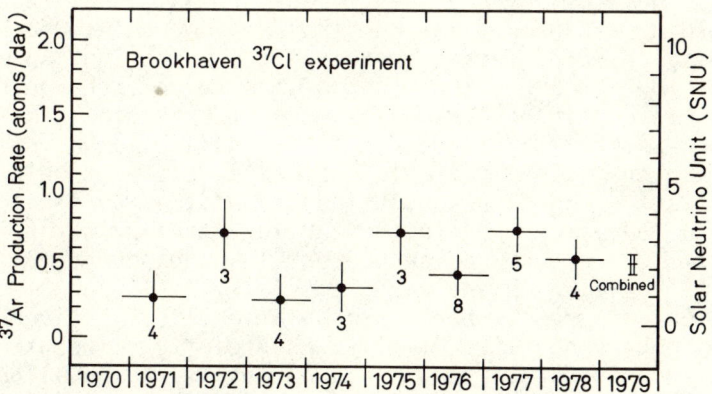

Fig. 3. Yearly averages of the ^{37}Ar production rate. The number given at each data point denotes the number of runs during that year.

There has been some discussion about the muon background which is subtracted. The number used above was obtained by measuring the cosmic ray muon signal with a smaller C_2Cl_4 tank at shallower depths and then extrapolating the results by means of the known intensity-depth relation and the known energy dependence of the photonuclear cross section for muons to the depth where the solar neutrino detector is located[2]. Fireman[3] has measured the ^{37}Ar production rate from potassium acetate by cosmic ray muons in the Homestake Mine at various depths underground. His results scaled for the Cl experiment give a cosmic ray muon background of 0.18+0.09 atoms ^{37}Ar per day, which subtracted from the observed 0.50+0.06 atoms per day would give 1.7+0.6 SNU. There remains, however, the question of whether the potassium acetate experiment really measures the same effect as the Cl experiment, because the reaction mechanism for the production of ^{37}Ar from potassium is different from that of ^{37}Ar from chlorine.

STANDARD SOLAR MODEL PREDICTION FOR THE ^{37}Cl EXPERIMENT

During the last few years Bahcall et al.[4] have worked on a new standard solar model calculation, and last year at the Neutrino '79 Conference in Bergen[5] a preliminary number of 7.5 SNU obtained by Bahcall and co-workers was reported. The value has now been fixed to 7.8+1.5 SNU. This result is significantly higher than the last published value[6,7]. The increase is due to recent revisions in some of the solar model input parameters. These latter changes will be discussed in the following.

The cross section for the basic pp-reaction is the only one of all relevant cross sections which cannot be measured in the laboratory, it has to be calculated. The Gamow-Teller coupling constant needed in these calculations is extracted from measurements of the neutron half-life. There has been a recent redetermination of this half-life[8] yielding a higher value and thus a lower GT coupling constant. This, in turn, decreases the pp-cross section. An additional decrease is due to a recalculation of the meson exchange corrections to this cross section[9]. A reanalysis[4] of the original $^3He+^4He$ cross section measurements[10,11] yielded for this reaction a lower cross section than that used in previous solar models (see below). The solar constant[12] and thus the luminosity of the sun is about 1.8% higher than the value used in the earlier standard solar model calculations. There are new solar elemental abundances compiled by Ross and Aller[13], which increase especially the C and O abundances by about 20% compared to earlier tabulations[14]. This increase results in a higher opacity. Finally, the opacity values themselves have been increased as a result of new opacity calculations at Los Alamos and Livermore[4].

Table 1 lists the resulting neutrino fluxes at the earth for the different neutrino sources[4]. Combined with Bahcall's cross section values[7] one obtains the capture rates in the last column of Table 1 giving a total capture rate of 7.8 SNU for the Cl experiment.

Table 1. Solar neutrino fluxes[4], cross sections[7] and capture rates for $^{37}Cl(\nu_e,e^-)^{37}Ar$.

Neutrino Sources and Energies [MeV]	Flux on Earth ϕ [cm^{-2}sec^{-1}]	Cross Section σ [cm^2]	Capture Rate $\phi\sigma$ [SNU]
p+p (0-0.42)	6.1 ×10^{10}	0	0
p+e$^-$+p (1.44)	1.5 ×10^8	1.54×10^{-45}	0.23
^{7}Be decay (0.86)	4.1 ×10^9	2.4 ×10^{-46}	0.98
^{8}B decay (0-14.06)	5.85×10^6	1.08×10^{-42}	6.32
^{15}O decay (0-1.74)	3.7 ×10^8	6.6 ×10^{-46}	0.24
^{13}N decay (0-1.19)	4.6 ×10^8	1.6 ×10^{-46}	0.07
		$\Sigma\phi\sigma$	7.8

COMPARISON OF ^{37}Cl EXPERIMENT AND THEORY

The measured rate of 2.2+0.4 SNU for the chlorine experiment has now to be compared with the result of the standard solar model calculation, 7.8+1.5 SNU. There is a clear discrepancy between experiment and theory, by more than a factor of three.

In Table 1 the important feature of the Cl-experiment can be seen: It is not sensitive to the bulk of the solar neutrinos, the almost model-independent pp neutrino flux, because the energy of these neutrinos is below the 814 keV threshold of the experiment. On the other hand, more than 90% of the signal comes from the ^{7}Be and ^{8}B branches which are known to be strongly dependent on the central temperature of the sun.

This fact has led to the invention of numerous so-called non-standard solar models in the last ten years or so, most of which are constructed to yield reduced internal temperatures and thus have lower ^{7}Be and ^{8}B neutrino fluxes. Details of these non-standard models will not be repeated here, see for instance the review by Bahcall[15]. We will only mention a new non-standard solar model calculation by Schatzman and Maeder[16] reported at this conference. They introduced into the model a turbulent diffusion, due to a mild turbulence which is caused by shear flow. This turbulence alters, in addition to the changes due to nuclear reactions, the ^{1}H, ^{3}He and ^{4}He abundances in the core. In particular, it increases the ^{1}H and ^{3}He abundances in the nuclear burning zone relative to the standard model leading to a lower central temperature. Depending on the diffusion coefficient used in the calculations, predicted capture rates down to less than 1 SNU for the Cl experiment can be obtained.

Two other possible explanations of the discrepancy that have been on the list for years but have not been taken too seriously have regained interest during the last one or two years. One has to do with the nuclear physics that goes into the models and the other one deals with the properties of the neutrinos.

^{3}He + ^{4}He reaction

It was already reported at the conference last year[5] that there are new measurements of the ^{3}He+^{4}He cross section at very low center-of-mass energies by Claus Rolfs' group in Münster[17]. In Fig.4 the low energy part of the cross section factor S for this reaction is plotted as a function of the center-of-mass energy. S contains the intrinsic nuclear properties of the reaction, it results from the measured cross section after elimination of the strongly energy-dependent terms (the 1/E factor and Coulomb barrier factor).

For nonresonant reactions, the S-factor is only a slowly varying function of energy. The measurements of Parker and Kavanagh[10] are indicated by open circles, solid circles are from Nagatani and co-workers[11], squares represent the new data by Rolfs' group. For the solar model calculations the measurements have to be extrapo-

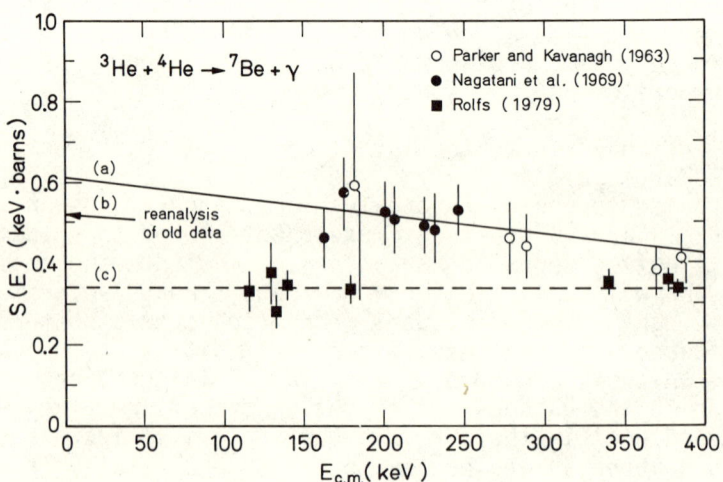

Fig. 4. Cross section factor S for the ^{3}He + ^{4}He reaction at low center-of-mass energies.

lated to almost zero energy because the relevant energies in the sun are in the neighbourhood of 10 kilovolts. The extrapolation is done by a polynomial fit through the data points (it should be noted that there are many more data points from Parker and Kavanagh and from Rolfs' group above 400 keV). Using the old data, the analysis by Nagatani et al.[11] yielded 0.61 keV·barns, which is the S-value adopted in all earlier standard solar model calculations (Fig.4, curve (a)). In the recent reanalysis of the same data, as discussed above, a value of 0.52 keV·barns was obtained[4] (Fig.4, (b)). Taking Rolfs' data, the result is 0.34 keV·barns[17] (Fig.4, (c)). Rolfs' measurements, however, are not absolute measurements but are normalized for this comparison to an absolute measurement by Parker and Kavanagh at 1.5 MeV. Rolfs[17] intends to perform a new absolute calibration, but this is not accomplished yet. There remains therefore the possibility that all his data points will move up or down.

In the past, one argument against Rolfs' S-value have been the theoretical calculations by Tombrello and Parker[18], which show an increase of S with decreasing energy. According to Rolfs[17] there are, however, new calculations that do not exhibit such a behaviour.

Using Rolfs' value of 0.34 keV·barns as the S-value for ^{3}He+^{4}He one arrives at a capture rate prediction for the Cl-experiment of 5.5 SNU instead of 7.8 SNU[4], not quite enough to agree with the Brookhaven result, but it is too early to draw any final conclusion. We have to wait for Rolfs' absolute calibration and for the results of two new measurements of the ^{3}He+^{4}He cross section being initiated at Caltech and in Brookhaven because of the confusion about this reaction after Rolfs' new measurements.

Neutrino oscillations

Neutrino properties, for instance neutrino oscillations[19], as possible explanations of the solar neutrino discrepancy have also been in discussion for years, but the discrepancy itself, if at all, has been the only experimental evidence for this possibility. This situation has changed quite drastically during the last few months. There is now at least one experiment with reactor antineutrinos which may have detected neutrino oscillations[20]. Let us therefore consider the consequences of neutrino oscillations for solar neutrino experiments.

Since the Cl-experiment (and also most of the proposed new experiments) are based upon the inverse β-decay, only electron type neutrinos can contribute to the useful interactions. Therefore, in the presence of neutrino oscillations, the probability of finding an electron neutrino after a certain path length in a beam that originally consisted only of electron-neutrinos is the important quantity. This probability, for the case of only two different neutrino types, is given by

$$P(\nu_e \to \nu_e) = 1 - \frac{\sin^2 2\Theta_M}{2} \left(1 - \cos 2\pi \frac{d}{L}\right).$$

$P(\nu_e \to \nu_e)$ depends on the mixing angle Θ_M, on the distance d between neutrino source and detector, here the sun-earth distance, and finally on the oscillation length L, which is given by

$$L[m] = \frac{2.5 E_\nu \text{ [MeV]}}{\Delta m^2 \text{ [eV}^2\text{]}}.$$

Here E_ν is the neutrino energy and Δm^2 the squared mass difference between the two neutrino states. Let us now consider the ratio R, the observed capture rate in a solar neutrino experiment in the presence of oscillations divided by the rate without oscillations. R is calculated as follows

$$R = \frac{\Sigma P(E)\phi(E)\sigma(E) + \int P(E)\phi(E)\sigma(E)dE}{\Sigma \phi(E)\sigma(E) + \int \phi(E)\sigma(E)dE},$$

where the sum is taken over the neutrino line sources (^{7}Be, pep) and the integral involves the continuum sources (pp, ^{8}B, ^{13}N, ^{15}O). For Δm^2 values less than 10^{-11} eV^2, R is near unity, the capture rate is thus only very little affected by oscillations. If Δm^2 is of the order of 10^{-10} to 10^{-11} eV^2, then L is in the range of the sun-earth distance and R will come out somewhere between 0 and 1. However, R will never be exact zero because of the broad energy spectrum of solar neutrinos (see Bahcall and Frautschi[21]). For a Δm^2 value near 10^{-8} to 10^{-9} eV^2, L is comparable to the sun-earth distance variation caused by the eccentricity of the earth's orbit. In principle, R now varies in the course of the year. However, this variation would be small for the Cl experiment because the fractional range in distance $\Delta R/R$ is smaller than the fractional range $\Delta E/E$ in the energy of the ^{8}B neutrinos[21]. Thus, for the main contribution to the ^{37}Cl capture rate such an effect is averaged out, it would be visible only for the small signal due to the monoenergetic ^{7}Be neutrinos. Nevertheless, there has been an attempt by Ehrlich[22] to extract an oscillation in this range out of the Brookhaven data, but the results of this analysis are very marginal. The data are probably also consistent with a non-varying ^{37}Ar production during the course of a year. Finally, if Δm^2 is larger than 10^{-8} eV^2, the oscillation length L becomes small compared to the size of the neutrino producing core in the sun. R will then even out to 0.5 for maximal mixing (or to a value between 0.5 and 1 if the mixing angle Θ_M is less than $\pi/4$) because now one averages (compared to L) over quite different source-detector distances.

If we now compare the Δm^2 values in the above discussion to the value indicated by the reactor antineutrino experiment[20] which is in the eV^2 range, it seems that for solar neutrino experiments the last case would apply. This implies that in a three neutrino world the lowest possible value of R would be 1/3 which is what is observed within the errors if we interpret the discrepancy between theory and experiment in the Cl experiment in terms of neutrino oscillations. The answer, however, whether this interpretation is correct cannot come from the chlorine solar neutrino experiment alone. One must consider simultaneously all available data including accelerator and reactor data. There are two recent papers[23,24] in which such an analysis in terms of neutrino oscillations in a three neutrino system is performed. The conclusion from both papers is that there exists at least one set of parameters (3 mixing angles, 2 squared mass differences) which is compatible with all experimental data, but is is clear that more experiments are needed to fix all the parameters to the desired degree. New solar neutrino experiments would be of great interest in this respect, and this is the topic of the next section.

PROPOSED NEW SOLAR NEUTRINO EXPERIMENTS

In Table 2 there is a list of these new possible experiments [25-33]. All of these except one are based upon the inverse β-decay. The exception is the last reaction listed in this table; it uses the elastic scattering of neutrinos from electrons. Because of background problems, one needs a high energy threshold, 7 MeV or so, in order to cut off all the Compton scattering events at lower energies, thus one would not be able to observe solar neutrinos other than from the ^{8}B branch. There have been two proposals[32,33] for experiments to detect the ^{8}B neutrinos via this reaction and these would be similar to an experiment at LAMPF where the cross section for the same reaction will be measured[32]. Experiments like these, because of their directional sensitivity, could answer the question whether the small signal observed in the Cl experiment is really due to solar ^{8}B neutrinos. A similarly high threshold is necessary in another direct counting experiment, where the relativistic electron emitted after inverse β-decay on deuterium is counted[31].

Table 2. Proposed new solar neutrino detectors.

Neutrino Reaction	Half-life of Product	Threshold Energy [MeV]	References
^{205}Tl$(\nu_e,e^-$IT$)^{205}$Pb	1.6×10^7 y	0.062	(25)
^{71}Ga$(\nu_e,e^-)^{71}$Ge	11.4 d	0.236	(26)
^{81}Br$(\nu_e,e^-$IT$)^{81}$Kr	2.1×10^5 y	0.512	(27,28)
^{7}Li$(\nu_e,e^-)^7$Be	53 d	0.862	(29)
^{115}In$(\nu_e,e^-2\gamma)^{115}$Sn	D.C.	0.120	(30)
d$(\nu_e,e^-$p$)$p	D.C.	~ 5	(31)
$e^-(\nu_e,\nu_e)e^-$	D.C.	~ 7	(32,33)

All other experiments in Table 2 are sensitive to a larger part of the solar neutrino spectrum as can be seen from the threshold energies listed. In Table 3 it is shown how the different solar neutrino sources contribute to the total capture rate for the different experiments. The rates were calculated from Bahcall's cross sections[7] and the neutrino fluxes obtained with the new standard solar model[4]. It is obvious that for reaction thresholds below 420 keV the bulk part of the total capture rate comes from the pp neutrinos, this is the case for the Ga, In and Tl experiments. For the Br experiment 11.5 SNU of the total rate of 16.5 SNU are due to ^{7}Be neutrinos. Finally, the Li experiment has its largest contributions from both pep and ^{8}B neutrinos.

The important feature of the three experiments sensitive to the pp flux is that the expected capture rate is almost independent of the solar model even with most of the non-standard models. This follows simply from the fact that one can calculate the pp neutrino flux without use of a detailed model just from the assumption that the energy generation in the sun is in equilibrium with its present luminosity. The outcome of a pp neutrino experiment would thus answer the question whether the discrepancy in the Cl experiment is due to problems with solar models (or nuclear cross sections used in these models) or is due to the properties of neutrinos. In the following we will focus only on those three experiments on which, as far as we know, work was in progress during the last two years. These are the Ga and the In experiments, both pp neutrino detectors, and the Br experiment.

Table 3. Expected capture rates (in SNU) for the standard solar model[4,7].

Target	Threshold [keV]	Neutrino Source						Total
		pp	pep	^{7}Be	^{8}B	^{13}N	^{15}O	
^{7}Li	862	0	9.0	3.9	18.1	1.9	8.4	41.3
^{37}Cl	814	0	0.23	0.98	6.32	0.07	0.24	7.8
^{71}Ga	236	65	2.4	26.3	∼ 2	2.5	3.5	102
^{81}Br+)	512	0	1.3	11.5	∼ 1	1.1	1.6	16.5
^{115}In	120	534	9.6	119	∼ 5	11.5	14.8	694
^{205}Tl	62	439	5.1	74	∼ 2	7.4	8.4	536

+) ^{81}Br capture rates calculated with log ft = 4.88 (Bennett et al.[34])

Bromine Experiment

Let us start with the Br experiment in which a natural Br deposit has to be used as a target because of the long halflife of the product nucleus ^{81}Kr with 210,000 years. The ^{81}Br-^{81}Kr scheme is shown in Fig. 5. Two years ago at the Neutrino '78 Conference[1], there were two important questions open: (a) what is the cross section for neutrino capture in ^{81}Br leading to the ^{81}Kr isomer and (b) how serious would background reactions be which also produce ^{81}Kr? Both questions have been answered in the meantime. Bennett and co-workers from Princeton[34] have detected the very weak electron capture decay branch of the ^{81}Kr isomer and have determined the log ft value to be 4.58. This means that the cross section for neutrino capture in ^{81}Br is on the more favourable side of the previously estimated range. On the other hand, after some of the important quantities for the background reactions were measured in Brookhaven and in Heidelberg[35], it now seems clear that a Br solar neutrino experiment is not feasible with presently known Br deposits. The deposit which was chosen as a possible target for the experiment, the Ronnenberg sylvinite at about 900 m depth underground in the Salzdethfurt potassium salt mine (Germany), contains 0.2 to 0.3% Br. What prevents us from using this deposit as a solar neutrino detector is the uranium content, which in all measured samples is higher than 3 ppb. A uranium content of less than 1.5 ppb would be necessary to reduce the contribution of α-particles to the ^{81}Kr production rate to less than 10% of the solar neutrino signal. Alpha-particles can produce ^{81}Kr by the ^{35}Cl(α,p) reaction followed by ^{81}Br(p,n)

The requirement of less than 1.5 ppb uranium is, however, not unrealistic. There exist potassium salt deposits with known U content below 1 ppb, but they are not deep enough to serve as a solar neutrino detector. A second difficulty arises from the high content of krypton dissolved in the potassium salt ($\sim 10^{-10}$ scc/g). This implies a ^{81}Kr/^{82}Kr isotope ratio as low as 10^{-12} in the mass spectrometric determination of ^{81}Kr, which,

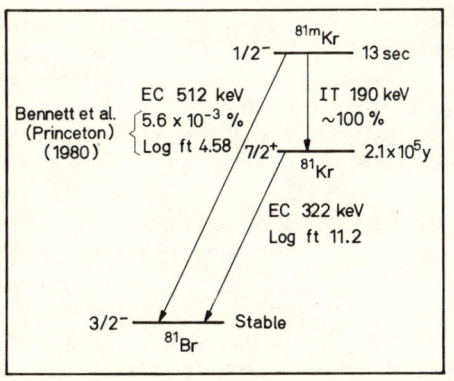

Fig. 5. The ^{81}Br - ^{81}Kr scheme.

with the present techniques, is not measurable.

Gallium Experiment

Next we will report on the progress made in the Ga experiment since the Neutrino '79 Conference in Bergen[36]. There is an international collaboration[37] with colleagues from Brookhaven, Philadelphia, Princeton, Rehovot and Heidelberg working on a pilot experiment with the aim of demonstrating that all steps in this experiment, from the extraction of the Ge to the counting of ^{71}Ge, are feasible at the level necessary for a full scale experiment. Neutrino capture in ^{71}Ga leads to ^{71}Ge, which then decays back to ^{71}Ga by electron capture with a halflife of 11.4 days. Fifty tons of Ga as $GaCl_3$ solution are needed for 1 neutrino capture per day (only pp and pep neutrinos). For the extraction a few mg Ge carrier are added and, in the presence of HCl excess, the Ge forms $GeCl_4$, a rather volatile compound which can be swept out of the solution by helium purge. For counting, the $GeCl_4$ will be reduced to GeH_4 by means of $NaBH_4$ and finally the GeH_4 will be introduced into a proportional counter to observe the ^{71}Ge decay.

Since the beginning of May, the 2.8 m^3 tank of the pilot experiment has been filled with 4.65 tons of $GaCl_3$ solution, equivalent to 1.26 tons of Gallium. Fig. 6 shows a plot of the very first Ge extraction out of the tank. The amount of (stable) Germanium present in the tank is plotted versus the extraction time. The Ge content of 200 mg, initially present in the $GaCl_3$ solution, was reduced to 0.7 mg after a 32 hour extraction. It follows that (under these circumstances) a 95% extraction yield can be obtained in 17 hours of extraction time. In the meantime many more extraction runs have been performed, adding only a few mg of Ge carrier to the $GaCl_3$ solution in the tank. Most of these gave extraction yields of better than 90%.

The ^{71}Ge counting procedure was described in detail at the Neutrino '79 Conference[36]. In the meantime, progress in the counter background was made: the present background count rates are almost sufficient for a full scale experiment. Thus, the outcome of the pilot studies is very promising so far, and a full scale Ga experiment could in principle be initiated soon.

Feasibility studies for a Ga solar neutrino experiment are also under way at the Nuclear Research Institute of the Academy of Sciences, Moscow, USSR. Experiments with a detector of 300 kg Ga metal have been performed in order to examine the extraction technique for Ge from large amounts of Ga metal[38]. Thus far, these investigations have been used to derive a lower limit for the lifetime of the ^{71}Ga nucleus against the charge-nonconserving decay ^{71}Ga → ^{71}Ge+(anything). For such an experiment the 300 kg detector was placed in an underground position (20 m.w.e.) in order to reduce

the cosmic ray production of
^{71}Ge. After about three ^{71}Ge
halflives Ge was extracted from
the Ga metal using a technique
analogous to that described by
Bahcall et al.[26]. ^{71}Ge was
finally counted as GeH$_4$ in a
small (5 mm diameter) proportional counter. An overall extraction yield of 80% was
achieved. The 90% confidence
lower limit obtained for the
^{71}Ga lifetime with respect to
the above process is 2.3x10^{23}
years. There are plans for a
solar neutrino experiment with
several tens of tons of gallium.
This full scale detector would
be installed in the deep-
foundation underground laboratory of the Baksan Valley Neutrino Observatory, North Caucasus[39]. The construction of
this underground site which
will be shielded by 5000 m.w.e.
of rock is expected to be
completed within the next
few years.

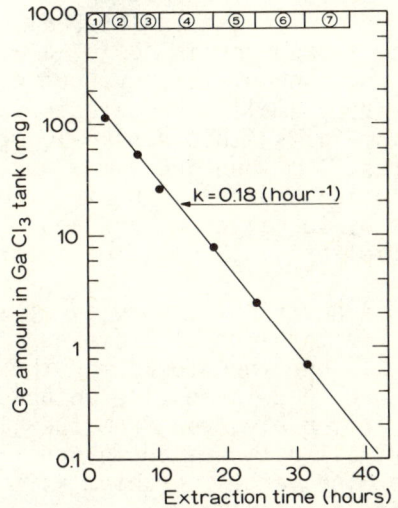

Fig. 6. Results of the very first Ge extraction out of the pilot tank (see text).

Indium Experiment

This experiment, first suggested by Raghavan[30], is now being studied in a collaboration of Raghavan (Bell Lab.) and Deutsch (MIT)[40]. The experiment is unique in so far as it is a low thresh-old detector (pp neutrinos) and at the same time a direct counting experiment. It is based on neutrino capture in ^{115}In leading by electron emission to a metastable state in ^{115}Sn, which then decays with a halflife of 3.3 µs to the ^{115}Sn ground state via emission of two gamma rays. In principle, the In detector is also a neutrino energy spectrometer: measuring the electron energy yields the ener-gy of the incident neutrino. There are, however, two disadvantages connected with this experiment: The log ft value of the ν capture transition is not known; it must be estimated. Secondly, there are serious background problems arising from the β-decay of ^{115}In to the ^{115}Sn ground state: the same amount of In needed for 1 ν capture per day from pp and pep neutrinos (3.7 t) gives rise to 10^{11} elec-trons per day from the ^{115}In β-decay. To reduce this background, Raghavan's basic idea was to use the delayed triple coincidence e$^-$-γ-γ and to divide the detector into small cells, counting the electron and the 116 keV gamma ray in one cell, and the 498 keV gamma ray in a neighbouring one. Random triple coincidences, however, require these single cells to be very small, even smaller than the 116 keV γ ray range. The present detector design is therefore based on the fact that the 116 keV transition proceeds in 50% of all cases via internal conversion, and the triple coincidence looked for is now $e^--e^--\gamma$. This procedure looses, of course, 50% counting efficiency, but now the single detector cells can be made very small. The detector will be sandwich-like, consisting of Indium-foils (thickness 5-10 mg/cm^2) alternated with layers of fibres, 1 m long and 1 mm in diameter. These fibres serve as scintillation counters, and the whole stack will be viewed from the side by so-called microchannel plates (position-sensitive photomultipliers), 15 cm in diameter.

A neutrino capture event will be characterized as follows: 614 keV total energy somewhere in the detector serves as a trigger signal, one then looks for the electron emitted after the neutrino capture in delayed time coincidence and local coincidence with the conversion electron from the 116 keV transition. By the end of this year, all data needed in order to build a prototype will be available. The largest problem is to find a fibre material with sufficient light collection.

COLLAPSING STAR NEUTRINOS

By the time, a massive star has exhausted its nuclear fuel in the core, the temperature and density there have become high enough to permit the electron capture reaction $p(e^-,\nu_e)n$. This neutronization produces a ν_e burst of a few milliseconds duration, followed by the collapse of the core because the electron degeneracy pressure is now removed. The subsequent cooling of the compressed core lasts for a couple of seconds and proceeds via emission of $\nu_e\bar{\nu}_e$, $\nu_\mu\bar{\nu}_\mu$ and $\nu_\tau\bar{\nu}_\tau$ pairs. The mean energies of the neutrinos emitted in both stages are of the order of 10 MeV. These are roughly the predictions resulting from current models of collapsing stars[41-44]. However, the exact features (fluxes of the different ν types, their energy, the time structure of the bursts) differ from model to model. In fact, these quantities may be determined experimentally in the near future by the detectors described below.

There are several existing detectors in the world which are in principle able to detect neutrinos from stellar collapses. The Homestake Long Range Neutrino Detector[45] is a 500 ton water Cerenkov counter located in the Homestake Gold Mine in Lead, South Dakota (USA) at a depth of 4,400 m.w.e. It surrounds the Brookhaven Cl solar neutrino detector. Then there are two liquid scintillation counters in the Soviet Union, a 100 ton detector at 600 m.w.e. in a salt mine near Artyomovsk, Ukraine[46] and a 330 ton detector built in a tunnel at 850 m.w.e. in the Baksan Valley Neutrino Laboratory,

North Caucasus[47]. A third liquid scintillation detector is being built at present by an Italian-Russian collaboration[48] in the Mont Blanc tunnel at a depth of 4270 m.w.e. 12 tanks with 15 tons of liquid scintillator are already running. The final configuration will be completed in 1981 and will consist of 48 tanks with 60 tons of liquid scintillator.

In addition, there are now several detectors being built or planned which are designed for the detection of proton decay[49] and some of which will also be sensitive to collapsing star neutrinos. For instance the largest one of these is the Michigan-Irvine-Brookhaven 8000-ton water Cerenkov detector[49,50].

Let us assume that in a collapsing star event taking place near the center of our galaxy ($\sim$10 kpc distance) 2×10^{53} ergs are emitted in form of ν_e's and 2×10^{52} ergs are emitted as $\bar{\nu}_e$'s, both with $\sim$10 MeV mean energy. This results in a neutrino fluence at the earth of 10^{12} ν_e/cm^2 and 10^{11} $\bar{\nu}_e/cm^2$. The ν_e can be detected either by the inverse β decay reaction (as in the case of solar neutrino detectors, and, in fact, solar neutrino experiments are in principle also sensitive to collapsing star events), or via the elastic scattering reaction on electrons. The $\bar{\nu}_e$ may be detected by the Cowan-Reines reaction $p(\bar{\nu}_e,e^+)n$.

The H_2O Cerenkov counters are sensitive to both kinds of neutrinos. They detect the relativistic electron or positron produced in either reaction[45], whereas the liquid scintillation counters look for the $\bar{\nu}_e$ component only, making use of the delayed coincidence between the positron and the 2.2 MeV γ-ray emitted after the neutron from the primary reaction is captured by a proton[46].

In Table 4 we compare the ability of different installations to detect the collapsing star event assumed above. The comparison includes 3 of the solar neutrino detectors, a 100-ton liquid scintillator (for instance the Artyomovsk detector) and a 500-ton H_2O Cerenkov detector (for instance the Homestake apparatus). Cross sections were taken from refs.[7,45,51] (the values for ^{71}Ga and ^{115}In were multiplied by a factor of two to allow for transitions to excited states in ^{71}Ge and ^{115}Sn). The last column in Table 4 gives the number of atoms (for the Cl and Ga experiments) or the number of counts (for direct counting experiments) produced by the collapsing star neutrino burst. Whether the obtained signal is detectable or not depends mainly on the background of the system. Since the Cl and Ga experiments integrate over the total exposure time, the collapsing star signal has to compete with the "background" signal produced by solar neutrinos, which is gradually building up during the exposure. The number of solar ν-produced atoms present at saturation is given in parentheses in the last column of Table 4. It is clear that neither experiment would be able to see the event, unless the distance from the collapsing star is smaller or, in the

Table 4. Comparison of different collapsing star neutrino detectors (see text)

Detector	Number of target atoms or electrons ($\times 10^{-30}$)	Reaction	Cross section ($10^{-42} cm^2$)	Number of produced atoms or counts
610t C_2Cl_4	2.2	$^{37}Cl(\nu_e,e^-)^{37}Ar$	2.7	5.9(21)
50t Ga	0.17	$^{71}Ga(\nu_e,e^-)^{71}Ge$	∼ 1	0.17(16)
	0.26	$^{69}Ga(\nu_e,e^-)^{69}Ge$	∼ 1	0.26(0.1)
3.7t In	0.019	$^{115}In(\nu_e,e^-2\gamma)^{115}Sn$	∼ 3	0.06
100t C_nH_{2n+2}	8.6	$p(\bar{\nu}_e,e^+)n(p,\gamma)d$	7.5	6.4
500t H_2O	170	$e^-(\nu_e,\nu_e)e^-$	0.17	29
	33	$p(\bar{\nu}_e,e^+)n$	7.5	25

case of the Cl experiment, the event happens just at the beginning of an exposure. In this case, however, an indication of the occurrence of such a collapsing star event by one of the "direct counting" detectors is required. Such kind of a "trigger" signal is also necessary if one intends to measure ^{69}Ge produced by neutrinos from a nearby(≲2 kpc) collapsing star with a 50t Ga detector. This follows from the short ^{69}Ge halflife (1.6 days). The advantage, however, is that there is almost no solar neutrino produced ^{69}Ge, in contrast to the ^{71}Ge case.

In the direct counting detectors the background count rate must be low enough to avoid the imitation of a neutrino burst by Poisson fluctuations of background counts. It seems that this condition is more or less fulfilled for the existing detectors[45-48]. However, because one cannot completely rule out the possibility that for a single detector a burst of pulses is produced by some electronic effect, it is desirable to observe the collapse neutrino signal worldwide and simultaneously in two or more of the existing detectors. In that case a neutrino burst such as assumed in Table 4, could clearly be identified, except for the In detector, which would not be able to observe events far beyond 1 kpc distance (due to its relatively small size).

Finally, there arises the question how often we can expect such a collapsing star event within our galaxy? Current estimates vary between 1 and, in the extreme case, 50 events per century depending on the methods used for estimating such numbers[52-54]. So, if we believe in the more optimistic numbers, we may really observe neutrinos from a collapsing star in the near future.

ACKNOWLEDGEMENTS. I am grateful to many colleagues for helpful discussions and suggestions in the course of preparing this review; especially I would like to thank J.N. Bahcall, B.T. Cleveland, R. Davis, I. Dostrovsky, T. Kirsten, K. Lande, J. Ray, and J.K. Rowley.

REFERENCES.

1. R. Davis Jr., J.C. Evans, B.T. Cleveland, Proc.Int.Neutrino Conf. Purdue, 53(1978); R. Davis Jr., priv. communication (1980).
2. A.W. Wolfendale, E.C.M. Young, and R. Davis Jr., Nature, Phys. Sci. 238, 130(1972).
3. E.L. Fireman, Proc.Int. Cosmic Ray Conf., Kyoto, Vol.12 (1979).
4. J.N. Bahcall, S.H. Lubow, W.F. Huebner, N.H. Magee Jr.,A.L. Merts, M.F. Argo, P.D. Parker, B. Rozsnyai, and R.K. Ulrich, Phys.Rev. Lett. 45, 945(1980).
5. D.N. Schramm, Proc.Int. Neutrino Conf., Bergen, Vol.1, 503(1979).
6. J.N. Bahcall, W.F. Huebner, N.H. Magee Jr., A.L. Merts, and R.K. Ulrich, Astrophys.J. 184, 1(1973).
7. J.N. Bahcall, Rev.Mod.Phys. 50, 881(1978).
8. J. Byrne, J. Morse, K.F. Smith, F. Shaikh, K. Green, and G.L. Greene, Phys.Lett. 92B, 274(1980).
9. C. Bargholtz, Astrophys.J.Lett. 233, L161(1979).
10. P.D. Parker and R.W. Kavanagh, Phys.Rev. 131, 2578(1963).
11. K. Nagatani, M.R. Dwarakanath, and D. Ashery, Nucl.Phys. A128, 325(1969).
12. C.Fröhlich, in: The Solar Output and its Variation, O.R. White, Ed., Colorado Associated University Press, Boulder, 93(1977).
13. J.E. Ross and L.H. Aller, Science 191, 1223(1976).
14. J.N. Bahcall and R.K. Ulrich, Astrophys.J. 170, 593(1971).
15. J.N. Bahcall, Space Sci.Rev. 24, 227(1979).
16. E. Schatzman and A. Maeder, Contribution to this conference.
17. C. Rolfs, Some Problems in Experimental Nuclear Astrophysics, Int. Workshop VII Hirschegg (1979), and priv. communication(1980).
18. T.A. Tombrello and P.D. Parker, Phys.Rev. 131, 2582(1963).
19. S.M. Bilenky and B. Pontecorvo, Phys.Rep. 41, 225(1978).
20. F. Reines, H.W. Sobel, and E. Pasierb, Evidence for Neutrino Instability, preprint(1980), submitted to Phys.Rev.Lett.
21. J.N. Bahcall and S.C. Frautschi, Phys.Lett. 29B, 623(1969).
22. R. Ehrlich, Phys.Rev. D18, 2323(1978).
23. A. De Rújula, M. Lusignoli, L. Maiani, S.T. Petcov, and R.Petronzio, A Fresh Look at Neutrino Oscillations, CERN-Report TH.2788(1979).
24. V. Barger, K. Whisnant, D. Cline, and R.J.N. Phillips, Possible Indications of Neutrino Oscillations, Preprint COO-881-135(1980).
25. M.S. Freedman, C.M. Stevens, E.P. Horwitz, L.H. Fuchs, J.L.Lerner, L.S. Goodman, W.J. Childs, and J. Hessler, Science 193,1117(1976).
26. J.N. Bahcall, B.T. Cleveland, R. Davis Jr., I. Dostrovsky, J.C. Evans Jr., W. Frati, G. Friedlander, K. Lande, J.K. Rowley, R.W. Stoenner, and J. Weneser, Phys.Rev.Lett. 40, 1351(1978).
27. R.D. Scott, Nature 264, 729(1976); Proc.Conf. Status and Future Solar Neutrino Research, G. Friedlander,Ed., BNL 50879, Vol.1, 293(1978).
28. W. Hampel, Annual Report, MPI Kernphysik Heidelberg, 158(1976); T. Kirsten, Proc.Conf. Status and Future Solar Neutrino Research, G. Friedlander,Ed., BNL 50879, Vol.1, 305(1978).
29. J.K. Rowley, Proc.Conf. Status and Future Solar Neutrino Research

G. Friedlander, Ed., BNL 50879, Vol.1, 261(1978).
30. R.S. Raghavan, Phys.Rev.Lett. 37, 259(1976).
31. A.M. Fainberg, Proc.Conf. Status and Future Solar Neutrino Research, G.Friedlander,Ed., BNL 50879, Vol.2, 93(1976).
32. H.H.Chen, Proc.Conf. Status and Future Solar Neutrino Research, G.Friedlander,Ed., BNL 50879, Vol.2, 55(1976).
33. K. Lande, Proc.Conf. Status and Future Solar Neutrino Research, G. Friedlander,Ed., BNL 50879, Vol.2, 79(1978).
34. C.L. Bennett, M.M. Lowry, R.A. Naumann, F. Loeser,and W.H.Moore, Feasibility of ^{81}Br as a Solar Neutrino Detector,Preprint(1980).
35. J.K. Rowley, B.T. Cleveland, R. Davis Jr., W. Hampel,and T. Kirsten, The Present and the Past Neutrino Luminosity of the Sun, submitted to Proc.Conf. Ancient Sun, Boulder (1979).
36. T. Kirsten, Proc.Int.Neutrino Conf., Bergen, Vol.2, 452(1979).
37. R. Davis, B.T. Cleveland, G. Friedlander, S. Katcoff, J.K. Rowley, J. Weneser (Broohaven National Laboratory); T.Kirsten, W. Hampel, G. Heusser, M. Hübner, J. Kiko, O.A. Schaeffer, R. Schlotz (Max-Planck-Institut f. Kernphysik, Heidelberg); I. Dostrovsky, Y. Eyal (Weizmann Institute of Science, Rehovot); J.N. Bahcall (Institute for Advanced Study,Princeton); K.Lande, W. Frati, R.I. Steinberg (Univ.of Pennsylvania, Philadelphia).
38. A.A. Pomanskii, Atomnaya Energiya, 44, 376(1978) (in Russian).
39. I.P. Barabanov, E.P.Veretenkin, V.N. Gavrin, Y.I. Zakharov, G.T. Zatsepin, G.Y. Novikova, I.V. Orekhov, and M.I.Churmaeva, Test of the Electric Charge Conservation Law, Preprint(1980) (in Russian).
40. R.S. Raghavan, private communication (1980).
41. D.N. Schramm, Proc.Int. Neutrino Conf., Baksan, Vol.1,131(1977)
42. G.E. Brown, Comments Astrophys. Space Phys. 7, 67(1977).
43. V.M. Chechetkin, V.S. Imshennik, L.N. Ivanova,and D.N.Nadyozhin, in: Supernovae, D.N.Schramm, Ed., D. Reidel Publ.Comp., Dordrecht, 159(1977).
44. T.J. Mazurek, Proc.Int.Neutrino Conf., Purdue, 125(1978).
45. M. Deakyne, W. Frati, K. Lande, C.K. Lee, R.I. Steinberg, and E. Fenyves, Proc.Int. Neutrino Conf., Purdue, 887(1978).
46. V.I. Beresnev, R.I. Enikeev, G.T. Zatsepin, P.V. Korchagin,A.S. Mal'gin, O.G. Ryazhskaya, V.G. Ryassny, and L.N. Stepanets, Proc.Int. Neutrino Conf., Purdue, 881(1978).
47. A.E. Chudakov and O.G. Ryazhskaya, Proc.Int. Neutrino Conf., Baksan, Vol.1, 155(1977).
48. P. Galeotti, O.G. Ryazhskaya, private communication (1980).
49. M. Goldhaber, Review Talk given at this conference.
50. J. LoSecco, private communication (1980).
51. W. Hampel, On the ^{69}Ge Production by Collapsing Star Neutrinos in a Ga Solar Neutrino Experiment, unpublished (1980).
52. Y.P. Pskovsky, Proc.Int. Neutrino Conf., Baksan,Vol.1,149(1977).
53. G.A. Tammann, in: Supernovae, D.N.Schramm, Ed.,D. Reidel Publ. Comp., Dordrecht, 95(1977).
54. B.M. Tinsley, in: Supernovae, D.N. Schramm, Ed., D. Reidel Publ. Comp., Dordrecht, 117(1977).

LOW-ENERGY NEUTRINOS IN ASTROPHYSICS

DISCUSSION (Chairman: M. Shapiro)

A. GARFINKEL, Lafayette: Please give us the theoretical value including errors presented at the Bergen Conference for the standard solar model.

W. HAMPEL, Heidelberg: Schramm in his summary talk at the Bergen Conference gave a value of 7.5 SNU which was a preliminary result from the new standard solar model calculation by Bahcall and others and which agrees well with the number published now 7.8 ± 1.5 SNU.

P. NEMETHY, Berkeley: What might be the time scale for a full-scale Gallium experiment?

W. HAMPEL, Heidelberg: In a few months we will write a progress report on the 1.5 ton pilot experiment and this report will then be submitted to the funding agencies. In case that there would be no problem with funding, we would acquire the necessary amount of 50 tons of Gallium in steps of 10 or 15 tons per year during the next 3 or 4 years.

A. BODEK, Rochester: Have there been any calculations of the effect of a possible small neutrino magnetic moment on the solar neutrino puzzle (via precession in a magnetic field)?

L. WOLFENSTEIN, Pittsburgh: Ruderman and I looked at this possibility assuming a large magnetic field inside the sun which changes the right-handed $\bar{\nu}_e$ to a non-interacting left-handed $\bar{\nu}_e$. Because of the index of refraction effect $\bar{\nu}_e)_R$ and $\bar{\nu}_e)_L$ have different effective masses in the sun and so with present astrophysical limits on the magnetic moment, an unreasonable field would be required.

E. FIORINI, Milano: Are there any proposals to measure neutrino capture on nuclei at the LAMPF beam dump where the neutrino energies are in the range of ^8B decay? Such experiments would be worthwhile in checking the reactions and detector systems used in detection of solar neutrinos.

T. BOWLES, Los Alamos: There is a proposal by Davis to measure neutrino capture on Cl at LAMPF, but this experiment doesn't appear feasible at the present time. There are no other proposals at present for neutrino reactions at LAMPF except for the ν_e-e elastic scattering experiment now being mounted. However, there are several proposals for experiments at LAMPF to be carried out when the storage ring comes on line in 1985.

NEW PARTICLES

THIS REVIEW IS DEDICATED TO THE MEMORY OF
E.H.S. BURHOP

Paul Musset

CERN, Geneva, Switzerland

INTRODUCTION

The content of this talk is centered on some of the new results recently obtained in the study of charmed and beautiful particles produced in neutrino, hadron and photon beams. The emphasis will be on neutrino reactions and on visual techniques, mainly those which the decay length of these short-lived particles is measurable.

Substantial progress was made during the last year[1] on the properties - mass and lifetime - of the charmed baryon states. Several examples of the decay of the F meson in emulsion are now available, and the question of the lifetime of the D meson, i.e., the difference between the D^0 and the $D^{\pm}$ lifetime is now well settled. We will also report on first evidence of associated charm production in hadron beams.

On the other hand, the problem of beauty is far from being settled. The Goliath experiment recently accumulated more statistics and in other experiments limits on the beauty cross-section by hadrons were obtained through the study of 3μ events. The indication for beauty which might have been obtained in antineutrino beams will be discussed.

CHARMED BARYONS

The Λ_c mass

At the last International Symposium on lepton and photon interactions[2] a new value obtained at SLAC for the Λ_c mass, i.e.,

(2.285 ± 0.006) GeV, and the value obtained in other experiments (15', photon experiment at SLAC, the ISR and the 7'), i.e., (2.257 ± 0.010) GeV were not in excellent agreement.

The Aachen-Bonn-CERN-Munich-Oxford experiment[3] using BEBC filled with hydrogen and equipped with a two-plane EMI studied the $\Delta S = -\Delta Q$ events produced by a ν_μ wide band beam. Four of these events were defined through a 3C- fit, and the corresponding background shown to be small ($\lesssim 0.2$ events).

Two of these four events are interpreted to be due to a Λ_c decay into the $pK^-\pi^+$ final states, with masses of 2.285 ±0.005 and 2.280 ± 0.003 (Fig. 1). No mass combination can produce a D meson mass and the other "wrong" mass combinations are 2.220 and 2.490 GeV, i.e., far from the Λ_c mass.

This study resulted in a precise determination of the Λ_c mass, in agreement with the highest of the two previous sets of measurements.

The Tohuku-IIT-Maryland-Stony Brook-Tufts Collaboration[4] made use of the 15' filled with deuterium and equipped with a two-plane EMI to study ν_μ interactions.

The channels $\Lambda\pi^+$ and $K^0 p$ were investigated and the invariance masses calculated in the two cases. The distributions (Fig. 2) show peaks. Nevertheless, since the angular distribution of the baryon Λ (resp K^0) in the $\Lambda\pi^+$ (resp $K^0 p$) system shows evidence for a backward peaking, a cut at $\cos\theta_\Lambda < -0.75$ (resp $\cos\theta_\phi < -0.90$) was made to check that the mass peaking was not due to the backward peaking of the baryon. On the contrary, the peak remained and the background was lowered.

The signal was found to be (8 ± 3.5) events in the first channel and (11 ± 4.5) events in the second channel, i.e., a total of (19 ± 5.7) events, which is a 3.3 standard deviation effect.

The mass obtained for this peak is $M(\Lambda_c) = (2.275 \pm 0.010)$ GeV, again in good agreement with the highest value of the two previous sets of measurements. The experiment was also able to provide a measurement of the branching ratio $(\Lambda_c \to \Lambda\pi^+)/\Lambda_c \to K^0 p) = 0.51 \pm 0.39$.

The Lifetime of the Λ_c

The following results on the lifetime of the Λ_c, as well as other results on charmed particle lifetimes, were obtained in three hybrid emulsion experiments that we now briefly describe.

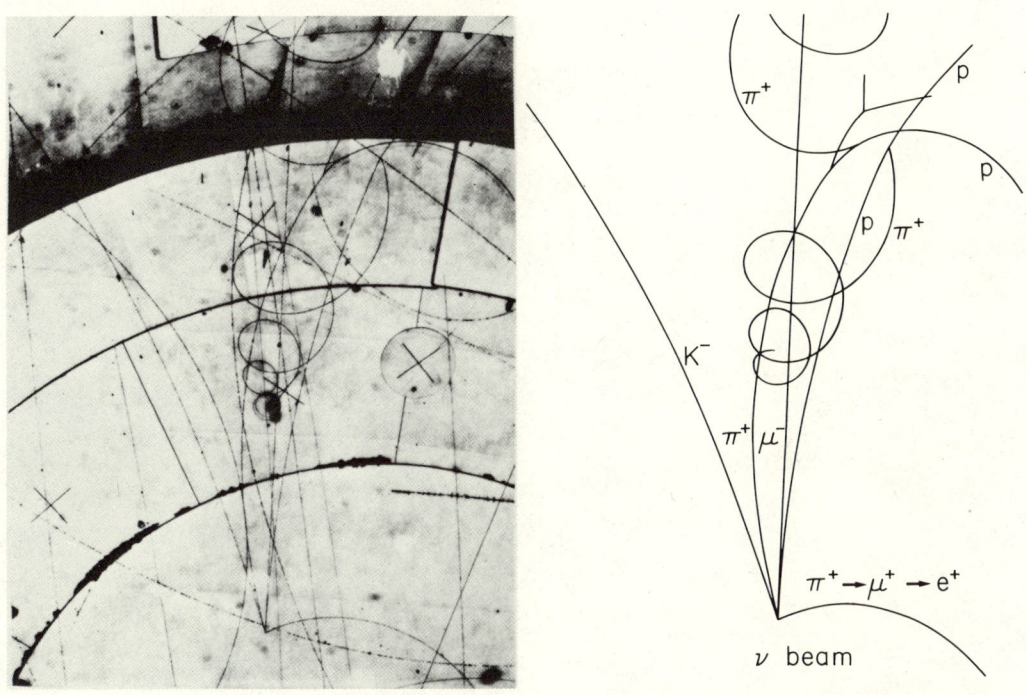

Fig. 1

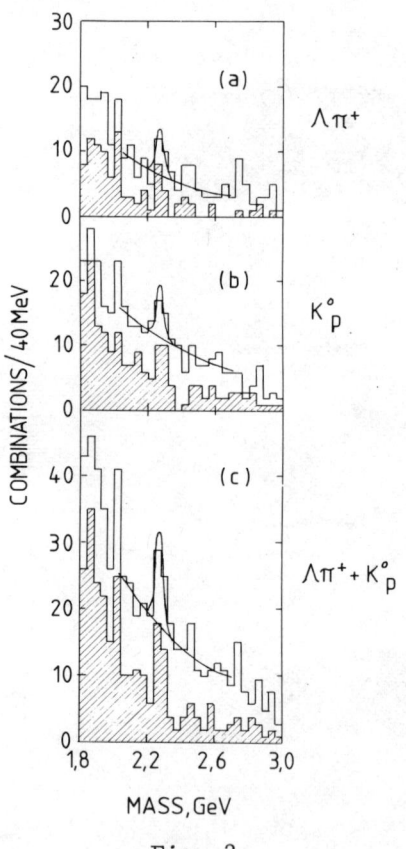

Fig. 2

NEW PARTICLES

The WA17 experiment[5] (Fig. 3) was performed in the CERN SPS wide band neutrino beam by the Ankara-Brussels-CERN-Dublin-London-Open University-Pisa-Rome-Turin Collaboration. A total of 31.5 ℓ of emulsion was used. Stacks of 600 μ pellicules, each of 1.5 ℓ, are assembled in front of BEBC which is filled with hydrogen. The emulsion pellicules are parallel to the beam. A veto coincidence counter encloses the emulsion stack and is coupled to the external muon identifier of BEBC. A total of eight charmed candidates were found out of the 169 charged current events.

The E-531 experiment[6] (Fig. 4) was performed in the FNAL wide band neutrino beam by the Aichi-FNAL-Kobe-Seoul-Montreal-Nagoya-Ohio-Okayama-Osaka-Ottawa-Tokyo-Toronto-Yokohama Collaboration. A total of 23 ℓ of emulsion was used, divided equally into two parts, one with 600 μ pellicules parallel to the beam, the other with 2 × 330 μ pellicules on a 70 μ polystyrene support perpendicular to the beam. Downstream a spectrometer, a time-of-flight hodoscope, a lead glass counter, a calorimeter and a muon identifier provide information for the analysis of the secondaries.

The photon-omega experiment[7] (Fig. 5) was performed in a tagged photon beam with energies from 20 to 80 GeV. Emulsion pellicules of 600 μ thickness are exposed at small (5°) angle to the beam and the experiment makes use of the Ω spectrometer, including a Cerenkov counter, a lead glass shower counter, a veto counter and hodoscopes. The analysis of the test experiment is in progress and the present results are obtained with only 2% of the data.

The WA17 experiment obtained one Λ_c event which decayed into $pK^-\pi^+$ where the p is identified by scattering and the K^- by momentum measurement and by ionization measurement in the emulsion. The $K^-\pi^+$ mass is found to be consistent with the mass of a K^{*0}. The decay time of this event is $t = (7.3 \pm 0.1)10^{-13}$s.

The E-531 experiment obtained four Λ_c events decaying into the following modes with the corresponding decay times:

$$\Lambda_c^+ \to \underline{\Lambda^0 \pi^+ \pi^- \pi^+} \qquad t = 3.58 \cdot 10^{-13}\text{s}$$

$$\Lambda_c^+ \to \underline{\Lambda^0 \pi^+ \pi^- \pi^+} \qquad t = 0.54 \cdot 10^{-13}\text{s}$$

$$\Lambda_c^+ \to \underline{p\pi^+ K^-} (\pi^0) \qquad t = 0.70 \cdot 10^{-13}\text{s}$$

$$\Lambda_c^+ \to \underline{p\pi^+ \pi^-} (K^0) \qquad t = 0.73 \cdot 10^{-13}\text{s}$$

where the secondaries underlined are positively identified. A likelihood method gives a lifetime of $\tau(\Lambda_c) = (1.14 ^{+0.90}_{-0.44})10^{-13}$s for these events.

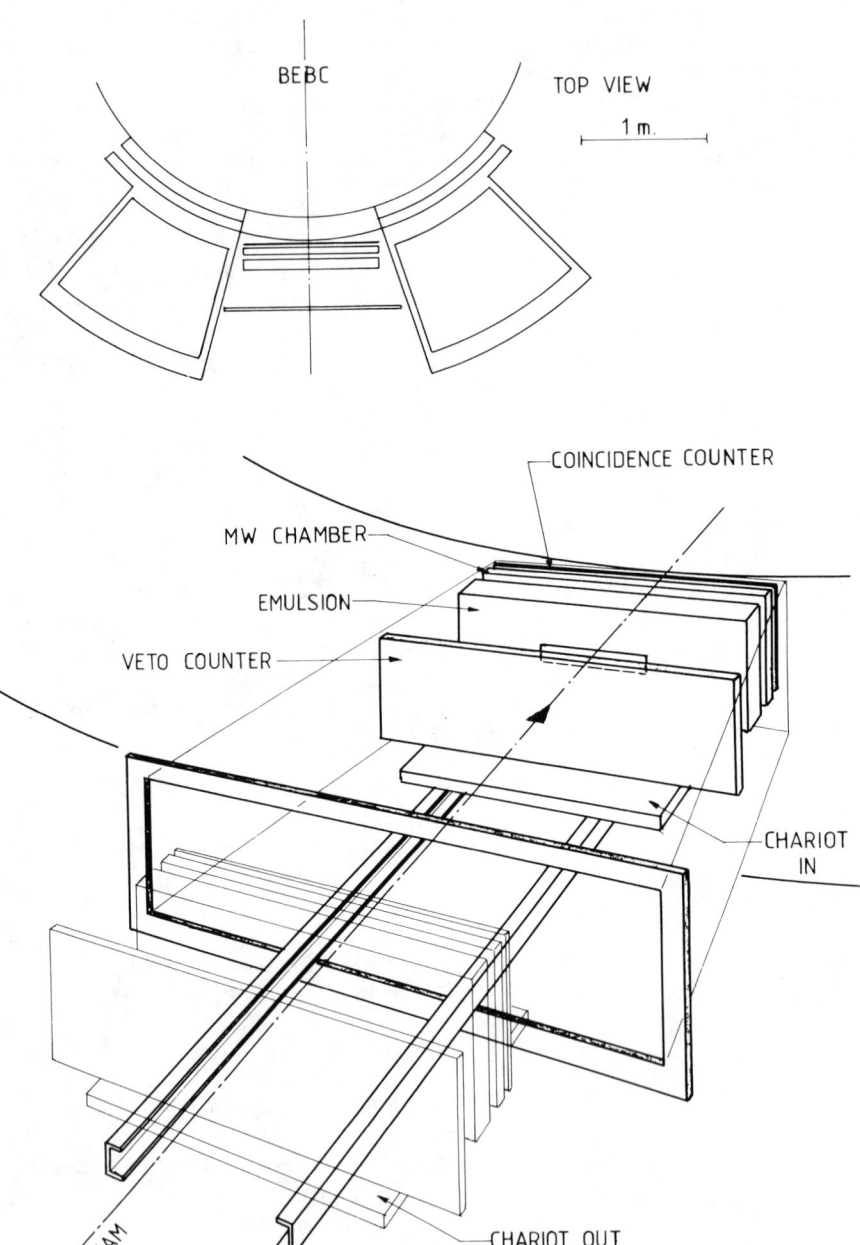

Fig. 3

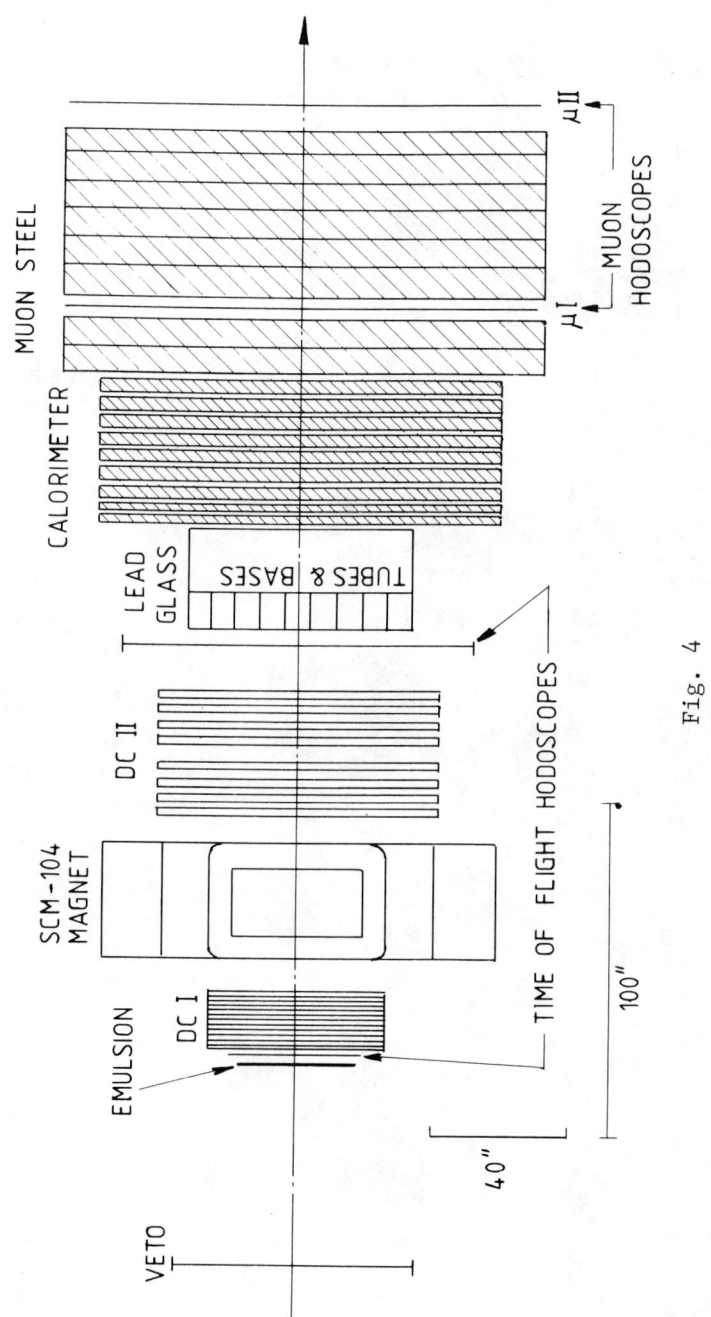

Fig. 4

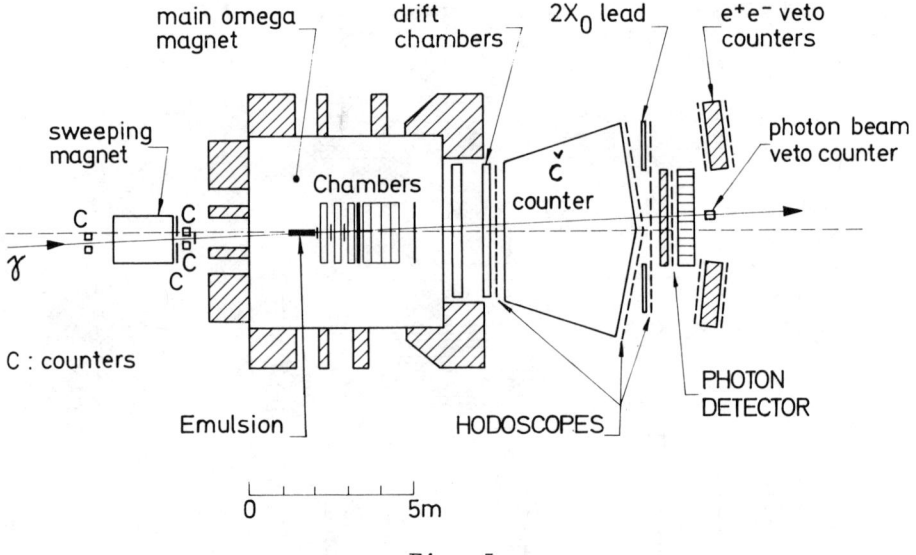

Fig. 5

NEW PARTICLES

The γ-Ω experiment has presently one Λ_c^+ decaying into $\underline{\Lambda^0 \pi^+}$ with a decay time $t = (0.59 \pm 0.02) 10^{-13}$ s.

In view of the limited statistics, these results can be considered to be in reasonable agreement.

Observations of a Σ_c^+

Observations of the Σ_c^{++} were previously reported[8]. The first observation of a Σ_c^+ is reported by the Bari-Birmingham-Brussels-CERN-LPNHE-Rutherford-Saclay-London Collaboration[9] using BEBC filled with a hydrogen-neon mixture and equipped with a H_2-TST. The chamber is in a neutrino wide band beam at the CERN SPS. A restricted fiducial volume is defined to allow for precise measurements of secondaries.

The very clear candidate for a Σ_c^+ decay (Fig. 6) has a μ^- identified by the two-plane EMI. A K^- is identified because the $\bar{p}$ hypothesis is excluded by the measurement of a δ-ray and because the π^- hypothesis is excluded by kinematics at interaction. A V^0 is produced in the interaction of this track. A π^+ is identified because the p and the K^+ hypotheses are excluded by track lengths and momentum measurement. A second π^+ is identified since the K^+ hypothesis is excluded by kinematics at interaction. Finally, a γ-pair is detected, the invariant mass of which is 140 ± 7 MeV.

The mass of the Σ_c^+ is measured to be

$$M(\Sigma_c^+) = (2.457 \pm 0.004) \text{ GeV}$$

Note that a precise measurement of the mass of the Λ_c is also provided by the same event

$$M(\Lambda_c) = (2.290 \pm 0.003) \text{ GeV}$$

We now can combine the four new and precise values of the Λ_c mass to obtain

$$M(\Lambda_c) = (2.285 \pm 0.003) \text{ GeV}$$

where the error is calculated taking into account not only the quoted errors of the various results, but also their internal dispersion. It is in very good agreement with the highest of the two values obtained in the previous measurements.

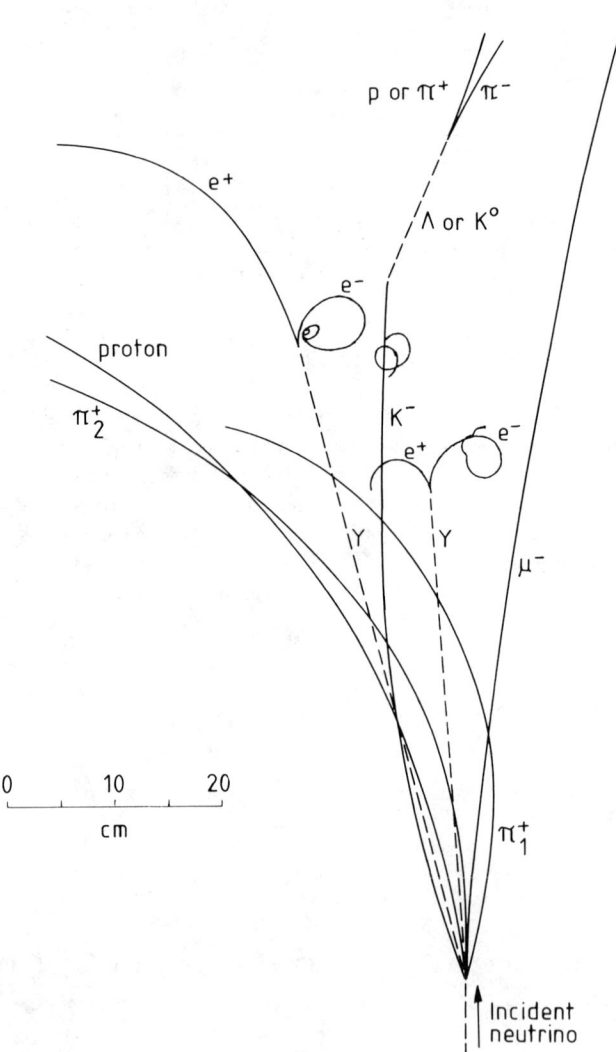

Fig. 6

NEW PARTICLES

CHARMED MESONS

Observation of the F Meson

The situation of the F meson was not completely settled at the 1979 International Symposium on Lepton and Photon Interactions[2]. A few examples of an $F \to \eta\pi$ decay were observed at DESY, whereas no confirmation, but no contradiction either, was obtained at SLAC.

A first indication of the direct observation of an F^+ decay in emulsion was recently obtained by the Kansas-FNAL-Serpukhov-Moscow-Crakow-Dubna-Seattle experiment[10] using cryogenic-sensitive emulsion inserted into the 15' chamber filled with deuterium and exposed to a wide band neutrino beam at FNAL.

A clean vertex (Fig. 7) is seen in the emulsion with three secondaries and a 3C-fit is obtained for the mass of the $F^+(2.030$ GeV) for an $F^+ \to \pi^+\pi^+\pi^-\pi^0$ decay. When the mass is not constrained, a 2C-fit gives $M(F) = (2.017 \pm 0.025)$ GeV, with the mass of one of the 3π system $M(\pi^+\pi^-\pi^0) = (0.808 \pm 0.020)$ GeV compatible with the ω hypothesis. Unfortunately, the interpretation is made more difficult by the fact that one of the tracks originated at the decay vertex misses the chamber, and three tracks detected in the bubble chamber are not seen in emulsion. Furthermore, an e^+e^- pair is well identified but the second γ is seen only as an e^+ track. The fit to the F is the best fit, whereas the Λ_c hypothesis leads to a poor fit. In the F hypothesis, the decay time of the event is $t = 1.5 \cdot 10^{-13}$ s.

Two other good candidates for the F decay were obtained in the neutrino emulsion-hybrid experiment E 531 at FNAL, described above[6]. They both are 3-prong events, and in the two cases a Cabibbo favoured D decay hypothesis is excluded by the identification of secondary particles.

The first event (Fig. 8) is identified as a an $F^- \to \pi^+\pi^-\pi^-\pi^0$ decay. The first two $\pi^\pm$ are identified by time-of-flight and the γ-rays of the π^0 are measured in the lead glass counter and checked to balance the p_T at the decay vertex. The mass is $M(F) = (2.026 \pm 0.056)$ GeV.

The second event is identified as an $F^+ \to K^+\pi^-\pi^+\bar{K}_0$ decay. The first particles are identified by time-of-flight, and the $\bar{K}_0$ hypothesis is made for the hadron particle measured in the calorimeter. The mass is $M(F) = (2.089 \pm 0.121)$ GeV.

A likelihood method for these two events gives a lifetime $\tau = (2.24 ^{+2.78}_{-1.05})10^{-13}$ s. Note that for these two events, a Cabibbo suppressed D decay, i.e., involving one and only one strange particle, has a probality less than 10^{-4}.

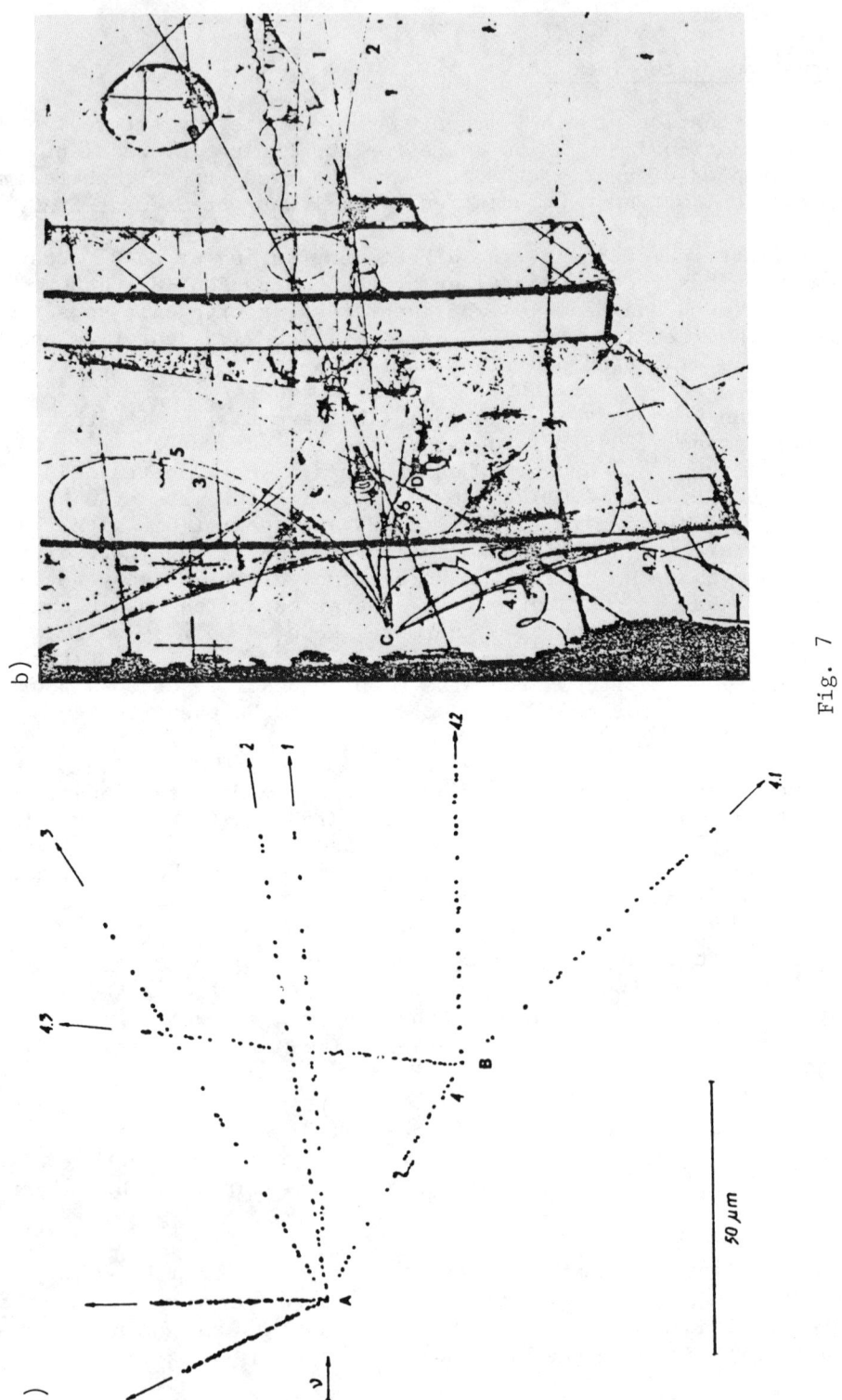

Fig. 7

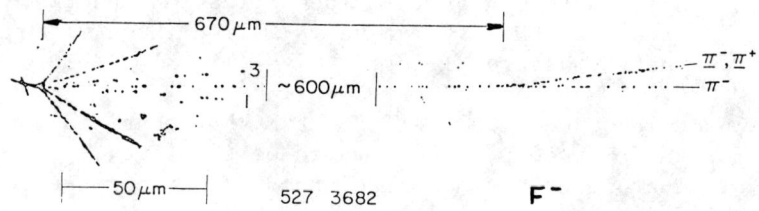

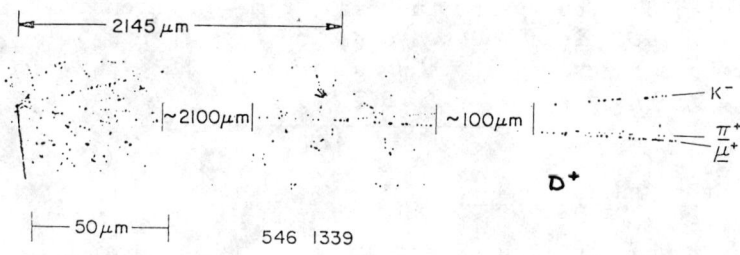

Fig. 8

Lifetime of the Charged D Meson

The question of the lifetimes of the D meson, and mainly the question of the ratio of the neutral D to the charged D meson lifetimes, was not completely settled until now[1]. Very interesting results have been obtained on the subject by the above mentioned hybrid emulsion experiments.

The neutrino experiment E-531 at FNAL[6] has obtained five good candidates for the charged D decays, which split into four positive D's and one negative D. They are all 3-prong events. One of the events is completely unambiguous, whereas another one is ambiguous with an F. The three other candidates are identified because the F hypothesis is excluded by particle identification and the Λ hypothesis is excluded by the mass values. A likelihood method for these five events gives the lifetime $\tau_\pm = (10.3 \,{}^{+10.3}_{-4.1}) 10^{-13}$ s.

The neutrino experiment WA17 at CERN[5] has obtained four candidates for the charged D^+ decays. Two of them are 3-prong events and two are 1-prong events. A likelihood method for the four events gives the lifetime $\tau_+ = (2.5 \,{}^{+2.2}_{-1.1}) 10^{-13}$ s.

Finally, the γ-Ω experiment[7] at CERN recently obtained a first D^- candidate, a 3-prong event, the decay time of which is $t = 1.00 \times 10^{-13}$ s. Since the two neutrino experiments both give their likelihood functions it is possible to combine the results for the lifetime (Fig. 9) $\tau_\pm = (5.5 \,{}^{+3.5}_{-1.2}) 10^{-13}$ s.

Lifetime of the Neutral D Meson

The neutrino experiment E-531 at FNAL[6] has a total of seven neutral decay events. Note, nevertheless, that a neutral baryon hypothesis is not excluded for three of them and that one of the decays is definitively a semi-leptonic decay. A likelihood method gives, for these seven events, a lifetime $\tau_0 = (1.00 \,{}^{+0.52}_{-0.31}) 10^{-13}$ s. (with a probability of 1% for a single lifetime).

The neutrino experiment WA17[5] at CERN has three neutral decay events which all include neutral secondaries. The likelihood method gives for the lifetime $\tau_0 = (0.53 \,{}^{+0.57}_{-0.27}) 10^{-13}$ s.

Finally, the γ-Ω experiment[7] at CERN recently detected two events. One is a 2-prong and the other is a 4-prong event (Fig. 10). This adds to a first 4-prong event detected in a previous experiment. They have, respectively, the decay times 0.4, 0.86 and 0.23 in 10^{-13} s units.

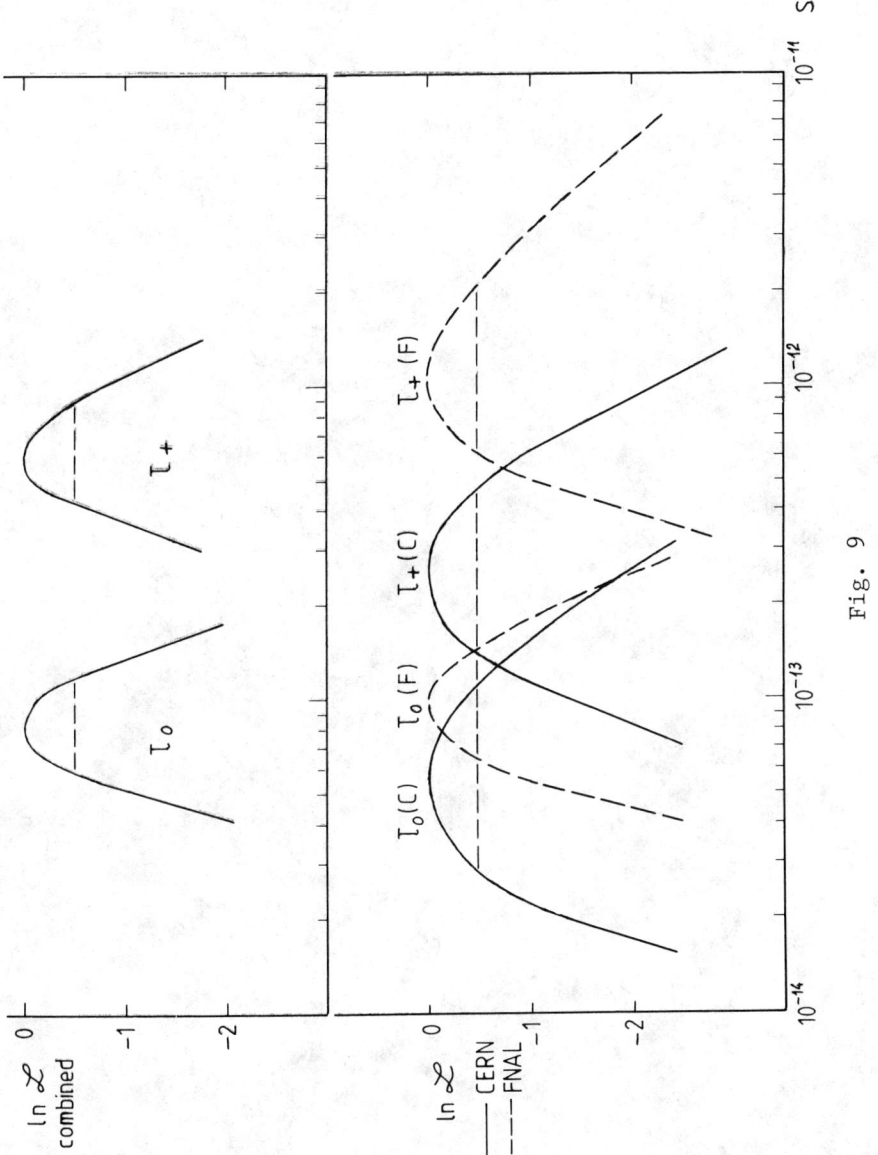

Fig. 9

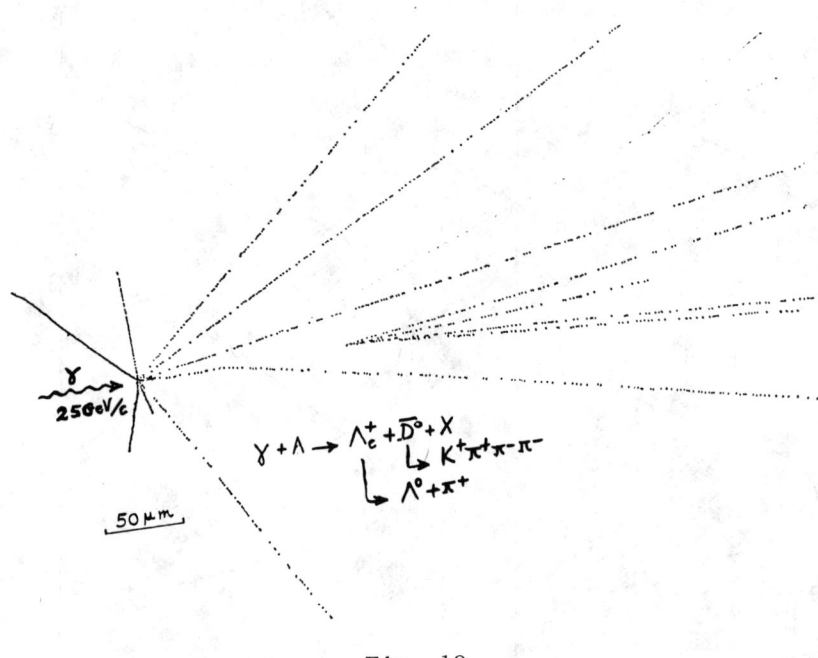

Fig. 10

The data of the two neutrino experiments can be combined to give (Fig. 9) the lifetime $\tau_0 = (0.8 ^{+0.3}_{-0.15}) 10^{-13}$ s. Not enough information is available to combine also the events from the γ-Ω experiment but we estimate that their contribution would bring the average to a value of τ_0 around $.7 \times 10^{-13}$ s. It is then clearly established that the lifetimes of the D_0 meson is substantially shorter (i.e., $(7 ^{+4}_{-2} {}^5)$ times) than the lifetime of the $D_\pm$ meson.

DIRECT OBSERVATION OF ASSOCIATED CHARMED PARTICLE PRODUCTION IN HADRON INTERACTION

The LEBC-EHS Experiment

The aim of this experiment[11] is to observe and measure charmed particle decays produced by a 340 π^- GeV beam in a high resolution bubble chamber[11]. The experiment is performed by the Brussels-CERN-Oxford-Padova-Rome-Rutherford-Trieste Collaboration.

The LEBC is a one-litre hydrogen chamber with a 20 cm diameter and a 4 cm depth. It is pulsed at 40 Hz frequency. The depth of focus is reduced to 5 mm to obtain a good optical quality. The bubble density is ~ 70/cm, the bubble size 40 to 50 μ and the resolution is 30 μ. This is obtained by selecting particles which enter the chamber in a one millisecond window around the minimal pressure by means of an interaction trigger. A flash delay of 200 μs is used. This trigger produces an event rate of 2.3 ev/μb in a first run.

In this study, the only substantial contamination is the strange particle production; mainly associated charmed particle production will be discussed (Fig. 11).

The quality of the events is characterized by the three parameters: L, length of decay; x, distance of the decay vertex to the beam track; and y, distance of an inclined decay track to the primary vertex. With the criteria: L > 2 cm and x > 0.6 cm, a total of 20 events were obtained, of which eight are estimated to be background. If the additional criterion, y < 0.1 cm is applied, 14 events remained, of which 3.9 are estimated to be background. According to a charged and neutral tentative classification, the 20 events split into 5 CC events (backg. 0.8), 10 C events (backg. 4.0) and 5 VV events (backg. 3.4).

The conclusion of the authors is that they have seen a signal for associated charm production. The cross-section obtained is dependent on the assumptions made for the lifetime. With

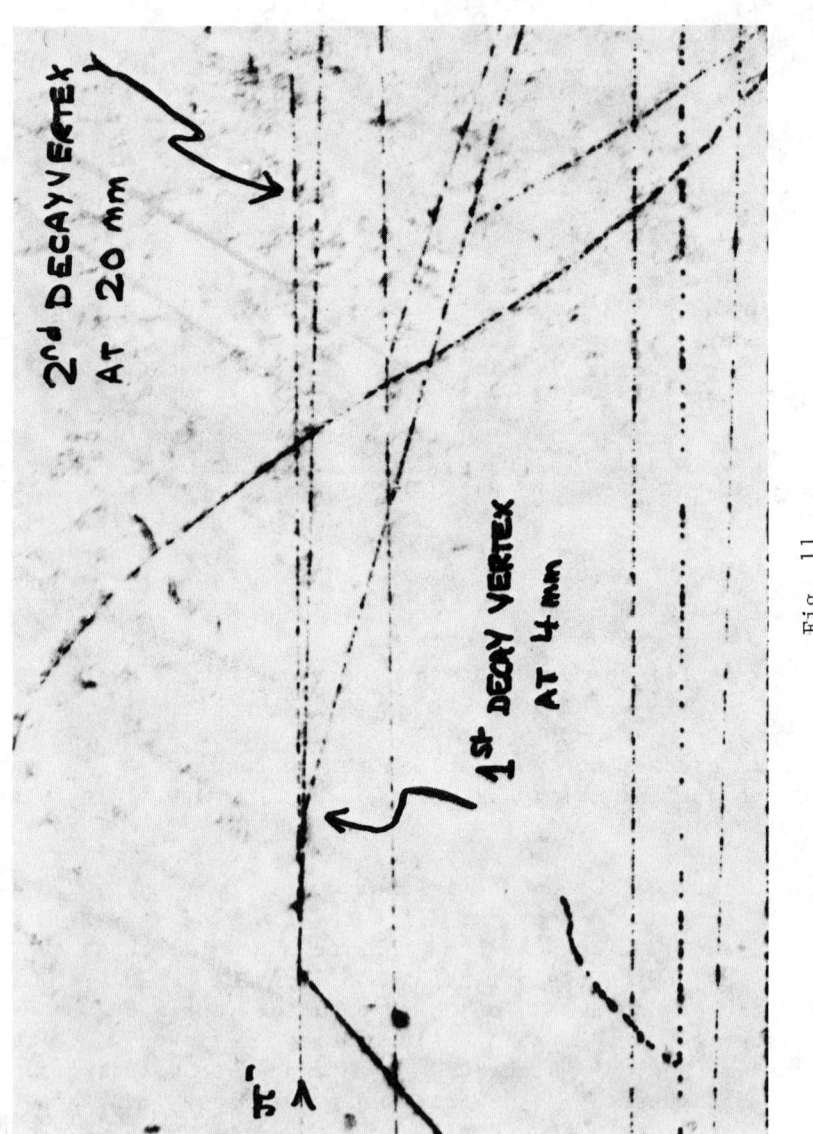

Fig. 11

$\tau = 5 \cdot 10^{-13}$ s they obtain $\sigma \sim 40 \mu b$. This number is corroborated by the study of 3-prong events in which the background is small because of the long lifetime of the $K^{\pm}$ and its small branching fraction into 3 prongs. From eight events, of which two are estimated to be background, a cross-section $\sigma \sim 35 \mu b$ is obtained, in agreement with the previous estimate. Let us remark that this cross-section is sensitive to charged D because it was just shown that the neutral D lifetime is too small to lead to a high detection efficiency. Note that the LEBC experiment will be pursued with the EHS spectrometer. Also a recent experiment with BIBC a propane bubble chamber, has been recently achieved at the CERN SPS.

The Streamer Chamber Experiment

Another piece of information on associated charmed particle production is produced by an experiment done in a high pressure (24 atm.) neon (90%)-hydrogen (10%) streamer chamber placed in a 350 GeV proton beam at FNAL[12]. The chamber is 4 cm long along the beam and has a width of 3 cm and a height of 0.45 cm. The streamer size is typically 50μ, and the track width 150 to 200μ, which leads to a precision of 40μ. The streamer chamber is followed (Fig. 12) by a muon identifier. A forward dead cone results in a 27% acceptance.

The experiment makes use of two triggers, one of which is the muon trigger, from which the geometrical criteria on the photograph produce a signal of 10 events with a background of 2.1 ± 0.4 events. The second trigger is simply an interaction trigger, which produces only one event with a background of 0.5 ± 0.1 events. From the signal the authors can estimate the production mass correction, assuming a given value for the lifetime. For $\tau = 5 \cdot 10^{-13}$ s they obtain $\sigma \sim 40$ to $60 \mu b$. Note that this result is of the same order of magnitude as the result obtained with pions, and is also in agreement with the result obtained in beam-dump experiments.

SEARCH FOR BEAUTY IN HADRON INTERACTIONS

The Goliath Experiment

This experiment[13], in which a 150 - 175 GeV π^- beam is used at the Goliath spectrometer, last year showed a 4σ signal, i.e., a 200 nb cross-section in the combination of the two channels $\psi K^0 \pi^{\pm}$ and $\psi K^- \pi^+$. The experiment was done by the Saclay-Imperial College-Southampton-Indiana Collaboration.

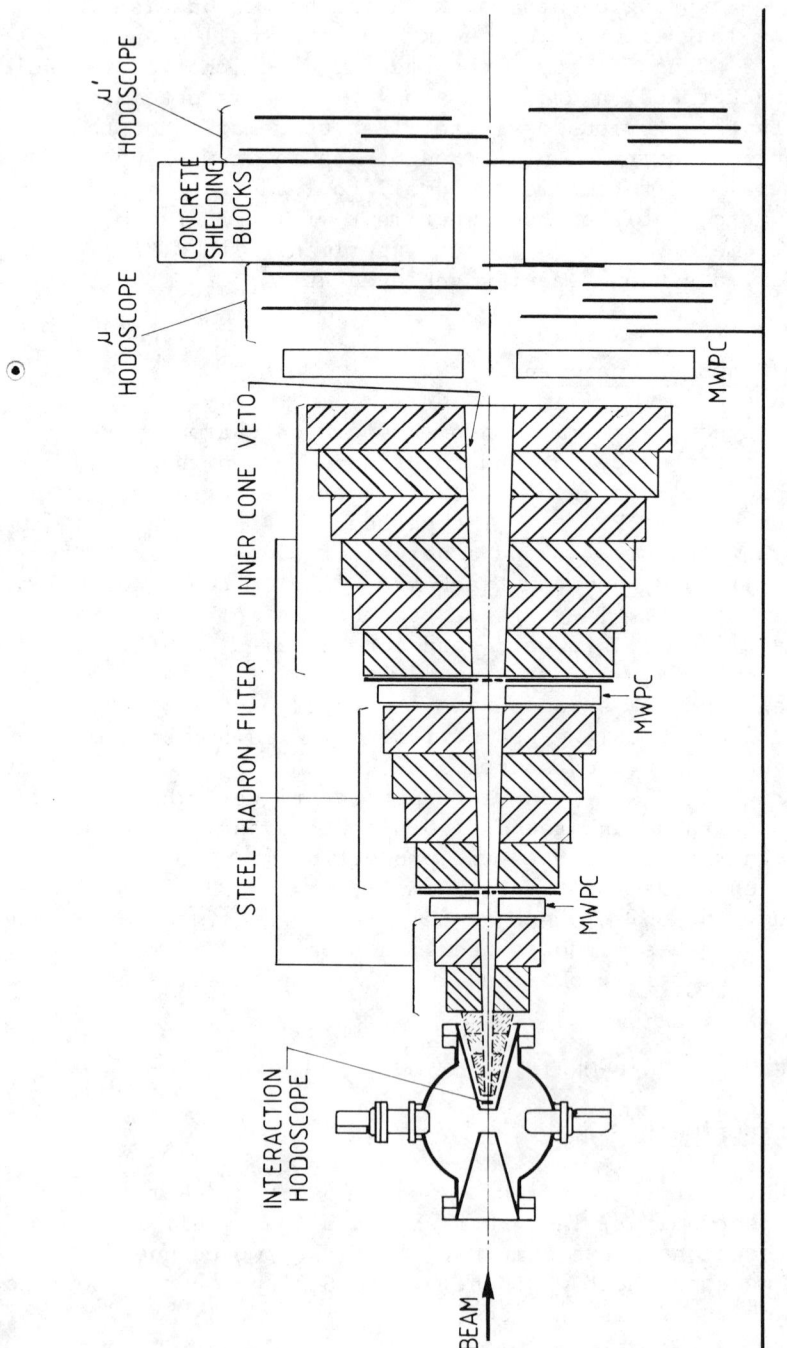

Fig. 12

NEW PARTICLES

The data taken during 1980 can produce about five times the previous statistics. A preliminary analysis does not confirm the existence of the signal, and the authors expect to set upper limits for the beautiful particle production in both the channels ψKπ and ψK, the last one being the most sensitive one.

Limits on Beauty Production from 3μ Events

A limit on beauty production has recently been obtained in a 225 GeV π⁻ beam at FNAL in the Chicago-Illinois-Princeton experiment[14]. The principle is to look to one beautiful particle decaying into ψX with ψ → 2μ, and the other beautiful particle having a third muon in its decay directly, or in the decay of a charmed particle secondary.

In order to make a straightforward comparison with the first Goliath results, they assume a 1% branching fraction for B → ψKπ and a 3% branching fraction for B → ψX. These two estimates originate from the same model. An additional assumption is that the branching fraction of B into muons is 18%, a generally accepted value.

From a sample of 65,900 ψ, a total of 487 show a third muon in the final state. In order to diminish the background, a cut in the transverse momentum p_T of the third muon is made and the corresponding efficiency ε for the detection of beauty is done by a Monte Carlo method. The results are summarized in the table:

p_T (GeV)	ε (%)	events	cross-section limits
1.4	1.6	4	11 nb
1.5	1.3	2	9 nb
1.6	1.1	1	7.5 nb

This result is in clear disagreement with the first Goliath result.

Another low limit on the associated beautiful particle production is obtained in proton interactions at 400 GeV in the Stanford-Caltech dump experiment at FNAL[15]. Protons interact on iron and can produce B pairs which subsequently decay into muons, with secondary charmed particles decaying into muons. The ψX decay mode can also lead to muons. The analysis was done by searching for 2μ and 3μ in the final state, and the lowest limit is obtained in the 3μ channel (ψμ). To lower the background, a p_T cut is made on the third muon. With one event left with p_T > 1.4 GeV, the limit obtained on the cross-section is 50 nb. Note that this limit is of the same order as the

limit obtained with pions and is smaller than the first Goliath
obtained with pions.

We conclude that there are now clear indications that the
associated production of beautiful particles by hadrons is much smaller
than the first experimental results led us to hope. However, in a
similar situation let us remember that the charmed particle production near threshold is not well accounted for by first order QCD
calculations, so that the beautiful particle production by hadrons
is certainly still an open question. The observation of the decay
lifetime of beautiful particles and their decay modes remain one
of the rare possible approaches to the measurement of the Cabibbo
mixing angles.

SEARCH FOR BEAUTY IN ANTINEUTRINO INTERACTIONS

It is well known that beautiful particles are expected to be
produced in antineutrino reactions through a beauty-changing charged
current. We report here about <u>possible</u> indications of beautiful
particle production in two experiments. These were obtained through
the analysis of dimuon events.

In the first experiment[16] done by the Bari-Birmingham-Brussels-
LPNHE-Rutherford-Saclay-University College Collaboration in the
BEBC equipped with an H_2-TST and filled with neon-hydrogen, a total
of 32 $\mu^+\mu^-$ events were detected, whereas 4 $\mu^+\mu^+$ and 5 $\mu^-\mu^-$ are
associated to background.

In order to separate ν from $\bar\nu$ produced events, a diplot with
the variables p^+/p^- and p_T^+/p_T^- is used where $p^\pm$ and $p_T^\pm$ are respectively
the total and transverse momentum of the positive and negative muon.
In the region $p^+/p^- + p_T^+ + p_T^- > 0$, where the Monte Carlo predicts
a 90% $\bar\nu$ selection, 21 events are found. Once the background, mainly
from charged pions and kaons, is subtracted 16.6 ± 5.6 events are
attributed to $\bar\nu$ interactions. The average number of strange particles per event is 1.3 ± 0.4 in good agreement with the GIM model.
Nevertheless, a bump of 8 events is seen at W (invariant hadronic
mass) with values between 5 and 6 GeV, whereas 0.7 events are
expected. These events are grouped in the region $1 < p_T < 1.3$ GeV
(Fig. 13).

Since a tentative explanation of this grouping of events might
be the production of beauty it is interesting to compare these
results with those obtained in another experiment[17] done by the
Berkeley-FNAL-Hawaii-Seattle-Madison Collaboration which makes use
of the quadrupole triplet beam at FNAL and of the 15' chamber equipped
with a two-plane EMI and filled with neon-hydrogen.

In this beam the high energy part is enhanced so that 1/2 (or 1/4) of the events have energies larger than 70 (or 115) GeV. The separation of $\bar{\nu}$ events is made by detecting the muon with the largest transverse momentum to be the primary muon. The Monte Carlo calculation showed that the $\bar{\nu}$ selection is correct 91% of the time. A total of 5.7 $\mu^+\mu^-$ are selected that way and Fig. 14 shows that their W spectrum is peaked around W-values of 5 GeV, much more than the $\mu^+\mu^-$ events due to neutrino interactions are. Despite the fact that the authors do not claim to draw any conclusion, do these facts lead to the conclusion that beauty is produced?

There are at least two objections. The first comes from the fact that the beauty hypothesis would most probably also imply the observation of other decay modes and, in particular, those leading to $\mu^+\mu^+$ events, for which other experiments put severe limits at this level of sensitivity. The second objection is that a bumping in W may be the result of two effects of purely kinematical origin. The first effect is that a selection of dimuon events indeed select charmed particles, i.e., heavy particles that tend to reduce the rate of low W-values. The other effect, when comparing antineutrino with neutrino reactions, is that the very high W-values are expected not to be favoured in antineutrino reactions because of the known y-dependency of the cross-section. All these considerations certainly prevent us from drawing any conclusions about the antineutrino production of beauty.

Note that in this second experiment, the ratio of neutrino produced dimuon events to one muon event is $R^{-+} = (0.37 \pm 0.10)10^{-2}$ in agreement with other experiments[18]. We also wish to mention a very recent result obtained with SKAT at Serpukhov[19] which, out of a total of 13 μ^-e^+ events, selected six events with $p_e > 400$ MeV (background 0.45 events) from which they extracted a similar number $R^{-+} = (0.35 \pm 0.14)10^{-2}$.

CONCLUSIONS

The past year has brought several important results in the study of the charmed baryon and meson states, their mass values and their lifetimes. Beautiful particles - and truthful particles even more - are very far from being established as unbound states.

ACKNOWLEDGEMENTS

We would like to acknowledge very useful discussions with P. Bareyre, P. Borgeaud, P. Bosetti, B. Conforto, C. Fischer, H. Fristch, G. Ringland, L. Montanet, K. Niu, N.W. Reay, G. Snow and W. Venus.

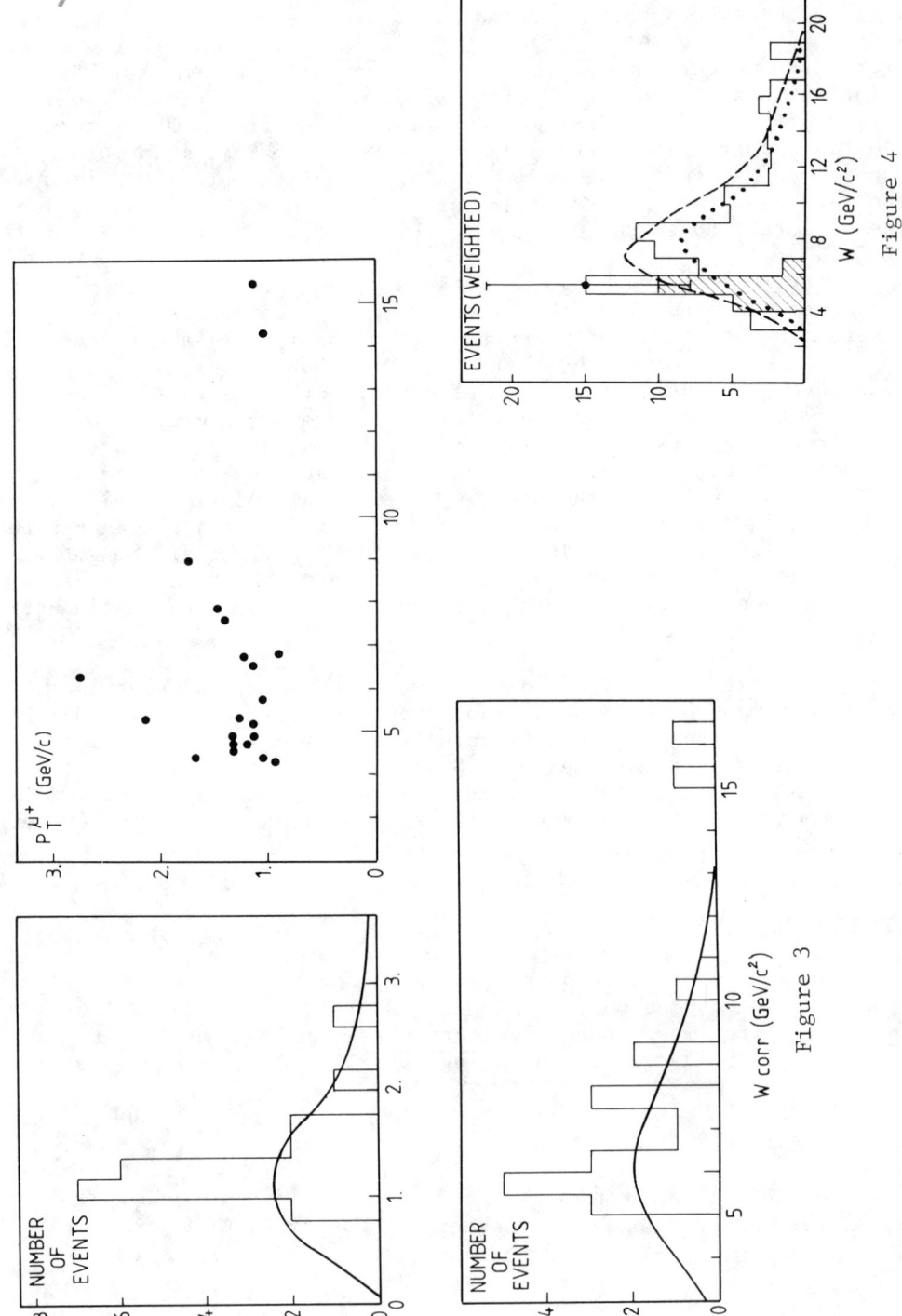

Figure 3

Figure 4

REFERENCES

1. For recent reviews on short-lived particles see:
   J. Sacton, XIV Rencontre de Moriond, Les Arcs, March 1979;
   M. Conversi, EPS International Conference on High Energy Physics, Geneva, July 1979;
   L. Voyvodic, International Symposium on Lepton and Photon Interactions at High Energies, Fermilab 1979.
2. A. Kernan and V. Lüth, International Symposium on Lepton and Photon Interactions at High Energies, Fermilab 1979.
3. J. Blietschau et al., to be published.
4. T. Kitagaki et al., to be published.
5. D. Allasia et al., submitted to Nucl. Phys.
6. U. Ushida et al., to be published.
7. M.I. Adamovitch et al., Phys. Lett. 89B (1980) 427; and to be published.
8. C. Baltay et al., Phys. Rev. Lett. 42 (1979) 1721.
9. M. Calicchio et al., to be published in Phys. Lett.
10. R. Ammar et al., Fermilab/Pub.80/39.
11. W. Allison et al., to be published in Phys. Lett.
12. J. Sandweiss et al., Fermilab/Pub. 80/16.
13. R. Barate et al., CERN EP/79-113.
14. R.N. Coleman et al., COO-3072-105.
15. A. Diamant-Berger et al., to be published.
16. N. Armenise et al., to be published in Phys. Lett.
17. H.C. Ballagh et al., Fermilab/Pub. 79/59.
18. V.A. Korotkov et al., SERP E107 and communication to this conference.
19. See, C. Matteuzzi, International Conference Neutrino '79, Bergen 1979.
20. Reviews of charmed particle production by hadrons can be found in:
    M. Boratav, 7th International Winter Meeting on Physics, Segovia 1979;
    W.M. Geist, EPS International Conference on High Energy Physics, Geneva 1979;
    F. Halzen, Madison preprint;
    F. Muller, Cargèse International School of Physics 1979;
    Y. Muraki, Topical Workshop on Forward Production of High Mass Flavourså Collège de France 1979.

DISCUSSION (Chairman: D. Cundy)

N.W. REAY, Columbus: Comment on the clustering of ν events with W: we (E-531) show the same effect with charm events, so the clustering in W does not necessarily indicate the existence of B-meson production. The question is: do you have any comments on the assumed large size of $B \rightarrow \psi + \chi$ in obtaining the Fermilab limit on B production in pion beams?

P. MUSSET, CERN: The limit does not depend on the $B \rightarrow \psi + \chi$ branching ratio, but so does the Goliath result, so assuming the B.R. is high, is useful in comparing the results of the two experiments.

WEAK CHARGED CURRENTS: HADRON FINAL STATES

B. Saitta

Department of Nuclear Physics
University of Oxford
Keble Road
Oxford OX1 3RH

INTRODUCTION

Bubble chamber experiments provide most of the data on hadrons produced in neutrino interactions, although results from large detectors using electronic techniques [1] have recently become available. Table 1 summarises the experiments and detectors whose results, presented at this conference or recently published, are discussed in this paper. Data from bubble chambers on hadron final states benefit from a 4π detection efficiency and accurate momentum resolution for charged particles, but suffer from the fact that neutral hadrons are not detected and therefore the energy of the incoming neutrino is not accurately known. "Energy correction methods", essentially based on transverse momentum balance [2], allow E_ν to be estimated. In the quark model the reactions which take place are particularly simple and for ν and $\bar{\nu}$ beams incident on nucleons, with the sea quarks and the Cabibbo angle neglected, involve only one quark flavour; explicitly we have (Fig. 1a)

$$\nu_\mu\, d \to \mu^-\, u$$
$$\bar{\nu}_\mu\, u \to \mu^+\, d \tag{1}$$

The study of final state hadrons could therefore provide information on the properties of the quark (u or d) that fragments into the observed hadrons. However the fact that the real scattering is on nucleons and not on quarks (Fig. 1b) complicates this simple scheme: consequences and problems are extensively discussed below.

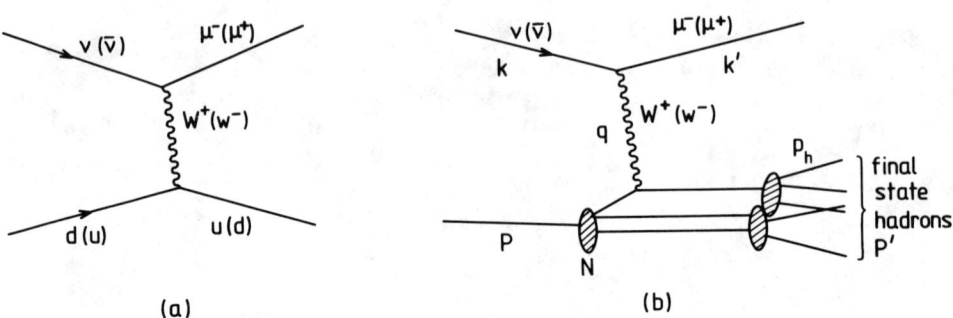

Fig.1 νN and $\bar{\nu} N$ charged current interactions in the parton model.

In section 1 the general features of the hadronic final states, multiplicities and charge distributions, are presented. Section 2 is a brief summary of the most recent data on non-strange vector meson production. A significant amount of information is available on the single particle inclusive cross section: the properties of energy and transverse momentum distributions are presented and discussed in sections 3 and 4 respectively and constitute the main part of this review. Section 5 presents the latest results on jet analysis and on searches for hard gluon bremsstrahlung effects.

The relevant variables are introduced in each section, here we define the usual kinematical quantities (fig.1b)

$$q = (k - k') \qquad Q^2 = (-q^2) = 2E_\nu E_\mu (1-\cos\theta_\mu)$$

$$x = \frac{Q^2}{2P \cdot q} \qquad y = \frac{P \cdot q}{P \cdot k}$$

$$W^2 = Q^2(\frac{1}{x} - 1) + M^2$$

1. MULTIPLICITY AND CHARGE DISTRIBUTIONS

Results on the mean charged multiplicity and multiplicity distributions come mainly from H_2 and D_2 experiments [3,4,5,6].

It is known, from hadron-hadron collisions, that the average multiplicity rises with increasing centre of mass energy. This seems to be true also in neutrino scattering as illustrated in fig. 2 where the mean charged multiplicity, $<n_{ch}>$ is plotted as a function of W^2.

HADRON FINAL STATES

Table 1. <u>Experiments from which results are presented</u>

Detector	Beam	Collaboration
15' H$_2$	ν	Berkeley-Fermilab-Hawaii-Michigan
15' H$_2$	$\bar{\nu}$ (ν)	Argonne-Carnegie Mellon-Purdue
15' D$_2$	ν	Illinois-Maryland-Stony Brook-Tohoku-Tufts
15' Ne/H$_2$	$\nu,\bar{\nu}$	Fermilab-IHEP-ITEP-Michigan
15' Ne/H$_2$	$\nu,\bar{\nu}$	Berkeley-Fermilab-Hawaii-Seattle-Wisconsin
BEBC Ne/H$_2$	$\nu,\bar{\nu}$	Aachen-Bonn-CERN-Demokritos-London-Oxford Saclay
BEBC H$_2$	ν	Aachen-Bonn-CERN-Munich-Oxford
GGM	ν	CERN-Milano-Orsay
SKAT Freon	ν	IHEP-IH(DDR)
CDHS	ν	CERN-Heidelberg-Dortmund-Saclay

Beyond the resonance region ($W^2 > 4$ GeV^2) the data are consistent with a linear dependence on $\ln W^2$; the results of this type of fit are summarised in table 2.

It should be stressed that no Q^2 dependence at fixed W is observed[6]. A completely similar behaviour characterises νN data obtained from an exposure of the bubble chamber SKAT[7] to a low energy neutrino beam, though the heavy mixture filling the chamber makes the analysis less certain.

The behaviour of higher moments of the multiplicity distribution, namely dispersion D and correlation parameter $\bar{f_2}$, defined as

$$D = (<n^2> - <n>^2)^{1/2}$$

$$\bar{f_2} = <n(n-1)> - <n>^2 = D^2 - <n>$$
(2)

is used to compare the neutrino data to other classes of processes

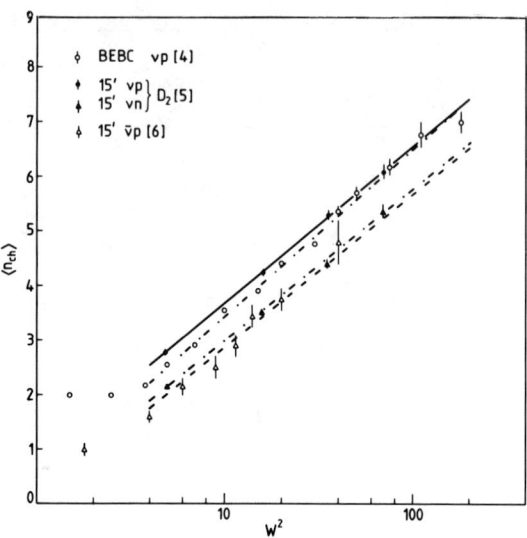

Fig. 2. Average charged multiplicity, $<n_{ch}>$, measured in νp, νn and $\bar{\nu}p$ scattering as a function of W. The lines are the result of fits of the type $a + b \ln W^2$

Table 2. Results of fits of the type $<n_{ch}> = a+b \ln W^2$, to the average charged multiplicity measured in ν and $\bar{\nu}$ scattering.

Reaction	a	b	Reference
νp	–	1.35 ± 0.15	3
νp	0.37 ± 0.05	1.33 ± 0.02	4
νp	0.80 ± 0.10	1.25 ± 0.04	5
νn	0.21 ± 0.10	1.21 ± 0.04	5
$\bar{\nu}p$	0.06 ± 0.06	1.22 ± 0.03	6

for example pp collisions and $\bar{p}p$ annihilations.

The Wroblewski equation[8], i.e. a linear relation between the dispersion D and the average multiplicity $<n>$, holds for ν scattering on both protons and neutrons[4,5]. Fig.3a shows D as a function of the average multiplicity from the 15' D_2 experiment and fig.3b D^- as a function of the average negative multiplicity from the BEBC νH_2 experiment.

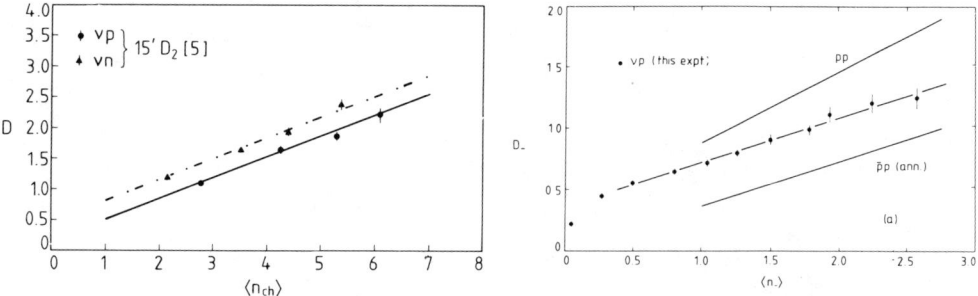

Fig.3 a) Dispersion D as a function of $<n_{ch}>$ for νp and νn scattering from ref.5. b) D^- versus $<n_->$ from ref.4. Also shown are the fits to pp and $\bar{p}p$ data. The lines are the results of fits of the type $D = K + A <n>$ The following values were obtained for the slope parameters:
$A_{\nu p} = 0.34 \pm 0.01$ (ref.5), $A_{\nu p} = 0.36 \pm 0.03$ (ref.4),
$A_{\nu n} = 0.34 \pm 0.01$ (ref.5)

The values of the slope parameters $A_{\nu p}$, $A_{\nu n}$ *, given in fig.3, are similar to that found in $\bar{p}p$ annihilation; this resemblance between the neutrino data and the class of annihilation processes is also supported by the behaviour of the correlation parameter f_2^{--} with $<n_->$ [4] (fig.4).

An asymptotic relation between multiplicity and dispersion has been discussed in the literature[9], relating D and $<n>$ to the multiplicity and dispersion of clusters of $(mass)^2 \simeq Q^2$, but it seems that the energy of the neutrino events is still too low to allow these types of ideas to be tested[5].

New interesting measurement come from studies of multiplicity differences of positive and negative hadrons travelling forward or backward in the hadronic centre of mass frame.

In the parton model description of neutrino scattering, the observed hadrons are the fragmentation products of the scattered quark or of the spectator diquark system. The problem of the separation between target and current fragments does not have a simple solution and various possibilities are discussed below(section 3). If we assume, for the time being, that this separation is achieved in the hadronic centre of mass system, then the forward ($x_F > 0$) and backward ($x_F < 0$) multiplicities would be related to the fragmentation of the

---

* It should be recalled that $D = 2D^- = 2D^+$, and if $D = a + b<n>$ then $D^- = a^- + b<n_->$ and $D^+ = a^+ + b<n_+>$ i.e. the slope parameter is the same.

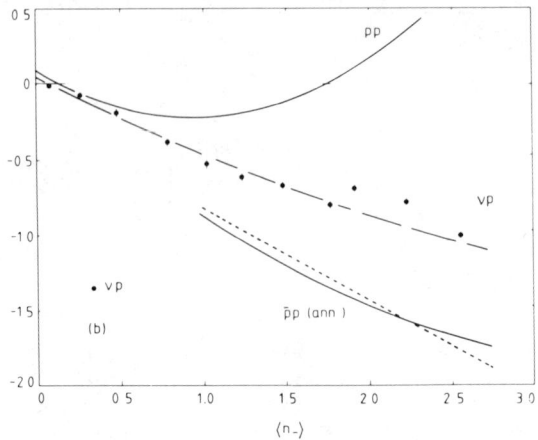

Fig.4 The correlation moment f_2^{--} as a function of $\langle n_-\rangle$ from ref.4. Also shown are pp and $\bar{p}p$ data. The line going through the data points is the result of a quadratic fit to the relation $f_2^{--} = A + B\langle n_-\rangle + C\langle n_-\rangle^2$.

struck quark (u for νscattering, d for $\bar{\nu}$) and of the diquark system (uu for ν, ud for $\bar{\nu}$) respectively (see fig.5).

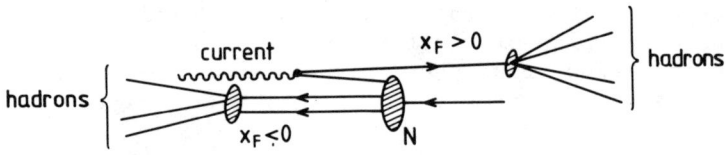

Fig.5 Ideal separation, in the current-nucleon centre of mass frame, between the fragments of the struck quark ($x_F>0$) and those of the spectator quarks ($x_F<0$)

For neutrino scattering the mean multiplicity forward is higher than backward, and this is due to positive hadrons at low W^2 (<16 GeV^2) and negative hadrons at high W^2, as shown in fig.6 where the difference $\langle Q\rangle = \langle n_+\rangle - \langle n_-\rangle$, which is the net average charge, is given as a function of W^2 for each hemisphere. The data are from the BEBC νH_2 experiment[4].

The relation between the net forward charge $\langle Q\rangle_F$ and the charge of the fragmenting quark, e_q is not a direct one. It has been

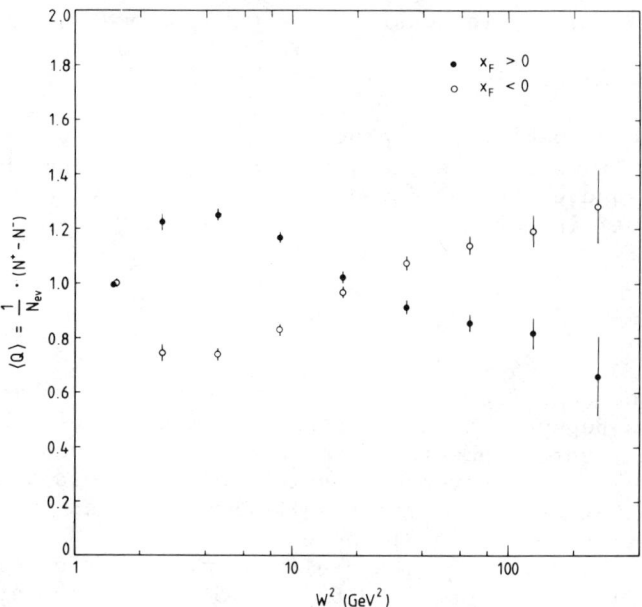

Fig.6 Average net charge in the forward and backward hemispheres as a function of W^2, from ref.4.

shown[10,11,12] that if the production of baryons is neglected, the average charge of a hadron jet initiated by a quark of charge e_q is

$$<Q>_q = e_q - \sum_i \gamma_i e_i \qquad (3)$$

where γ_i is the probability of creating a quark-antiquark pair of the type i. It is easy to show that the absolute value of the quark charge cannot be determined in a study of jets consisting of mesons only; the difference $<Q>_j - <Q>_k = e_j - e_k$ of quark charges can however be measured. Making the minimum assumptions that $e_u - e_d = e_u - e_s = 1$ and that $\gamma_u + \gamma_d + \gamma_s = 1$, equation (3) for a meson jet initiated by a u or a d quark becomes

$$<Q>_u = 1 - \gamma_u$$
$$<Q>_d = -\gamma_u \qquad (4)$$

The net charge is therefore independent of the quark charge and only the probability γ_u is measured. The situation changes if the production of baryons is considered. Eq. 4 becomes now

$$<Q>_u = (1-P)(1-\gamma_u) + P(3e_u - 4/3) \qquad (5)$$

where P is the probability of producing a baryon*; for P=1/3 $<Q>_u = e_u$. It is now clear why the relation between the net forward charge $<Q>_F$ and e_q is not direct. But consistency tests can still be carried out; in particular the difference $\Delta Q = <Q>_F^\nu - <Q>_F^{\bar\nu}$ should satisfy

$$\Delta Q = e_u - e_d \qquad (6)$$

As can be seen from fig.6, the average forward charge depends on W and this dependence can be assigned to an overlap between target and current fragments which is expected to vanish as 1/W in the limit $W \to \infty$[13]; the asymptotic value for ΔQ should satisfy the relation (6). The experimental result from a νN and $\bar\nu N$ experiment at Fermilab[14] is shown in fig.7; the extrapolation $1/W \to 0$ leads to a value of 0.98±0.15 for ΔQ. The preliminary values for $<Q>_F$ extrapolated to $W \to \infty$, obtained by the 15' D_2 collaboration[15], seem to be the same for νp and νn scattering (fig.8). This could rule out some simple models, where, for instance, the nucleon is mainly produced backward and the remaining charge is randomly distributed[4]. Such a type of model would in fact give $<Q>_F = 0.75$ for νp, $<Q>_F = -0.25$ for $\bar\nu p$ and $<Q>_F = 0.25$ for νn.

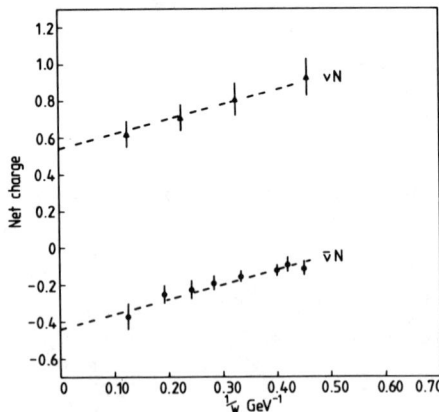

Fig. 7. The net forward charge as a function of 1/W for νN and $\bar\nu N$ interactions, from ref. 14.

---

*Note that eq.5 is valid only if the fragmenting quark is assumed to combine with the quark pairs uu, ud, us, dd, ds, ss, with equal probability.

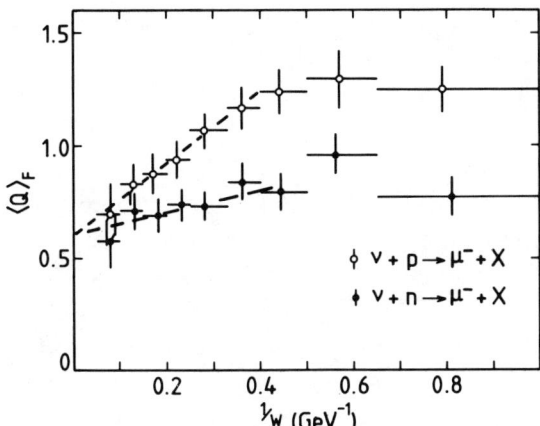

Fig.8 Preliminary result from the 15' D_2 experiment[15]. The net forward charge versus 1/W separately for νp and νn interactions. The lines are drawn through the points to guide the eye.

2. VECTOR MESON PRODUCTION

The way a quark fragments into hadrons should be determined experimentally; the detected particles could be directly produced during the fragmentation process or be the results of decays of heavier resonances. It is important to determine the amount of resonance production to test and eventually tune specific fragmentation models, and also because the production of a large fraction of heavier particles would affect the interpretation of some of the properties of the final state hadrons (e.g. transverse momenta). The results presented here, from νp, νD_2 and $\bar{\nu} p$ experiments[16,17,18] concern ρ^0 production rates as a function of various relevant kinematical quantities like Q^2, W, P_T. If valence quark processes dominate the production of vector mesons, then on the basis of the quark content ($\rho^0 = \frac{1}{\sqrt{2}}(u\bar{u} - d\bar{d})$) one would expect that

$$\frac{d\sigma(\nu N \to \mu^- \rho^0 + X)}{d\sigma(\nu N \to \mu^- + X)} = \frac{d\sigma(\bar{\nu} N \to \mu^+ \rho^0 + X)}{d\sigma(\bar{\nu} N \to \mu^+ + X)} \qquad (7)$$

and also a similar relation (the integral form of eq. 7) for the production rates ρ^0/ev. Table 3 and fig. 9 support the validity of this assumption. Slight discrepancies between the ν and $\bar{\nu}$ experiments are found if the ratios ρ^0/leading π (π^+ in ν and π^- in $\bar{\nu}$) are studied as functions of the centre of mass rapidity $Y_R = \frac{1}{2} \ln \frac{E + P_L}{E - P_L}$ and P_T.
In particular the sudden increase in ρ^0/π^- observed in the $\bar{\nu} p$ experi-

ment[18] in the current fragmentation region ($Y_R>0.5$) has no equivalent in the D_2 experiment (see Table 4).

Table 3. Production rates ρ^o/ev measured in νp, νD_2 and $\bar{\nu} p$ experiments

Reaction	Reference	<W>	ρ^o/ev
νp	16	~ 5	.21±.04
νp	17	~ 5	.21±.08
νn	17	~ 5	.18±.06
$\bar{\nu} p$	18	~ 3.4	.21±.03

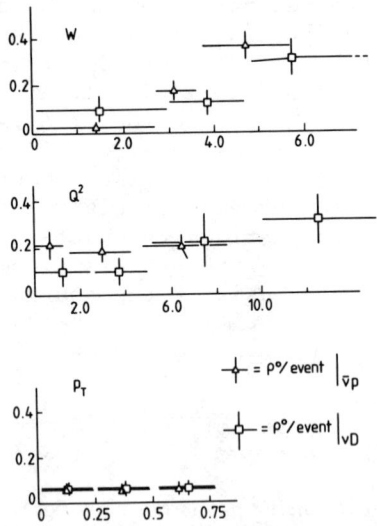

Fig.9. ρ^o/event from ν and $\bar{\nu}$ experiments (refs. 17 and 18) as functions of W, Q^2 and p_T.

Table 4. Ratios ρ^0/leading π as functions of centre of mass rapidity Y_R and P_T.

	$Y_R<-0.5$	$-0.5<Y_R<0.0$	$0.0<Y_R<0.5$	$Y_R>0.5$
$(\rho^0/\pi^+)_{\nu D}$	0.03±0.02	0.09±0.02	0.12±0.04	0.07±0.02
$(\rho^0/\pi^-)_{\bar{\nu} p}$	0.16±0.06	0.12±0.05	0.13±0.04	0.31±0.04

	$P_T<.25$	$.25<P_T<.50$	$P_T>.50$
ρ^0/π^+	0.05 ±.01	0.06±0.01	0.13±.03
ρ^0/π^-	0.09 ±.03	0.09±0.03	0.33±.09

3. FRAGMENTATION FUNCTIONS

A considerable amount of both theoretical and experimental work has been devoted to the study of the properties of the single-particle inclusive cross sections. In this section energy distributions of secondary hadrons produced in the process (sketched in Fig. 1b) $\overset{(-)}{\nu} N \to \mu^{-(+)} + h + X$ are discussed in some detail.

The differential cross section to produce a hadron of the type h can be written as a sum of three generalised structure functions[19]:

$$\frac{d\sigma_{\nu(\bar{\nu})}^h}{dx\,dz\,dQ^2} = \frac{G^2}{2\pi}\left[(1 - \frac{x_0}{x})F_1^h + \frac{x_0^2}{2x^2} F_2^h \mp (\frac{x_0}{x} - \frac{x_0^2}{2x^2}) F_3^h\right] \quad (8)$$

with $F_i^h = F_i^h(x,z,Q^2)$ where we have used the relation $Q^2 = 2ME_\nu xy$ and defined $x_0 = Q^2/2 ME_\nu$. In ref. 19 $z = \frac{P_h \cdot P}{P \cdot q}$ is the fraction of the parton energy carried by the hadron. The functions F_i^h are products of parton densities and fragmentation functions and are such that

$$\sum_h \int_0^1 dz\, z\, F_i^h(x,z,Q^2) = F_i(x,Q^2) \quad (9)$$

and the F_i are related to the ordinary structure functions of deep inelastic scattering F_i by $(F_1, F_2, F_3) = (2F_1, F_2/x, F_3)$. In the naive parton model the functions F_i^h assume a simple factorised form and in particular for ν scattering on protons if antiquarks and Cabibbo angle are neglected, we have $F_1^h = F_2^h = -F_3^h$ and

$$\frac{d\sigma_{\nu p}^{h}}{dxdz\,dQ^2} = \frac{G^2}{\pi} d(x) D_u^h(z) \tag{10}$$

Important consequences of eq. (10) are the <u>factorisation</u> in x and z of the cross section and the fact that the <u>fragmentation functions</u> $D_u^h(z)$ do not depend on Q^2 (or on x).

a) QCD Modifications

This sample picture is altered in QCD by the order α_s diagrams (fig. 10)[19,20,21].

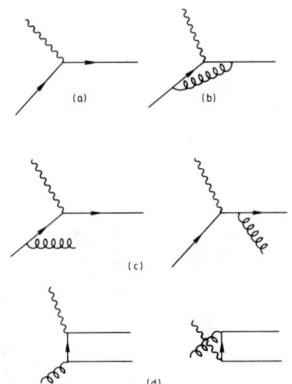

Fig.10. Feynman graphs contributing to the process current + N → H + anything.

The gluon emission before and after the absorption of the current would cause the fragmentation functions to have a Q^2 dependence $(D(z) \rightarrow D(z,Q^2))$ given by the Altarelli-Parisi evolution equations; for the fragmentation functions of a quark q we have for example:

$$\frac{\partial D_q^h(z,Q^2)}{\partial t} = \frac{\alpha_s(t)}{2\pi} \int_z^1 \frac{dz'}{z'} \left[P_{qq}(\tfrac{z}{z'}) D_q^h(z',t) + P_{Gq}(\tfrac{z}{z'}) D_G^h(z',t) \right] \tag{11}$$

where $t = \ln Q^2/\Lambda^2$, P_{qq} and P_{Gq} are the Altarelli-Parisi functions[22]. The diagram of fig. 10b would cause a violation of factorisation. Detailed calculations can be found in ref. 19; here it is sufficient to recall that for νp scattering, neglecting the strange sea and Cabibbo angle, we have

HADRON FINAL STATES

$$F_{1,2}^h(x,z) = 2\int_x^1 \frac{dy}{y} \int_z^1 \frac{dz'}{z'}\left[(d\, D_u^h + u D_d^h)\tilde{F}_{qq}^{1,2} + (d+\bar{u})D_G^h \tilde{F}_{Gq}^{1,2} + (D_u^h + D_{\bar{d}}^h)\tilde{F}_{qG}^{1,2}G\right]$$

(12)

$$F_3^h(x,z) = 2\int_x^1 \frac{dy}{y} \int_z^1 \frac{dz'}{z'}\left[(-dD_u^h + \bar{u}D_d^h)\tilde{F}_{qq}^3 + (-d+\bar{u})D_G^h \tilde{F}_{Gq}^3 + (-D_u^h + D_{\bar{d}}^h)\tilde{F}_{qG}^3 G\right]$$

where $u, \bar{d}, G$ are quark and gluon densities (explicit dependence on y and t omitted) and $D_u, D_{\bar{d}}, D_G$ are quark and gluon fragmentation functions (explicitly dependent on z/z' and t). Expressions for the functions $\tilde{F}_{qq}^{1,2,3}, \tilde{F}_{Gq}^{1,2,3}$ are given by equations 58–61 of ref. 19.

The breaking of factorisation is better studied in terms of double moments in x and z of the functions F_i^h because of the property of double convolution integrals that if

$$F(x,z) = \int_x^1 \frac{dy}{y} \int_z^1 \frac{dz'}{z'}\, D(\tfrac{z}{z'})\, \tilde{F}(\tfrac{x}{y}, z)\, q(y) \qquad \text{then}$$

$$\int_0^1 dx\, x^{n-1} \int_0^1 dz\, z^{m-1} F(x,z) = \left(\int_0^1 dz\, z^{m-1} D(z)\right)\left(\int_0^1 x^{n-1}\int_0^1 z^{m-1}\tilde{F}(x,z)\,dxdz\right)$$

$$\times \left(\int_0^1 x^{n-1} q(x)\, dx\right) \qquad (13)$$

The QCD predictions are particularly simple for non-singlet (NS) moments. For fragmentation functions the difference $D_i^h - D_j^h$ $i \ne j$ is a non-singlet,[*] while for structure functions $xF_3^{\nu N}$ is a non-singlet. For this special combination (double non-singlet) the predictions for the double moments from eq. 12 are:

$$F_3^{NS}(n,m) = \frac{C_n}{\left[\ln q^2/\Lambda^2\right]^{d_n}} \left(1 + \frac{\alpha_s}{\pi}\frac{4}{3}f_{nm}^3\right)\frac{C_m}{\left[\ln Q^2/\Lambda^2\right]^{d_m}} \qquad (14)$$

$\underbrace{\qquad\qquad\qquad}_{Q^2 \text{ dependence of S.F.}}$ $\underbrace{\qquad\qquad\qquad}_{Q^2 \text{ dependence of frag. function}}$

---
[*] Note that by charge conjugation invariance $D_u^+ - D_{\bar{u}}^-$ is a non-singlet combination, irrespective of the nature of the hadrons involved.

with $d_k = \dfrac{4}{33-2f} \left[1 - \dfrac{2}{k(k+1)} + 4 \sum\limits_{2}^{k} \dfrac{1}{j} \right]$ (f=number of flavours)

and

$$2 f_{nm}^{(3)} = \sum_{1}^{n-1} \dfrac{1}{j} \sum_{1}^{m-1} \dfrac{1}{k} + \sum_{1}^{n+1} \dfrac{1}{j} \sum_{1}^{m+1} \dfrac{1}{k} + \dfrac{1}{m(n+1)} + \dfrac{1}{n(m+1)} - \dfrac{1}{nm} - \dfrac{1}{(n+1)(m+1)}$$

(15)

which expresses quantitatively the violation of factorisation*
Similar expressions can be obtained for other combinations.

b) <u>Higher Twist</u>

The Q^2 dependence and the breaking of factorisation is not a characteristic of QCD alone. The so-called "higher twist" terms, which are corrections $O(m^2/Q^2)$, would also lead to the same type of effects. Specific predictions are made by Berger[23] for hadrons produced in deep inelastic neutrino scattering by considering the dissociation of an off shell virtual quark into a pion plus anything (fig.11).

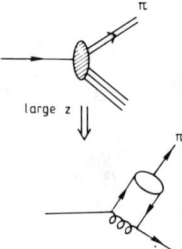

Fig.11. Dissociation of an off shell parton into a pion plus anything from ref.23.

For π^+ production in νN and π^- in $\bar\nu N$ scattering the cross section is expected to have the form (y dependent):

$$\sigma^\nu(z,y,Q^2) \propto (1-z)^2 + \dfrac{4(1-y)}{9} K_B^2/Q^2$$

$$\sigma^{\bar\nu}(z,y,Q^2) \propto (1-z)^2(1-y)^2 + \dfrac{4}{9}(1-y) K_B^2/Q^2$$

(16)

---
* Eq. 15 was obtained using the expressions for F_{qq}^3 of ref. 19. Slightly different answers would be obtained if the Breit frame is used as in Sakai[30]. In particular the equivalent of eq.15 is not symmetric in n and m.

HADRON FINAL STATES

where K_B is a parameter which includes all mass terms. Averaging eq. 16 over y, the following expression is obtained for the fragmentation functions:

$$D(z,Q^2) \propto (1-z)^2 + C <K_B^2> /Q^2 \qquad (17)$$

with $C=\frac{2}{9}(\frac{6}{9})$ for $\nu(\bar\nu)$. Eq. 17 should be valid only in the large z ($\gtrsim 0.5$) limit and for π^+ production by ν and π^- by $\bar\nu$.

c) Experimental Results

The theory does not provide a definite prescription on how to separate the fragments of the struck quark from those of the spectator quarks. Different reference systems have been suggested which may also enable target mass corretions to be taken into account [19,20].

It seems that the variable best suited to represent the fraction of the parton energy carried by the hadron and the frame in which the struck quark fragments can be isolated must be determined experimentally. Several variables and reference frames have been used in the different analyses:

Laboratory system variables: $z_L = \dfrac{P_h \cdot P}{P \cdot q + P^2} = \dfrac{E_h}{\nu + M}$; $z'_L = \dfrac{P_h\, P}{P \cdot q} = \dfrac{E_h}{\nu}$

c.m.s.: $z^* = \dfrac{2E_h^*}{W}$; E_h^* = energy in the c.m.s.

Breit frame: $z_B = -2\dfrac{P_h \cdot q}{Q^2}$; $\zeta = \dfrac{E^B + P_L^B}{q}$

In terms of any of these "z-variables" the fragmentation functions, which in general are functions of x and Q^2, are operationally defined as:

$$D^h(z,x,Q^2) = \dfrac{1}{N_{ev}(x,Q^2)} \dfrac{dN^h}{dz} \qquad (18)$$

where $N_{ev}(x,Q^2)$ is the number of events at that given x and Q^2 and dN^h is the number of hadrons of type h in the interval dz.

Depending on which frame is chosen for the analysis, the struck quark fragments are selected to be particles travelling forward in the centre of mass ($x_F > 0$), or forward in the Breit frame ($z_B > 0$), or with a considerable fractional energy in the laboratory ($z_L \gtrsim 0.2$). Some support for the separation $x_F > 0$ comes from the difference in the c.m.s. rapidity (Y_R) distributions of the positive and negative hadrons, where a clear step is visible at $Y_R=0$, but only at high W values ($\gtrsim 4$ GeV). This result was first obtained by the 15' $\bar{\nu}p$ collaboration[24] and then confirmed in ν and $\bar{\nu}$ experiments at CERN[4,25].

With a W cut, the z-distributions for positive and negative hadrons from the 15' $\bar{\nu}p$ experiment[24] do not exhibit any Q^2 dependence (fig.12) and the second moment of the fragmentation function ($<z>$) is rather independent of Bjorken x (fig.13)

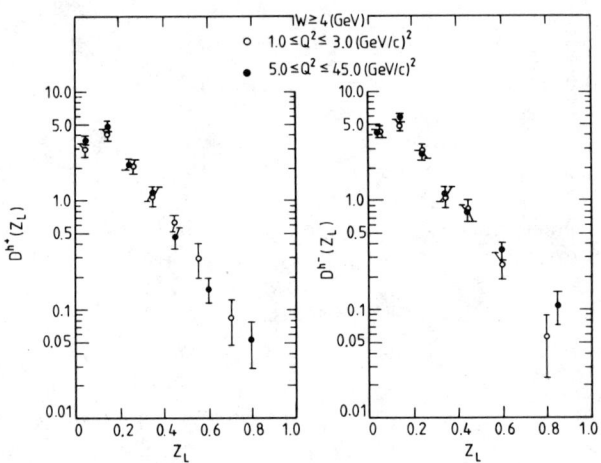

Fig.12. d-quark fragmentation functions in positive and negative hadrons for two different Q^2 regions (from ref.24)

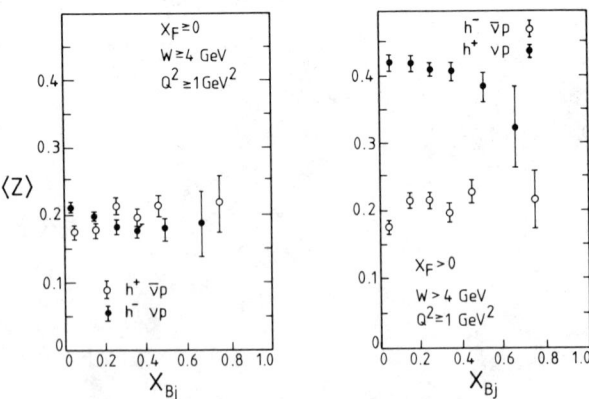

Fig.13. $<z>$ for positive and negative hadrons (W>4 GeV) as a function of Bjorken x, measured in νp and $\bar{\nu}p$ interactions.

Essentially the same trend is observed in the νp data of the ABCMO collaboration if identical cuts are applied[26,27], though there could be some indication of an x dependence of <z> for positive hadrons. The fact that <z> for positive hadrons in νp scattering is significantly higher than <z> for negative hadrons in ν̄p interactions, could indicate an excess of fast protons (those two quantities should be equal if only pions were considered), but this is beyond the scope of the present study.

A cut in W constitutes a severe limitation on the accessible x region (for instance, for E_ν=40 GeV, Q^2=2 GeV and W>4 GeV, only 0.027< x <0.116 is accessible). If this condition is relaxed and the whole W range accepted (above $Q^2_{min} \sim$ 1-2 GeV^2, effects due to the resonance region are considerably reduced), then the fragmentation functions show a significant Q^2 dependence, first shown by the BEBC νH_2 collaboration[28] and confirmed by the GGM νN SPS experiment[29]. The Q^2 evolution of the non-singlet moments (averaged over x)

$$D^{NS}(m,Q^2) = \int_0^1 z^{m-1} \left[D^+(z,Q^2) - D^-(z,Q^2) \right] dz \quad (19)$$

with $D^\pm(z,Q^2) = \dfrac{1}{N_{ev}(Q^2)} \dfrac{dN^\pm}{dz}$ is in agreement (surprisingly) with the predictions of the first order QCD calculation (fig.14).

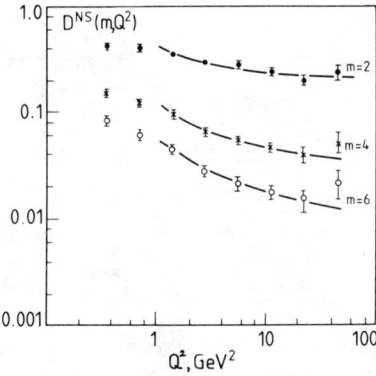

Fig.14. Even moments of the non-singlet combination $D^{NS}(m,Q^2) = D^+(m,Q^2) - D^-(m,Q^2)$ as a function of Q^2 from the BEBC νH_2 experiment[28]. The curves show the logarithmic dependence predicted by first order QCD.

A value of 0.52±.08 GeV for Λ is quoted in ref.28. The fragmentation functions exhibit also a significant Bjorken x dependence for Q^2<10 GeV^2, as illustrated in fig.15a, where $D^+(3,Q^2)$ from ref.28 is plotted versus x, and in fig.15b, where $D(z,Q^2)$ obtained by the GGM collaboration is shown for two x intervals at fixed Q^2.

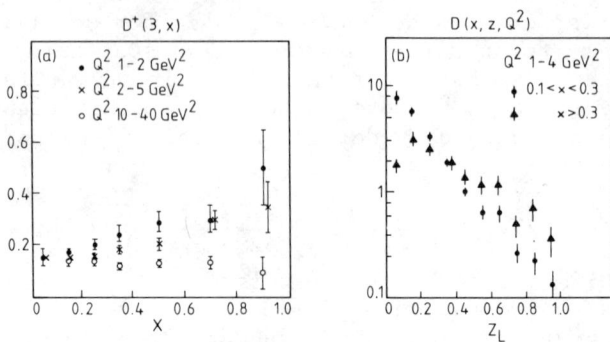

Fig.15. a) $D^+(3,Q^2)$ from the BEBC νH_2 experiment plotted versus x for 3 ranges in Q^2.
b) $D(z,x,Q^2)$ from ref.29 for two ranges in x at fixed Q^2.

In order to compare the predictions of next-to-leading order QCD with the experimentally observed breaking of factorisation, the Q^2 dependence of the non-singlet double moments of the fragmentation functions was investigated by the ABCMO collaboration[30]. More precisely, the normalised double moments, defined as:

$$D^h(n,m,Q^2) = \frac{\int_{E_{min}}^{E_{max}} \phi(E)\, dE \int_{x_1}^{1} x^{n-1} dx \int_{0}^{1} z^{m-1} dz \; \frac{d\sigma^h}{dx\,dz\,dQ^2}}{\int_{E_{min}}^{E_{max}} \phi(E)\, dE \int_{x_1}^{1} x^{n-1} dx \; \frac{d\sigma_{ev}}{dx\,dQ^2}}$$

(20)

where $x_1 = \frac{Q^2}{2M(E_\nu - E_\mu^c)}$; E_μ^c = minimum muon energy; E_{min} and E_{max} are the minimum and maximum ν energy, $\phi(E)$ = energy spectrum, $\frac{d\sigma_{ev}}{dx\,dQ^2}$ = cross section, are studied to divide out the Q^2 dependence of the structure functions (hence the denominator in eq.20) and to take care of the finite spectrum and energy of the beam (hence the integral ove the energy and the integral from x to 1 rather than from 0 to 1). The non-singlet combination $(D^+ - D^-)^1$ of the double moments defined above are shown in fig.16 for various choices of n and m, for hadrons going forward in the hadronic centre of mass system. The theoretical predictions were obtained by integrating the expression in eq.12, having assumed a parametrisation of the quarks and gluons density functions. The curves obtained are compared with the data in fig.16

and are normalised to coincide with the data points for n=5 at $Q^2 \approx 10$ GeV^2. (For low n values the predictions are not reliable[30] and therefore are not shown.)

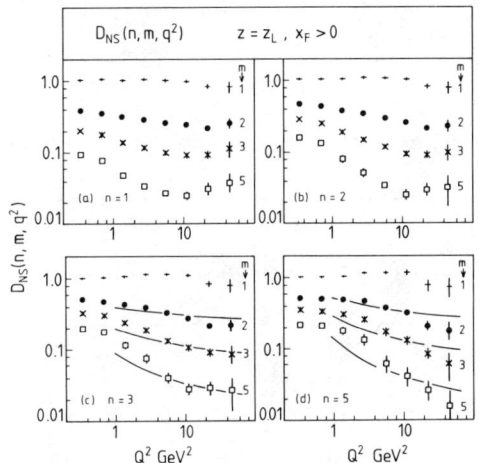

Fig. 16. Non-singlet double moments $D^{NS}(n,m,Q^2)$ for various choices of n and m plotted against Q^2. The solid lines are the predictions from next-to-leading order QCD.

The Q^2 dependence of the NS double moments defined by eq.20 is of the type

$$D^{NS}(n,m,Q^2) = \frac{C_m}{\left[\ln Q^2/\Lambda^2\right]^{d_m}} \left(1 + \frac{\alpha_s}{\pi} \frac{4}{3} K_{nm}\right) \qquad (21)$$

with d_m and K_{nm} calculable in the theory. At fixed n therefore there is one free parameter, C_m, for each m. Once the C_m have been determined, the n dependence is entirely predicted. The two ratios S_1 and S_2, first introduced by Sakai[20], defined as:

$$S_1(n_1, n_2, m) = \frac{D^{NS}(n_2, m)}{D^{NS}(n_1, m)} \qquad (22)$$

and

$$S_2(n_1, m_1; n_2, m_2) = \frac{S_1(n_1, n_2, m_1)}{S_1(n_1, n_2, m_2)} \qquad (23)$$

can be well determined experimentally (once correlations are taken into account) and would both be equal to 1 if factorisation holds. S_1 and S_2 have an absolutely predicted Q^2 behaviour since all the arbitrary constants cancel in taking the ratio of moments. S_1 and S_2 are shown in fig.17, together with the predictions for various choices of n_1, n_2, m_1, m_2. A violation of factorisation ($S_{1,2} \neq 1$) is observed in the range $2 < Q^2 < 16$ GeV^2, but the size and behaviour are not well described by QCD; it must be emphasised that this is a stringent test of the theory since there are no arbitrary constants left. Slight differences in the values of S_1 and S_2 but <u>not</u> in the trend are found if the variable z_B and the Breit frame are used[30].

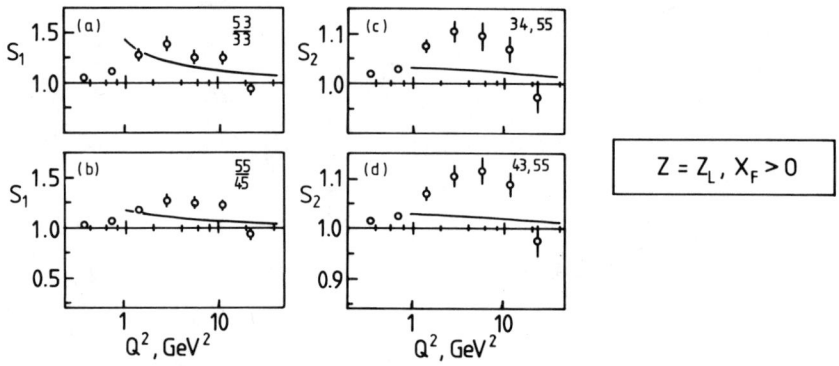

Fig.17. Ratios of double moments S_1 and S_2 (defined in the text) plotted versus Q^2. The lines are the predictions from next-to-leading order QCD.

As shown in fig.15b, the data of the CERN-Milano-Orsay collaboration[29], exhibit a violation of factorisation in x at low Q^2. This has been interpreted as a reflection of the presence in the inclusive

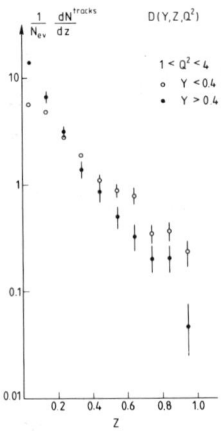

Fig. 18. $D(y,z,Q^2)$ from ref.29 for two ranges in y at fixed Q^2.

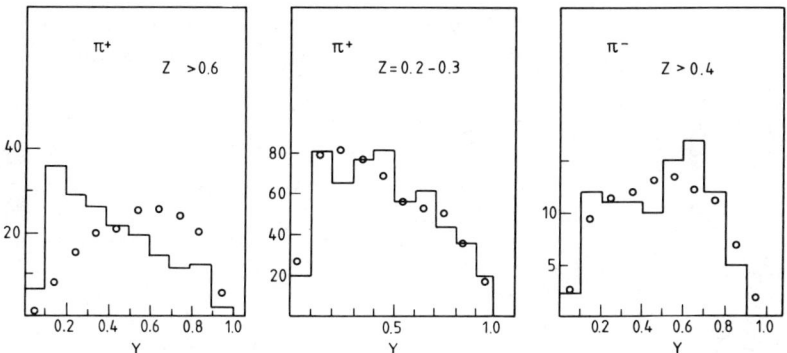

Fig.19. y-distributions at fixed z for positive and negative hadrons compared with Monte Carlo results (open circles) from ref.29.

cross section of y dependent terms, predicted by the model described in ref.23. A comparison of figs.15b and 18 shows a similar behaviour with x and y of the distributions at fixed (low) Q^2 and, furthermore, the y distribution at large z for positive hadrons suggest the presence of a term proportional to $(1-y)$ (eq.16). No discrepancy between the data and a Monte Carlo simulation, based on conventional Feynman and Field model for the fragmentation of the u-quark, was found for negative hadrons or low-z positive hadrons (fig.19). The higher twist contribution would of course affect the value of Λ obtained in the lowest order analysis of the NS fragmentation functions. Estimates for the value of the parameter $<K_B^2>$ in eq.17 have been given.

It has been suggested[31] that the difference between fragmentation functions of positive and negative hadrons should be used to eliminate the background from particles produced via a mechanism different from the one considered in ref. 23.

The CERN-Milano-Orsay collaboration obtained the preliminary value of $1.73 \pm .21$ for $<K_B^2>$ assuming $\Lambda = .140$ GeV from a fit to the Q^2 evolution of the NS moments[32]. If a form of the type

$$D^{NS}(m,Q^2) = C_m + A_m/Q^2 \qquad (24)$$

is assumed for the non-singlet moments and the only Q^2 dependence at large z (>0.5) comes from the higher twist terms of eq.17, then A could be easily related to $<K_B^2>$. The values of A_m obtained from a fit to the BEBC νH_2 published data (for z>.5)[33] are shown in fig.20 as a function of m. The solid and dashed lines were obtained by setting $K_B^2 = 2.4$ and $<K_B^2> = 2.9$ respectively* in eq.17. The fact that eq. 24 gives a good description of the data does not constitute in itself evidence for the presence of twist four terms.

*The values of K_B^2 depend on the functional form assumed for $D^{NS}(z,Q^2)$. In this analysis, $D^{NS}(z,Q^2) = c(1-z)^2 + 2<K_B^2>/Q^2$ was used.

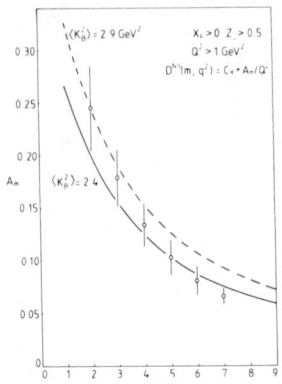

Fig.20. Coefficient A_m, from a fit of the type $D^{NS}(m,Q^2)=C_m+A_m/Q^2$ to the published BEBC νH_2 non-singlet moments, plotted versus m. The solid and dashed lines are the expected values from ref.23.

4. TRANSVERSE MOMENTUM

An increase with W^2 (and Q^2) of the average transverse momentum squared of hadrons produced in neutrino scattering was reported some years ago[34]. A similar rise observed in analyses of data on $e^+e^- \rightarrow$ hadrons[35], in conjunction with an excess of so-called "planar events" interpreted as evidence for hard gluon bremsstrahlung, has renewed the interest in the transverse properties of the final state hadrons.

The variables listed below are used in this section: (see Fig.21

$\vec{p}_T$: the component of the hadron momentum transverse to the current direction $\vec{q}$ (or to the total measured hadrons momentum vector $\vec{P}_H$).

p_n : the component of $\vec{p}_T$ out of the lepton scattering plane.

z_L, z^* : already defined in section 3.

$z_\pm = \dfrac{E_h^h \pm P_L^h}{E_H^H \pm P_L^H}$ where E_h, P_L^h (E_H, P_L^H) refer to a single hadron (all hadrons).

The direction $\vec{q}$ of the current corresponding to the exchanged virtual boson provides a "natural" axis with respect to which transverse momenta can be evaluated, though its determination could be affected in a systematic way by the uncertainty in estimating the energy of the incoming neutrino.

On the other hand p_n, the component of p_T out of the lepton plane, is determined using only the ν and μ directions which are accurately known; hence it is completely insensitive to the energy correction method employed. In the absence of any azimuthal asymmetry of dynamical origin

$$<p_n^2> = \frac{1}{2} <p_T^2> \qquad (25)^*$$

This relation is used in the analyses as a check: the validity of eq. 25 would imply that no systematic bias is introduced in the estimate of the current direction. For completeness, it should be said that p_T is relative to $\vec{q}$ in the BEBC νH_2[37] and relative to $\vec{P}_H$ in the 15' Ne/H_2[38] and BEBC Ne/H_2 [25] analyses.

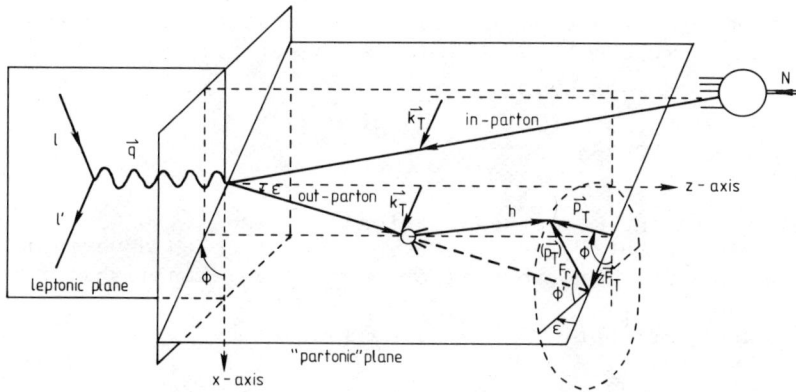

Fig.21. Transverse momentum of a secondary hadron in absence of perturbative processes, from ref.36.

The transverse momentum squared of the observed hadron, ignoring perturbative processes for the time being, is given by [37] (fig.21)

$$p_T^2 = z^2 K_T^2 (p_T^2)_{Fr} + 2 K_T (p_T)_{Fr} \cos\phi' + O(\varepsilon^2) \qquad (26)$$

where z is the fraction of the parton momentum carried by the detected hadron, K_T is the transverse momentum that the initial state parton had before absorbing the current, and $(p_T)_{Fr}$ is the transverse momentum relative to the direction of the fragmenting parton (for other symbols see fig. 21). Averaging eq. 26 over azimuthal angles one obtains

$$<p_T^2> = <z^2><K_T^2> + <p_T^2>_{Fr} \qquad (27)$$

---

* An azimuthal asymmetry would change this relation to

$$<p_n^2> = \frac{1}{2} <p_T^2> - \frac{1}{2} <p_T^2 \cos 2\phi>$$

The effect of gluon radiation by the incoming or outgoing parton is to add an extra contribution $\langle p_T^2 \rangle_{QCD}$, to the hadronic $\langle p_T^2 \rangle$ of eq. 27:

$$\langle p_T^2 \rangle = \langle z^2 \rangle \langle K_T^2 \rangle + \langle p_T^2 \rangle_{Fr} + \langle p_T^2 \rangle_{QCD} \qquad (28)$$

This perturbative part is calculable [39,40,21] or, more precisely, the transverse momentum that the parton acquires during the radiative process is calculable. The result is that:

$$\langle p_T^2 \rangle_{parton} \propto \alpha_s(Q^2) Q^2 f(x,y) \qquad (29)$$

The function $f(x,y)$ has very little y dependence and behaves approximately like $\frac{1-x}{x}$; the final result is therefore:

$$\langle P_T^2 \rangle_{parton} \approx K \alpha_s(Q^2) W^2 \qquad (30)^*$$

For the coefficient K, Altarelli gives a value of ~ 0.031 for almost all x [41]. Other calculations obtain $K \sim 0.04$ with a variation of about 15% with Q^2. Previous experiments have established that:

i) the distribution $\frac{d\sigma}{dp_T^2}$ is well described by $e^{-\alpha m_T}$ (where $m_T^2 = p_T^2 + m_\pi^2$) with $\alpha \sim 6$ GeV^{-1} [34];

ii) the average p_T, $\langle p_T^2 \rangle$, rises with increasing W^2 (and somewhat less with Q^2), particularly for high z_L ($\gtrsim 0.2$) [34,26].

Results recently obtained have clarified the situation and from an experimental point of view the following facts are now well established:

i) e^{-6m_T} is still a good parametrisation of $\frac{d\sigma}{dp_T^2}$, at least for $p_T^2 \lesssim 1-2$ GeV$^2/c^2$. This has been confirmed by data from the SKAT experiment [7].

---

*The Q^2 dependence of eq. 29 is distorted when integrating over x and y at fixed neutrino energy; eq. 30 holds for W^2 not close to the maximum value allowed by E_ν.

ii) The BEBC Ne/H_2 experiment with a threefold increase in the number of both ν and $\bar{\nu}$ charged current events has shown that ν and $\bar{\nu}$ data are in quantitative agreement with each other, as far as $<p_T^2>$ is concerned [25] (fig.22). This resolves an earlier uncertainty as to whether $<p_T^2>$ of hadrons produced by $\bar{\nu}$ increases with W^2.

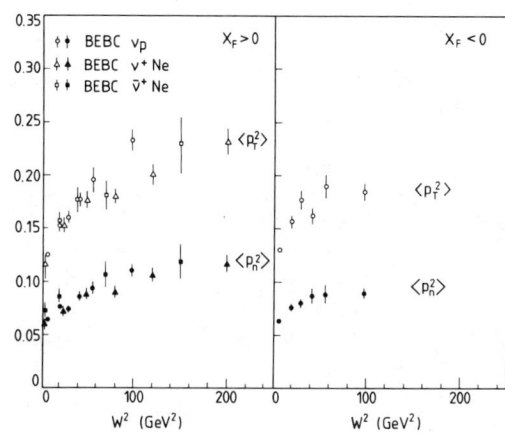

Fig. 22. $<p_T^2>$ and $<p_n^2>$ as a function of W^2 for particles with $x_F > 0$ and $x_F < 0$.

The rise of $<p_T^2>$ with W^2 is mainly associated with particles going <u>forward</u> in the hadronic centre of mass system. Fig. 22 shows data from the BEBC νH_2 collaboration on $<p_T^2>$ and $<p_n^2>$ separately for tracks with $x_F > 0$ and $x_F < 0$. The rate of increase with W^2 is higher forward than backward by a factor of about 3 [42]. This symmetry becomes more enhanced if a selection on z is made. This is illustrated in fig.23, where $<p_T^2>$ at fixed W is plotted versus z^* (the convention that z^* is negative for hadrons with $x_F < 0$ was adopted). The data exhibit a clear forward-backward symmetry at $<W> = 5$ GeV, while an evident asymmetry appears at larger W associated with high $|z^*|$ ($\gtrsim 0.3$). A similar conclusion is reached by the BFHSW collaboration on the basis of the behaviour of $<p_T^2>$ versus W^2 for values of $z_\pm > 0.2$ (fig.24).

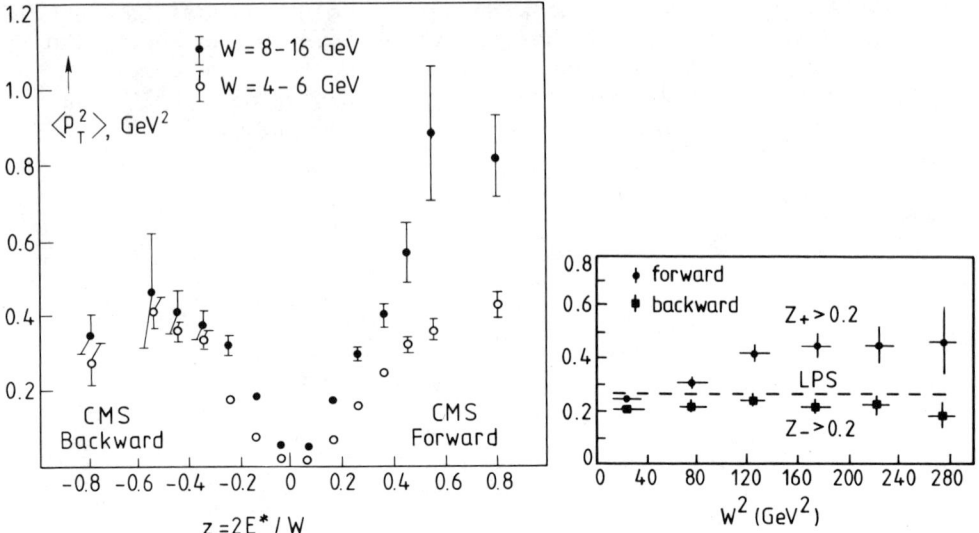

Fig. 23. $\langle p_{kT}^2 \rangle$ as a function of z^* for two different values of W, from the BEBC νH_2 experiment.

Fig. 24. $\langle p_T^2 \rangle$ as a function of W^2 for particles with $z_{\pm} > 0.2$. Preliminary from ref. 38.

An analysis of the dimuon event sample, relevant to the subject of transverse momentum, has been presented at this conference by the CDHS collaboration [1]. The production of a charmed particle, for instance via the reaction $\nu s \to \mu^- c$, and its subsequent semileptonic decay, would give an extra muon in the final state. The average transverse momentum squared of the "second muon" shows the same tendency to increase with W^2 as it does for hadrons, although some corrections should be applied to take into account the fact that the particle which is primarily produced is the parent charmed meson rather than the detected muon.

All the data (hadrons + muon from charm decay) are consistent with $\langle p_T^2 \rangle$ (or $\langle p_n^2 \rangle$) having a linear W^2 dependence but could equally well be described by the form A + BW. As an example, figs. 25a, 25b and 25c show the CDHS (preliminary), ABCDLOS and ABCMO results: A+BW and A+BW^2 would both give good fits.

The W range spanned by the present data does not allow an accurate determination of the power of the W dependence. The recent results from a muon scattering experiment at CERN [1] seem to agree very well with the ν data (fig. 26) (if the same cuts are applied) and could help in this respect.

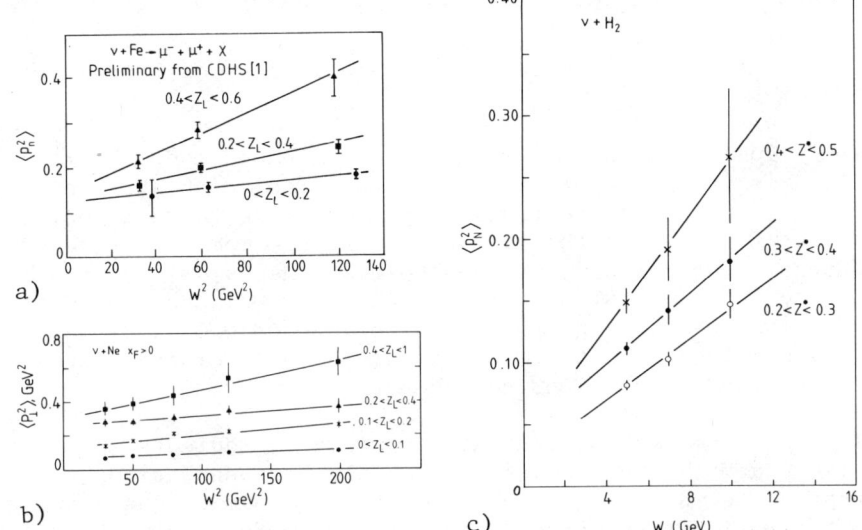

Fig. 25. a) $\langle p_n^2 \rangle$ of the μ^+ produced in the reaction $\nu N \to \mu^- \mu^+ X$ versus W^2 at fixed z. Preliminary result from ref.1.
b) $\langle p_T^2 \rangle$ of secondary hadrons with $x_F > 0$ as a function of W^2 for various z intervals from ref.25.
c) $\langle p_n^2 \rangle$ as a function of W for 3 values of z^* from the BEBC νH_2 experiment [37]. The lines are drawn to guide the eye. The form of the W dependence ($A+BW^2$ or $A+BW$) cannot be accurately determined since the W range is limited.

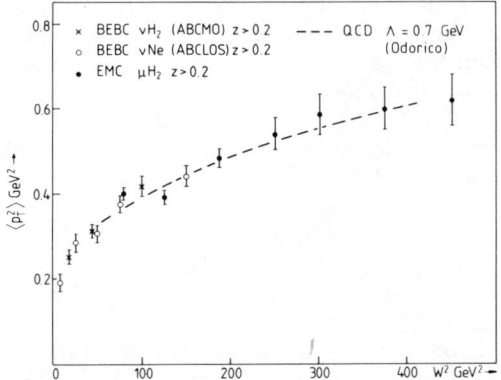

Fig. 26. $\langle p_T^2 \rangle$ as a function of W^2 for particles with $z_L > 0.2$ from neutrino [25,37] and muon [1] experiments. The dashed line is the prediction from a QCD based model [44].

Let us turn now to the comparison of the data with theoretical predictions. The $\frac{d\sigma}{dp_T^2}$ distributions obtained by the BFHSW group agree well, at high p_T^2, with the QCD predictions of Halzen and Scott [43].

The result of a calculation from a QCD based model [44] is shown in fig.26 and is in agreement with the data. The calculation assumes a value of 0.7 GeV for Λ and that non-perturbative terms in eq.28 contribute ~ 0.2 GeV^2 to the average transverse momentum squared of the hadrons. Furthermore, if the whole of the $<p_T^2>$ rise with W^2 is assigned to QCD effects, using eq.30 with K=0.031 and the relation $<p_T^2>_{QCD} = <z^2><p_T^2>_{parton}$, then a value of 0.48 ± 0.08 for the average $\alpha_s(Q^2)$ is obtained from the ABCMO data; the preliminary value of 0.4 was obtained by the CDHS collaboration [45].

It should be emphasised that necessary ingredients in these comparisons are a) the assumption that the W dependence of the non-perturbative terms is negligible, and b) the form of quark and gluon fragmentation functions used in the various model calculations. If forward-backward differences of p_T^2 are considered[37], as an attempt to cancel non-perturbative effects, then the agreement with theory becomes less encouraging. Furthermore, the higher twist model discussed in section 3 predicts a p_T-z correlation whose effect should be subtracted and models of fragmentation with a W dependence [46] of $<p_T^2>_{Fr}$ could explain the results.

5. GLUON SEARCHES

If the increase of $<p_T^2>$ with W^2 is due to the emission of a single hard gluon (and not to the emission of several gluons in succession or due to other mechanisms), the event should have a planar structure which reflects the "three body" (diquark system, quark and gluon) origin of the final state hadrons.

A detailed study of the event shape is performed by using the normalised eigenvalues Q_i (i=1,3)*(and the corresponding eigenvectors $\vec{n}_i$) of the momentum tensor [4]

$$M_{\alpha\beta} = \sum_{1}^{N} P_{j\alpha} P_{j\beta} \tag{31}$$

where N is the number of observed hadrons and p_j is the momentum of the j-th hadron is a given reference frame (here the c.m.s. is used). If the Q_i are ordered so that $Q_1 < Q_2 < Q_3$, then the sphericity S is

---

*Note that $\sum_{i=1}^{3} Q_i = 1$.

defined as $\frac{3}{2}(Q_1+Q_2)$ with $\vec{n}_3$ as sphericity axis and $A=\frac{3}{2}Q_1$ is the aplanarity which gives a measure of the "flatness" of the event.

The ideal two-jet event (ie. where no hard gluon has been radiated) would have A=S=0. This is never the case experimentally because the values of A and S are smeared by the parton hadronisation. Detailed predictions for the expectation values of these quantities when hard gluon radiation is included can be found in ref.48; here it is sufficient to mention that the gluon bremsstrahlung should lead to events with high S and low A (ie. a flat, wide jet).

In neutrino interactions knowledge of the current direction enables the principal axis of the event (sphericity or thrust*) to be imposed a priori. The momentum tensor of eq.31 is then constructed in a plane orthogonal to the current (or to the total hadron momentum vector), Q3 is then fixed, while Q_1 and Q_2 are given by diagonalising $M_{\alpha\beta}$. Using this method, the BFHSW group [38] defines the planarity P for the event as

$$P = \frac{Q_2 - Q_1}{Q_2 + Q_1} \qquad (32)$$

and the dispersion in transverse momentum as

$$\Pi_N = \frac{A}{\sqrt{N_F}} \sum_F (p_T - \langle p_T \rangle) \qquad (33)$$

where N_F is the number of forward hadrons and the sum is only over particles with $x_F>0$. A is chosen so to give unit width to the distribution.

Hard gluon bremsstrahlung effects can be detected a) by comparing the behaviour with W^2 and Q^2 of quantities like S, A (average values or differential distributions) with theoretical predictions and b) by searching for an excess of events showing a planar three-jet structure.

Figs. 27a, 27b and 27c show $\langle S \rangle$ and $\langle A \rangle$ from $\bar{\nu}H_2$, νH_2 and νD_2 experiments [49,37,50] as functions of W. It is seen that these quantities have a strong dependence on the multiplicity (as is easily understood). No discrepancy is found with expectations from Monte Carlo calculation without gluon effects included. This is perhaps as expected, since the fraction of events with a "hard gluon" is supposed to be small ($O(\alpha_s)$).

---

* See for instance ref.47 for definition.

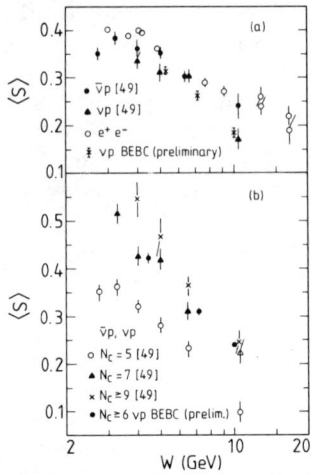
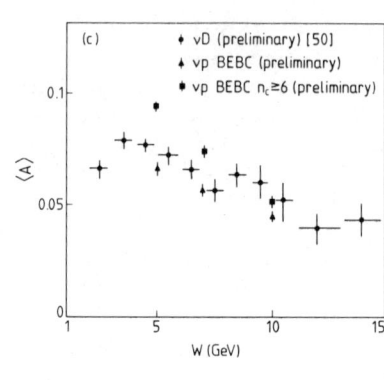

Fig. 27. <S> as a function of W from $\bar{\nu}p$ [49] and νp [37] experiments.
a) all multiplicities;
b) fixed multiplicity;
c) <A> versus W. Preliminary results from refs. 37 and 50.

Searches for an excess of planar (three-jet) events have been made by the ABCMO [37] and BFHSW [38] collaborations. The definitions of "planar" were different in the two cases: the selection S>0.25 and A<0.04 was used by the ABCMO group (as in analyses of e^+e^- data); Π_N>2.4 and P>0.6 by the BFHSW collaboration. The results obtained are summarised in Tables 5 and 6 respectively. Also indicated are the additional cuts made on multiplicity and kinematical variables. To ascertain whether there is an excess of planar events, the observed number should be compared with the estimates of the expected number. These were obtained either by rotating each track by a random angle about the current direction $\vec{q}$ or from Monte Carlo generated events.

The first method gave a result which does not differ from the observed number. It should be said however that the rotations about $\vec{q}$ would not destroy any real excess of planar events if the configuration of the hadron momenta in the plane orthogonal to $\vec{q}$ is such that one of the tracks has a p_T considerably larger than the others.

The Monte Carlo result seems to indicate some effect. However, S and A are sensitive to the multiplicity distribution and a finely tuned Monte Carlo is necessary to make a fair comparison.

The conclusions reached by the BFHSW are somewhat different. There is an excess of events with high Π_N (hence high p_T) and high planarity (phase space would favour the opposite). The energy fraction radiated in each angular region $d\theta_i$ (where θ_i is the angle between $\vec{q}$ and the momentum vector of a particular hadron projected into

Table 5. Number of events with given sphericity S and aplanarity A, from the BEBC νH_2 experiment (preliminary).

$$n_{ch} > 6, \quad Q^2 > 1 \text{ GeV}^2$$

	$<W>$ GeV	5	7	10
All S All A	observed	405	416	439
S >0.25 A <0.04	observed	58	35	26
	Rotation about $\vec{q}$	61	31	25
	Monte Carlo	83	31	15

Table 6. Number of events with given planarity P and dispersion Π_N. Preliminary result from the BFHSW collaboration.

$$Q^2 > 2 \text{ GeV}^2, \quad W^2 > 4 \text{ GeV}^2$$

P	$\Pi_N < 2.4$	$\Pi_N > 2.4$
0. - 0.2	96	2
0.2 - 0.4	304	1
0.4 - 0.6	378	7
0.6 - 0.8	359	15
0.8 - 1.0	233	15

the ($\vec{q},\vec{n}_i$) plane) shows a dip in the forward direction for θ_1 but not for θ_2, which is characteristic of a 3-jet structure. These facts could be explained by the emission of a single hard gluon, though other possibilities, like baryon production, are not definitely excluded.

6. SUMMARY

This summary contains some of the points which I think are reasonably well established, as well as some questions which still need to be answered.

- The moments, D and f_2, of the multiplicity distributions show a behaviour similar to that found in $p\bar{p}$ annihilation. No comparison with asymptotic predictions is possible at present energies.

- The extrapolation $W \to \infty$ of the net forward charge $<Q>_F$ allows consistency checks of the usual assignment of the charge for the quarks. In particular, the difference $<Q>_F^\nu - <Q>_F^{\bar{\nu}}$ is consistent with being 1 and $<Q>_F$ is the same for νp and νn scattering.

- The fragmentation functions and their moments show a Q^2 dependence and do not factorise in x and z if "low W" (<4 GeV) data are included.

- There is no Q^2 dependence of D(z) at high (or fixed) W, but in this case there are severe limitations in the accessible x range.

- The magnitude and the Q^2 dependence of the factorisation breaking are not well described by next-to-leading order QCD predictions.

- There is evidence for the presence of higher twist terms in fragmentation functions. If most (or all) of the Q^2 dependence is assigned to higher twist corrections, then the mass squared which characterises their magnitude is estimated to be ~ 2 GeV^2.

- $<p_T^2>$ increases with increasing W (both averaged over z and at fixed z); the rate of this rise is higher for particles going forward in the c.m.s. than for those going backwards. An accurate estimate of the form of the W dependence is not allowed by the W range available at present.

- The results are consistent with predictions of QCD based models, but some assumptions must be made about the non-perturbative terms (which appear to contribute significantly).

- The observed excess of planar events in the highest W region could be interpreted as evidence for hard gluon radiation. The results need confirmation however.

I am grateful to D H Perkins, G Myatt, P Grossmann and R Batley for a large number of discussions. I would like to thank C Matteuzzi (CERN) and G Snow (Maryland) for communications and discussions about their data. I am indebted to Richard Batley for his careful reading of the manuscript.

REFERENCES
1. European Muon Collaboration, J Aubert et al, CERN prep.EP-80-119. J Rander, CERN-Dortmund-Heidelberg-Saclay collaboration, contribution to this conference.
2. G Myatt, CERN ECFA Report 72/4, Vol 2: 117 (1972).
3. J Bell et al., Phys.Rev.D19:1 (1979)
4. P Allen et al., Acchen-Bonn-CERN-Munich-Oxford collaboration, paper submitted to this conference.
5. T Kitagaki et al., Illinois-Maryland-Stony Brook-Tohoku-Tufts collaboration, paper submitted to this conference.
6. M Derrick et al., Argonne-Carnegie-Mellon-Purdue collaboration, Phys.Rev.D 17:1 (1978).
7. V V Makeev, IHEP, IH (DDR) collaboration, paper submitted to this conference.
8. A Wroblewski, contribution to 1972 High Energy Physics Conference at FNAL.
9. A Amati, G Veneziano, Preprint CERN Th-2620.
K Konishi, Preprint RL 79-035.
A Basetto, M Ciafaloni, G Marchesini, Phys.Lett.83B:418 (1979).
10. G R Farrar, J L Rosner, Phys.Rev.D7:2747 (1973).
11. R Feynman, R Field, Nucl.Phys.B136:1 (1978).
12. P Grossmann, Oxford Bubble Chamber Group note, OXF80-PG1.
13. B Musgrave, Proceedings of Neutrino '79, Bergen, Vol 2:556 (1979).
14. J P Berge et al., Phys.Lett. 91B:311 (1980).
15. G A Snow, Illinois-Maryland-Stony Brook-Tohoku-Tufts collaboration, private communication.
16. J P Berge et al., Berkeley-Fermilab-Hawaii-Michigan collaboration, Preprint UH-511-368-79.
17. C C Chang et al., Illinois-Maryland-Stony Brook-Tohoku-Tufts collaboration, paper submitted to this conference.
18. M Derrick et al., Phys.Lett. 91B:307 (1980).
19. G Altarelli et al., Nucl.Phys. B160:301 (1979).
20. N Sakai, Phys.Lett. 85B:67 (1979).
J Sheiman, Nucl.Phys. B171:445 (1980).
21. R D Peccei, R Rückl, Nucl.Phys. B162:125 (1980).
Wu Chin-Min, Nucl.Phys. B167:337 (1980).

22. G Altarelli, G Parisi, Nucl.Phys. B126:298 (1977).
23. E L Berger, Phys.Lett. 89B:241 (1980).
24. M Derrick et al., Phys.Lett. 91B:470 (1980).
25. H Deden et al., Aachen-Bonn-CERN-Demokritos-London-Oxford-Saclay collaboration, paper submitted to this conference.
26. N Schmitz, Proceedings of the International Symposium on Lepton and Photon Interactions, Fermilab:359 (1979).
27. Aachen-Bonn-CERN-Munich-Oxford collaboration, private communication.
28. J Blietschau et al., Phys.Lett. 88B:281 (1979).
29. C Matteuzzi, CERN-Milano-Orsay collaboration, paper submitted to this conference.
30. P Allen et al., Aachen-Bonn-CERN-Munich-Oxford collaboration, paper submitted to this conference.
31. E L Berger, SLAC-PUB-2362 (1979).
32. C Matteuzzi, CERN-Milano-Orsay collaboration, private communication.
33. B Saitta, Oxford Bubble Chamber Group note, BS2-80.
34. P C Bosetti et al., Nucl.Phys. B149:13 (1979).
35. R Brandelik et al., Phys.Lett. 86B:243 (1979).
    D P Barber et al., Phys.Rev.Lett. 43:830 (1979).
    Ch Berger et al., Phys.Lett. 86B:418 (1979).
    W Bartel et al., Phys.Lett. 91B:142 (1980).
36. A Mendez et al., Nucl.Phys. B148:499 (1979).
37. P Allen et al., Aachen-Bonn-CERN-Munich-Oxford collaboration, paper submitted to this conference.
38. H C Ballagh et al., Berkeley-Fermilab-Hawaii-Seattle-Wisconsin collaboration, paper submitted to this conferenec.
39. A Mendez, Nucl.Phys. B145:199 (1978).
40. G Altarelli, G Martinelli, Phys.Lett. 76B:89 (1978).
41. G Altarelli, Proceedings of the 13th Rencontre de Moriond, Vol II: 395 (1978).
42. B Saitta, Proceedings of the 15th Rencontre de Moriond, 1980, to be published.
43. F Halzen, D M Scott, University of Hawaii preprint UH-511-386-80 (1980).
44. P Mazzanti et al., Phys.Lett. 81B:219 (1979).
    R Odorico, private communication.
45. J Rander, CDHS collaboration, private communication.
46. P Grossmann, private communication.
47. Sau Lan Wu, G Zobernig, Z.Physik C2:107 (1979).
48. P Binetruy, G Girardi, Nucl.Phys. B155:150 (1979).
    J Ranft, G Ranft, Phys.Lett. 82B:129 (1979).
49. M Derrick et al., Phys.Lett. 88B:177 (1979).
50. K Tamai, Illinois-Maryland-Stony Brook-Tohoku-Tufts collaboration, private communication via G Snow.

DISCUSSION (Chairman: D. Cundy)

H.H. BINGHAM, Berkeley: It seems to me that your data also support the proposition that there is an excess of planar events over the Monte Carlo prediction as W increases.

CHARGED CURRENT INCLUSIVE INTERACTIONS

F. Eisele

Institut für Physik der Universität Dortmund

Dortmund, Federal Republic of Germany

Summary

New results on total cross section and structure function measurements are reviewed. Substantial progress has been achieved in the determination of the flavour content of the nucleon. Scaling violations also at high values of Q^2 are now established by several experiments. The origin of these scaling violations is discussed.

I. INTRODUCTION

By summer 79 it had become evident that deep inelastic scattering of neutrinos is an excellent tool to measure the nucleon structure. Neutrino experiments separate quarks and antiquarks, they distinguish different flavours and they give high event rates in the high Q^2 region.

As a result of these old experiments we know a lot about nucleon structure:
- The shape and magnitudes of structure functions on isoscalar targets ($u + d$, $\bar{u} + \bar{d}$) are known.
- The basic predictions of the quark parton model have been verified.
- The Q^2-variation of F2 and XF3 is measured over a large Q^2-range ($1 \div 200$ GeV^2/c^2).
- Scaling violations are clearly established for $Q^2 \gtrsim 40$ GeV^2/c^2.
The pattern of these scaling violations agrees with that predicted in QCD and is quantitatively described by leading order QCD down to $Q^2 \gtrsim 1$ GeV^2/c^2. It was however not possible to conclude that perturbative QCD had been definitely tested. There is the worrying possibility that the observed scaling violations are predominantly

due to so called higher twist effects and that perturbative QCD effects are small in our Q^2-range.

More precise experiments over the largest possible Q^2-range were urgently needed and also stronger theoretical prejudice on higher twist effects.

At the time of this conference new experimental results on H2 and D2-targets are available to further separate the flavour content of the nucleon. High statistics experiments on isoscalar targets present new precise measurements of the Q^2 variation of the structure functions F2, XF3, $\bar{q}^{\bar{\nu}}$ and of the strange sea.

II. NEUTRINO INTERACTIONS ON PROTONS AND NEUTRONS

These experiments measure the flavour content of the nucleon and allow in principle the separate measurement of the quark densities $u(x)$, $d(x)$, $\bar{u}(x)$ and $\bar{d}(x)$.

Experimentally the cross section ratio $\sigma^{\nu n}/\sigma^{\nu p}$ is most easily accessible. Three groups have given new preliminary results. They are sumarized in table I together with published data.

Table I   Cross-section ratios $\sigma^{\nu n}/\sigma^{\nu p}$ and $\sigma^{\bar{\nu} n}/\sigma^{\bar{\nu} p}$

Experiment	target	E_ν [GeV]	$\sigma^{\nu n}/\sigma^{\nu p}$	$\sigma^{\bar{\nu} n}/\sigma^{\bar{\nu} p}$
ANL 12' [1]	D2	1.5 ÷ 6	1.95 ± .21	
BNL 7' [2]	D2	≤ 10	1.48 ± .17	old results
CERN GGM [3]	Propane/Freon	1 ÷ 10	2.08 ± .15	
BEBC-TST [4]	H2 /Ne	~ 44	1.83 ± .17	
15' IMSTT [5]	D2	~ 50	1.83 ± .15	
SKAT(IHEP) [6]	Freon	~ 10	1.44 ± .16	0.45 ± .05

All results are in reasonable good agreement and support the notion that the leading u-quark carries about twice the momentum of the d-quark. The expected value in the quark parton model is $\sigma^{\nu n}/\sigma^{\nu p} \sim 1.8$.

The real progress in this field however is the first measurement of the shape of the structure functions on neutrons and protons with reasonable accuracy. The IMSTT-collaboration[5] has studied neutrino interactions in a deuterium bubble chamber. Their analysis is based on ~ 2800 events $\nu n \to \mu x$ and 1900 events $\nu p \to \mu x$ which are separated by counting the number of charged tracks. Apart from statistics the main limitation of this experiment is due to substantial

CHARGED CURRENT INCLUSIVE INTERACTIONS

rescattering corrections of the order of 20 % which are hard to evaluate. Their x-distributions for νn and νp-scattering normalized to the published total cross-section for isoscalar targets are shown in fig. 1a. In terms of quark densities these distributions can be written as

$$\nu n: \quad \frac{1}{\sigma_o} \frac{d\sigma^{\nu n}}{dx} = x\,[u(x) + s(x) + \frac{1}{3}\bar{d}(x)\,] \approx F2^{\nu n}$$

$$\nu p: \quad \frac{1}{\sigma_o} \frac{d\sigma^{\nu n}}{dx} = x\,[d(x) + s(x) + \frac{1}{3}\bar{u}(x)\,] \approx F2^{\nu p}$$

such that νn measures essentially the distribution of the leading quark whereas νp measures the non leading down quark. The shape of $F2^{\nu n}$ agrees reasonably well with the published structure function $F2^{\bar{\nu}p}$ of the ACMP-collaboration[7] and with the shape as measured in electroproduction (dashed-lines)[8] in agreement with the predictions of the QPM. The leading quark has a harder momentum spectrum than the non leading quark as can be seen in fig. 1b, which gives the "non-singlet" structure function

$$F2^{\nu n} - F2^{\nu p} = x\,[u_v - d_v + \frac{2}{3}(\bar{u} - \bar{d})]$$

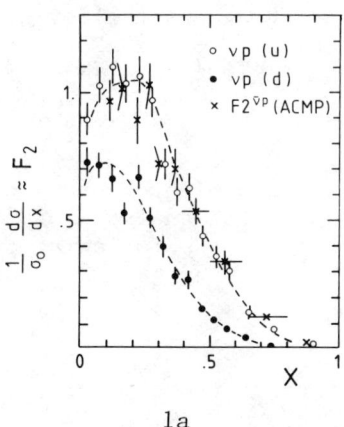
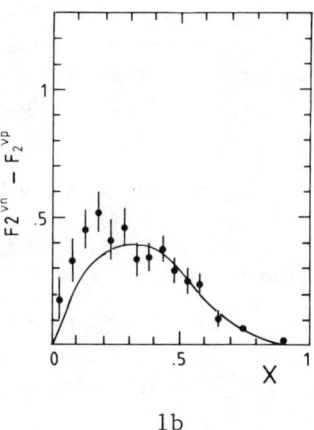

1a 1b

Fig.1a: Measured x-distributions of the IMST-collaboration for νn(o) and νp(●)-scattering compared to $F2^{\bar{\nu}p}$(x) and the shape measured in electroproduction (dashed lines).
Fig.1b: Non-singlet structure function $F2^{\nu n}-F2^{\nu p}$ compared to the shape of $F2^{ep}-F2^{en}$ (solid line).

A similar conclusion can be drawn from a result contributed by the ABCMO-collaboration[9] based on $\sim$ 3400 neutrino interaction in hydrogen with E > 20 GeV. This allows to determine $F2^{\nu p}(x)$ in the

Q^2-range 2 ÷ 18 GeV^2. A comparison with the e-p-data from SLAC-MIT[10] is shown in fig.2 which gives the ratio $F2^{\nu p}(x)/F2^{ep}(x)$. Outside of the sea region (x ≳ · 3) this is equal to the quark ratio $d(x)/u(x)$. This ratio decreases with increasing x compatible with (1-x) as suggested by Feynman[11]. At present the data is not precise enough to distinguish such a behaviour from the prediction based on QCD plus Su(6) that $d/u \to 1/5$ for $x \to 1$ [12].

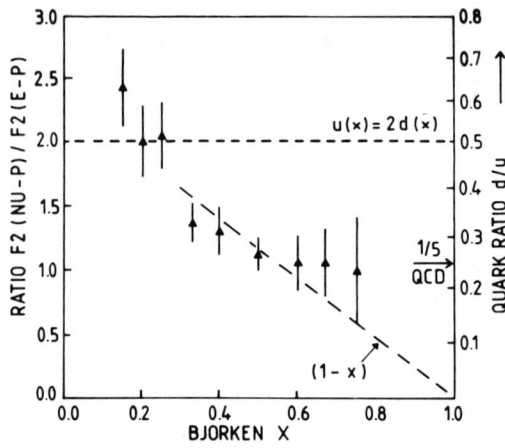

Fig.2:
Ratio of structure function F2 as measured in neutrino proton and electron proton scattering versus x.

Neutrino experiments offer the interesting possibility to test the charge symmetry of the sea, i.e. to test if $\bar{u} = \bar{d}$. Fig. 1b shows the structure function $F2^{\nu n} - F2^{\nu p} = x [u_v - d_v + \frac{2}{3}(\bar{u} - \bar{d})]$ compared to the shape of the structure function

$$F2^{ep} - F2^{en} \sim x [(u_v - d_v + 2(\bar{u}-\bar{d})].$$

The two distributions differ for small x which might be an indication that $\bar{d} > \bar{u}$. An independent check comes from the analysis of the y-distributions. The IMSTT-collaboration obtains the following preliminary results:

$$\int_0^1 x\bar{u}(x) \, dx = .017 \pm .007$$

$$\int_0^1 x\bar{d}(x) \, dx = .027 \pm .008 \quad \text{(stat. errors only)}$$

Again there is an indication that $\bar{d}$ may carry more momentum than $\bar{u}$. Before final conclusions can be drawn, it is necessary to study the effect of scaling violations in the comparison of electroproduction and neutrino data and to evaluate the influence of systematic errors.

The question of the charge symmetry of the sea remains a challenge to neutrino physics and can be best settled by studying antineutrino scattering on neutrons and protons.

III. TOTAL CROSS-SECTION MEASUREMENTS ON ISOSCALAR TARGETS

Four experiments have contributed new measurements to this conference. They are summarized in Table II.

Table II: New measurements of total cross-sections

Experiment	$<E_\nu$ [GeV]>	σ^ν/E	$\sigma^{\bar\nu}/E$ [10^{-38} cm^2/GeV^2]
SKAT(IHEP)[6]	10	0.70 ± .06	0.23 ± .05
BEBC(TST)[4]	14	0.64 ± .06	---
GGM(SPS)[13]	46	---	0.29 ± .03
CHARM[14]	20-200	.594±.027	0.292 ± .015

Figure 3 shows a graphical summary of existing data.[16] There is a good overall agreement and the new data fits well*). For antineutrino scattering all measurements are compatible with a constant value of $\sigma^{\bar\nu}/E$ independent of energy whereas for neutrinos the slope σ^ν/E shows a definite rise towards low energies. This behaviour is in agreement with the measured pattern of scaling violations. Predictions based on a QCD-parametrization of structure functions by Buras and Gaemers[15] are shown in fig.3 as dashed lines.

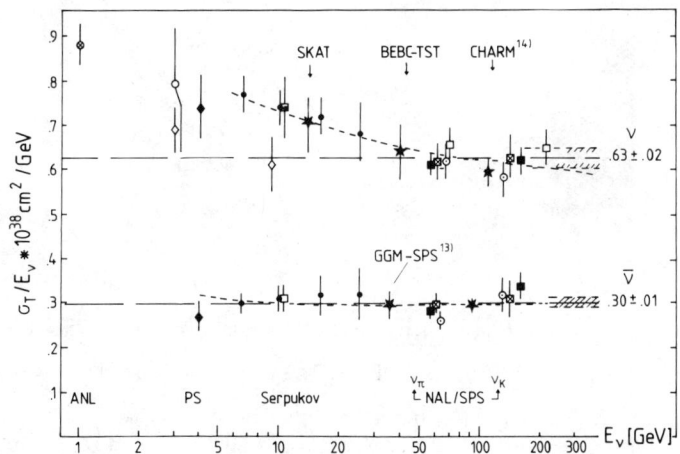

Fig.3: σ_T/E_ν for neutrino and antineutrino charged current total cross section as a function of neutrino energy. The dotted line is the prediction based on a QCD parametrization of structure functions. ANL ⊗ , GGM ◇◆, IHP ◉●, CRS ○, CITF ■ , CDHS ⊠, BEBC ⊙, CFRR □ [16]

*) This statement is no longer valid since in the meantime the CFRR-collaboration has presented new results to the Madison conference[17]: σ^ν/E_ν=.733±.07 and $\sigma^{\bar\nu}/E$=.37±.03 for 30<E_ν < 300GeV which are inconsistent with the other high energy measurements. Fortunately the main uncertainty for all experiments seems to be in the overall flux normalization whereas the $\bar\nu/\nu$ ratio and the energy dependence are known much better. For the measurement of structure function this leads to an overall scale error but does not effect the shape or Q^2-dependence.

The errors of the total cross section measurements mainly due to the flux uncertainties are still substantial and constitute a serious limitation for neutrino physics especially for the determination of structure functions. Substantial systematic improvements are necessary in future experiments. It is also clear, however, that this will be a difficult task.

IV STRUCTURE FUNCTIONS ON ISOSCALAR TARGETS (MAINLY Q^2-VARIATION)

The structure functions F2 and xF3 are easily obtained by adding or subtracting neutrino and antineutrino differential cross sections:

$$\nu + \bar{\nu} : \quad F2(x, Q^2) = q + \bar{q}$$
$$\nu - \bar{\nu} : \quad xF3(x, Q^2) = q - \bar{q} = q_{valence}$$

A direct measurement of the sea-distribution in the nucleon is possible by studying antineutrino interactions at high $y = E_h/E_\nu$ as illustrated in figure 4. It is clear that such a measurement requires very high statistics in an antineutrino beam.

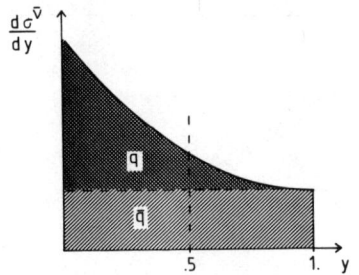

Fig.4: y-distribution for $\bar{\nu}$-scattering

The structure function which can be measured this way is

$$\bar{q}^{\bar{\nu}}(x, Q^2) = x(\bar{u} + \bar{d} + 2\bar{s})$$

It should be noted that the determination of F2 and $\bar{q}^{\bar{\nu}}$ suffers from substantial corrections due to our limited knowledge about the strange sea and especially due to a possible violation of the Callan-Gross relation which would imply that $R' = (F2-2xF1)/F2$ is non zero. At present the value of R' has to be assumed rather than measured in the determination of these structure functions.

Three experiments have contributed new structure function measurements to this conference. Their x-Q^2-range is illustrated in figure 5.

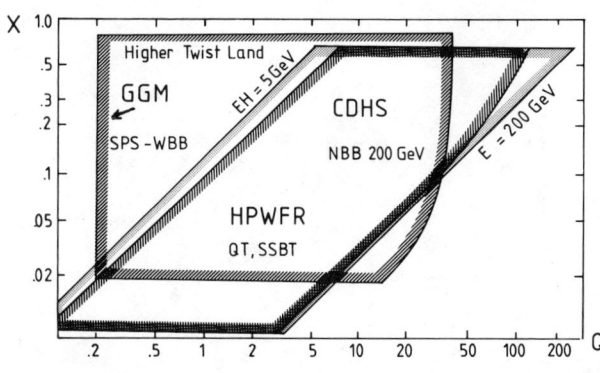

Fig. 5:
Range in x-Q^2 of three experiments to determine the nucleon structure functions

There are two high statistics counter experiments with very similar (x-Q^2)range. The HPWFR-collaboration[18] has accumulated 21000 neutrino events and 7000 antineutrino events in the Fermilab wide band beams (QT and SSBT). The analysis of the CDHS-collaboration is based on 65000 neutrino and 25000 antineutrino events obtained in the CERN narrow band beam. The high Q^2 region is kinematically limited by the maximal available energy, at low Q^2 these counter experiments are limited by their moderate hadron energy resolution which does not allow to use data with hadron energies below ~ 5 GeV. In a bubble chamber on the other hand low energy hadrons are well measured. This is the reason why the Gargamelle-SPS experiment[13] can fill up the region at large x and low Q^2. Their analysis is based on 3000 ν and 4000 $\bar{\nu}$-events from the CERN wide band beam.

The strucutre functions F2 and xF3 as measured by the HPWFR-collaboration are shown in figures 6a ÷ 6d (solid boxes) as a function of x and Q^2. For comparison the published data of the CDHS-collaboration[19] are also included (open boxes). At small values of x the structure function F2 shows a substantial rise with Q^2, around x = .1 there is scaling and for large x there is substantial shrinkage up to the highest values of Q^2. The same shrinkage is seen for the structure function xF3. (figure 6c). Thus scaling violations are clearly established by this experiment also at high Q^2 and the amount of scaling violations agrees very well with published data. There is however evidence for systematic differences in the shape of F2 between the HPWFR and the published CDHS data. The origin of these differences is presumably due to uncertainties in hadron calibration, flux ratios and resolution functions.

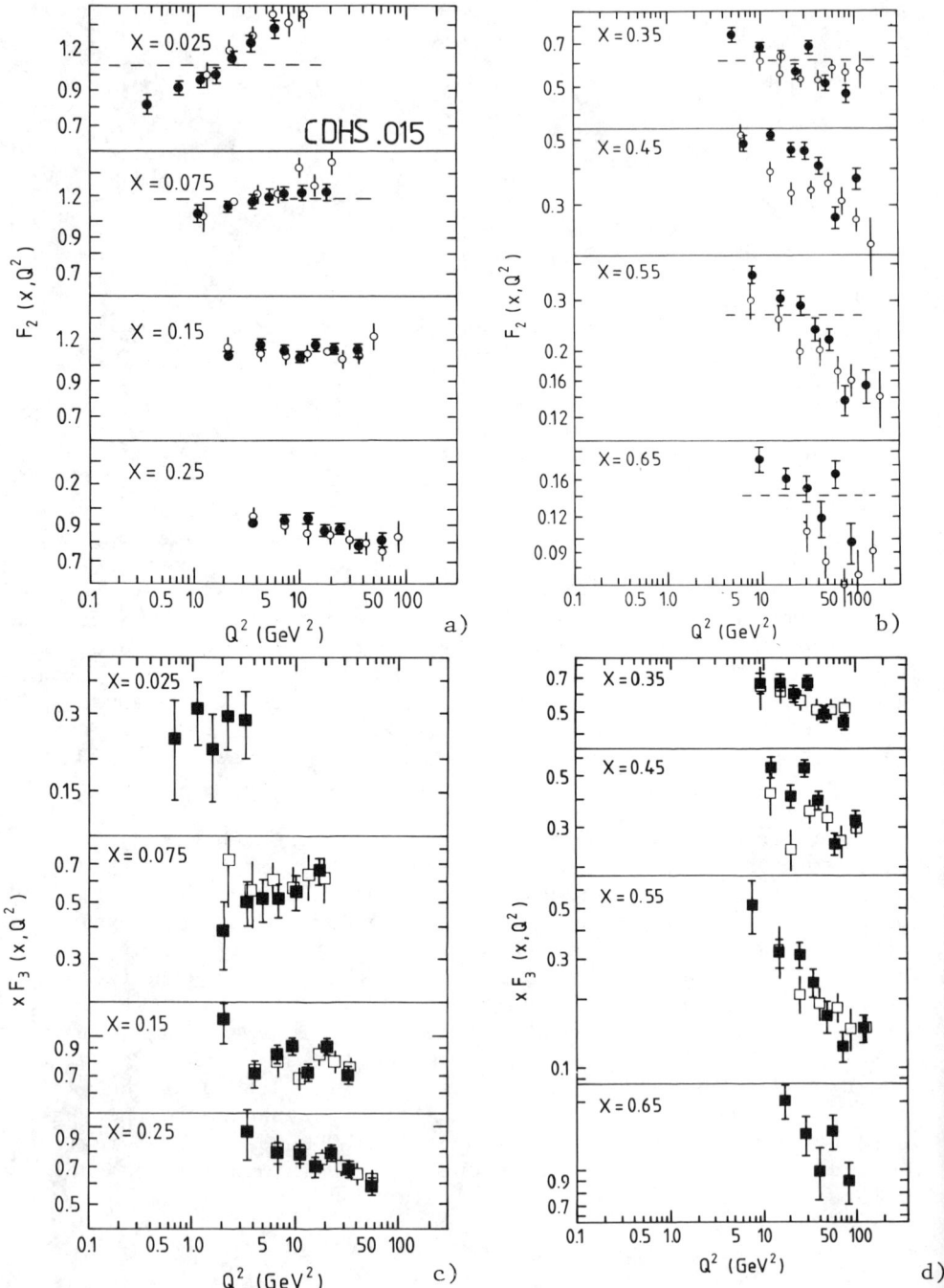

Fig. 6: Structure functions versus Q^2 for different values of x. Solid boxes: HPWFR-collaboration[18], open boxes: CDHS, published data[19] a), b) structure function F2, c) and d) structure function xF3

CHARGED CURRENT INCLUSIVE INTERACTIONS

Figure 7a) and 7b) show the preliminary data on F2 and xF3 of the GGM-SPS-collaboration[13] versus x and Q^2.*)

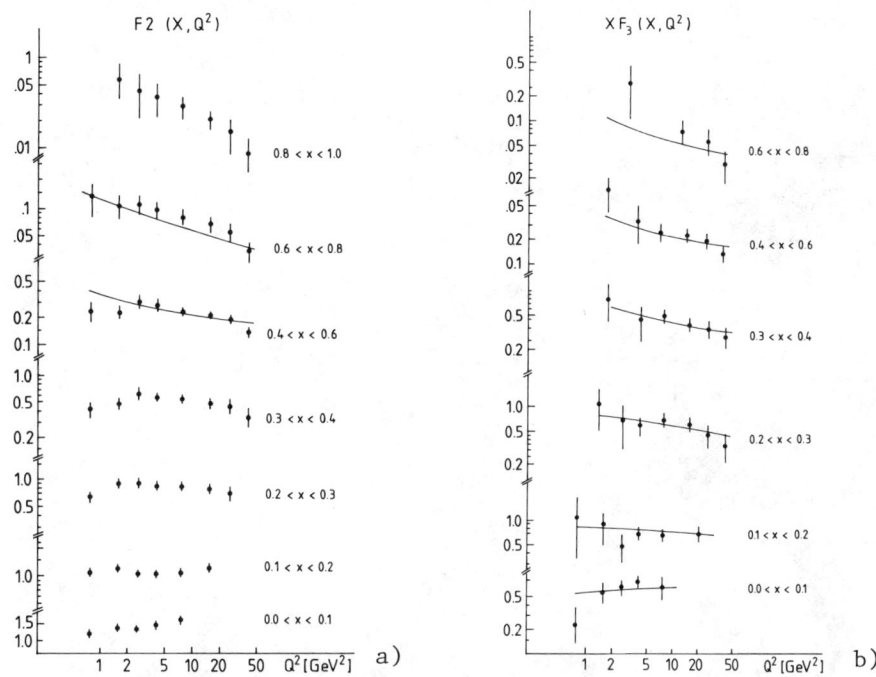

Fig.7: Structure function F2 and xF3 as measured by the GGM-SPS-collaboration. Solid lines: second order QCD-fit.

This data covers the maximal range in $\ln Q^2$ within one experiment with reasonably high statistics. The measurement of F2 fills in the hole at large x, low Q^2 which is not accessible to counter experiments and which may play an important role in separating out higher twist contributions. There is evidence for a shrinkage at large x above $Q^2 \sim 2$ GeV^2/c^2 in agreement with the other experiments.

The CDHS collaboration has contributed new measurements of F2 and xF3 which are based on four times higher statistics compared to the published results. The structure function F2 shown in figure 8a) is evaluated for a value of R $=\sigma_L/\sigma_T$ = o.1; effects of Fermi motion are not unfolded. The structure function shows a very significant rise for $x \leq 0.1$ by a factor of ~ 1.6. For large values of x there is a very strong shrinkage effect which extends to the highest values of Q^2. It should be noted that this data is definitely inconsistent with scaling even if one regards only data with $Q^2 > 20$ GeV^2/c^2.

*) These data differ somewhat from those presented at the conference due to a different normalization for the neutrino data which improved the agreement with other experiments.

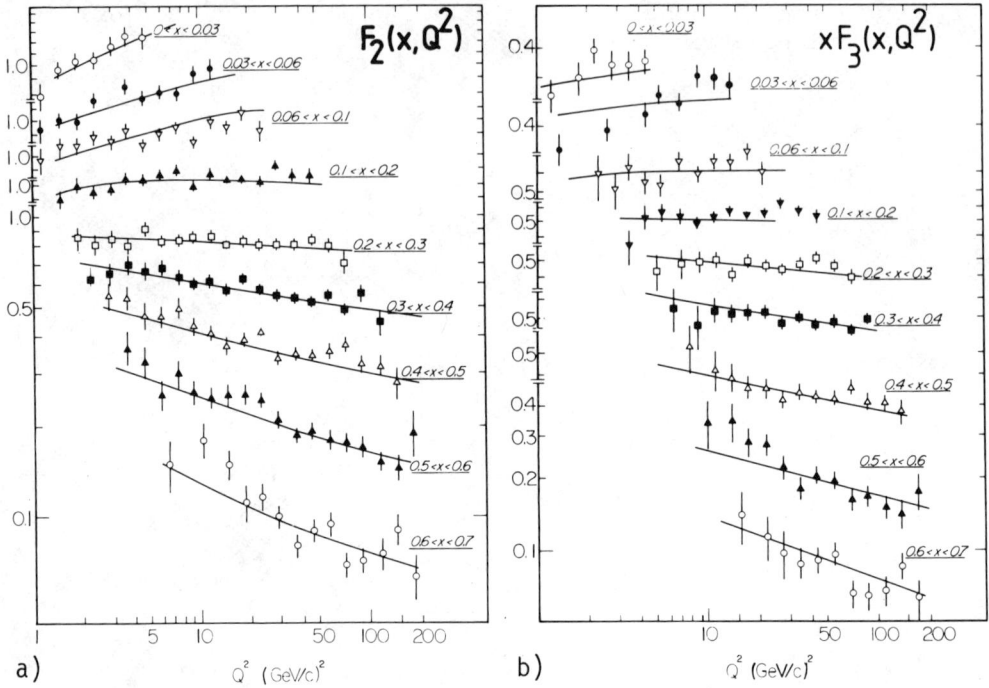

Fig.8: Structure function F2 and xF3 as measured by the CDHS collaboration. The solid lines are the result of a leading order QCD-fit with $\Lambda_0 = 0.5$. Both data sets have a scale error of ± 7 % due to the flux uncertaincies. $R = \sigma_L/\sigma_T = .1$

Figure 8 b) shows the measurements of xF3. For the first time the structure function is measured precisely also at small values of x. The data shows a moderate rise at small values of x and the same amount of shrinkage as for F2 at large x. The new data is in excellent agreement with the published data[19] and also with the results of the HPWFR collaboration.

The antiquark distribution

The antiquark distribution $\bar{q}^\nu(x, Q^2) = x [\bar{u} + \bar{d} + 2\bar{s}]$ has been measured for the first time as a function of x and Q^2 by studying the antineutrino scattering at high y. This is the most direct way to measure the sea-quark distribution in the nucleon and therefore very interesting from an experimental point of view. In the framework of QCD this measurement should allow a good determination of the gluon distribution since the relative effect of gluon bremsstrahlung for $\bar{q}^\nu$ is larger than for F2.

The preliminary result of the CDHS collaboration is shown in figure 9. The measurement is based on about 100000 $\bar{\nu}$- and 60000 ν-events with E > 20 GeV accumulated in the CERN wide band beam and 25000 $\bar{\nu}$- and 65000 ν-events from the narrow band beam. The wide band beam data has been normalized to the published total cross-section[19]. $R = \sigma_L/\sigma_T$ has been fixed to the value 0.1.

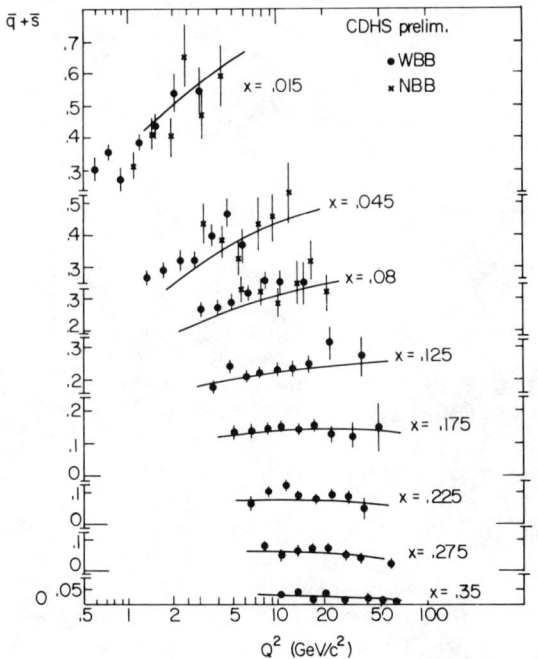

Fig.9:
Antiquark-distribution $\bar{q}^\nu = x[\bar{u}+\bar{d}+2\bar{s}]$ as measured by the CDHS-collaboration. Solid dots: data points obtained in the wide band beam. Crosses: points from narrow band beam. The solid lines are the prediction based on a QCD-fit to F2 and xF3 only.

The data shows a very strong rise with Q^2 for $x \lesssim .15$. For x-values above .2 there is indication of shrinkage. The total momentum carried by antiquarks shows a significant increase with hadron energy and therefore also with Q^2. The solid lines in figure 9 are the prediction based on a QCD-parametrization of F2 and xF3 only. The measurement of $\bar{q}^\nu$ therefore is an excellent agreement with F2 and xF3 both in magnitude and in the amount of scaling violations.

Summary: measurement of structure functions

The available measurements of F2, xF3 and $\bar{q}^{\bar{\nu}}$ form a consistent set of structure functions. This is further illustrated in figure 10 which shows measurements of all structure functions for a fixed range in Q^2. The measurements of F2 by the neutrino counter experiments agree well with each other and with the e-d structure function measured at SLAC multiplied by the QPM-factor 9/5.

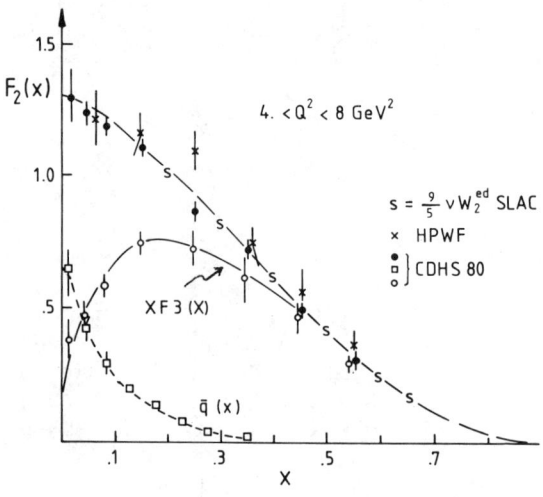

Fig.10:
The structure functions F2, xF3 and $\bar{q}^{\bar{\nu}}$ for a fixed band of Q^2. The lines are the result of a QCD fit to the CDHS data of F2 and xF3 only.

The Q^2-variation of all three structure functions has been measured rather accurately over a large Q^2-range. Scaling violations are clearly established by all experiments. The most important result is the observation of substantial shrinkage of F2 and xF3 by the two high statistics counter experiments up to the highest values of Q^2. This shows clearly that scaling violations are not a low Q^2 behaviour

How do these results compare with those of muon experiments? Within rather large normalization errors there is now reasonably good agreement in the overall shape and magnitude of F2 for all experiments. At least two muon experiments (MSUF[20] EMC[21]) find evidence of shrinkage at large x and high Q^2 ($Q^2 \gtrsim 20$ GeV^2/c^2). The amount of shrinkage however seems to be somewhat smaller than for the neutrino experiments resulting in values of $\Lambda_{L.O.} \approx .1$. Such a difference could be due to physics i.e. a weak propagator with a small mass of $m_W \approx 50$ GeV could explain it. It's more likely however that the existing differences are mainly due to experimental problems and just reflect the fact that the measurement of scaling violations is a very delicate job.

CHARGED CURRENT INCLUSIVE INTERACTIONS

The comparison of measurements from different experiments (neutrino and muon) and even from different data sets of the same experiment show very clearly that systematic errors exist which are sometimes large compared to the statistical errors. To make substantial progress in this respect will be very hard both for neutrino and for muon experiments and will take a long time. To my understanding the task is easier for neutrino experiments and I believe that the measurements presented at this conference have a high level of credibility.

Longitudinal structure function (Callan-Gross-relation)

The measurement of the longitudinal structure function q_L is of substantial theoretical interest but it seems to be a very difficult and depressing task. In neutrino physics we are used to measure the function $R' = (F2-2xF1)/F2 = q_L/F2$ which was found to be small and compatible with zero[22].

Probably the most respectable measurement of $q_L(x, Q^2)$ comes from ep and ed experiments of the SLAC-MIT-collaboration. They find a value of

$$R = \sigma_L/\sigma_T = R' + \frac{Q^2}{\nu^2}\frac{F2}{2xF1} = 0.21 \pm .1$$

which corresponds to $R' \approx .1 \pm .05$, with no significant dependence on x and Q^2. The error given is purely systematic. The interesting fact is, that R' is probably nonzero also at large x which is the only result up to now which indicates a disagreement with perturbative QCD and gives support to the idea of "higher twist dominance".

The CDHS-collaboration has made a new effort to measure R' as function of x and Q^2. Figure 11 shows their measurement of R' for different values of x averaged over Q^2.

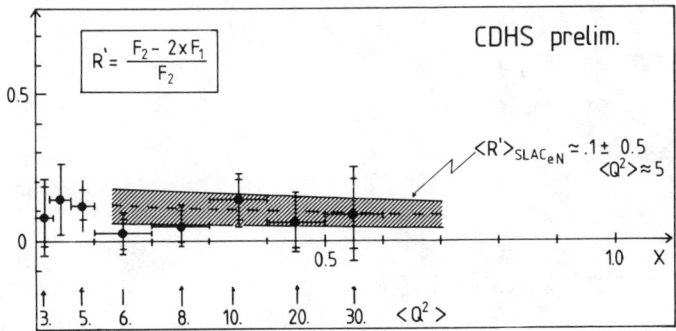

Fig.11: Measurement of R' versus x for the CDHS-collaboration. The outer error bars include the estimate of systematic errors.

The preliminary results can be summarized as follows: The data has statistical significance to measure R' versus x or versus Q^2. R' seems to be nonzero with a value R'= .1±.07 which is the same magnitudes as for SLAC but at four times larger values of Q^2. There is no evident dependence on x or Q^2.

Clearly the present knowledge on R' is very unsatisfactory and leads to an apreciable uncertainty in the measurement of F2 and $\bar{q}^{\bar{\nu}}$ as shown in table III.

Table III: F2(R = 0.1)/F2(R = 0.) for the CDHS experiment

Q^2[GeV]^2	x =	.08	.45	.65
1.4		1.01		
18		1.06	1.01	1.00
140			1.08	1.05

Strange sea: Q^2-evolution

Opposite sign dimuon events are known to be predominantely due to single charm production and subsequent semileptonic decay. The differential cross section on isoscalar target is

$$\frac{d^2\sigma(\genfrac(){0pt}{1}{-}{\nu} \to \genfrac(){0pt}{1}{-}{c})}{dxdy} \sim 2\cos^2\theta_c xs(x) + \sin^2\theta_c x[\genfrac(){0pt}{1}{-}{u} + \genfrac(){0pt}{1}{-}{d}]$$

Thus $\bar{\nu}$ induced $\mu^+\mu^-$-events are mainly due to the scattering off strange quarks whereas for ν-events a similar amount comes from valence and sea quarks.

The momentum distribution of strange quarks $x\bar{s}(x)$ as obtained from $\bar{\nu}$-events and of all sea quarks is very similar as can be seen in figure 12.

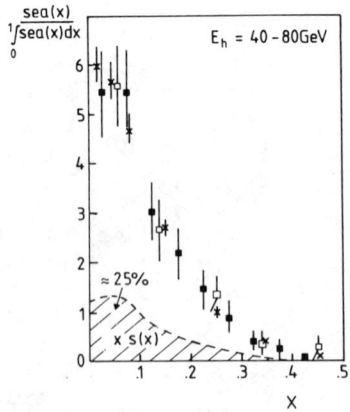

Figure 12: x-distribution of all sea quarks (✗) compared to that of the strange quarks:
(■) CDHS-collaboration (950 $\bar{\nu}N \to \mu^+\mu^- x$)
(□) HPWFR-collaboration ($\sim$190 $\bar{\nu}N \to \mu^+\mu^- x$)

The magnitude and Q^2-variation of the strange sea relative to the well known valence quark can be obtained from the x-distribution of neutrino induced $\mu^+\mu^-$-events. Figure 13 shows x-distributions based on 10000 events from the CDHS-collaboration for four different bins in $\nu = E_H + P_{\mu+}$. These distributions have been fitted to a functional form $F(x) = \alpha x s(x) + (1-\alpha)q_V(x)$ to obtain the relative amount of scattering off sea-quarks (mainly strange quarks) and valence quarks.

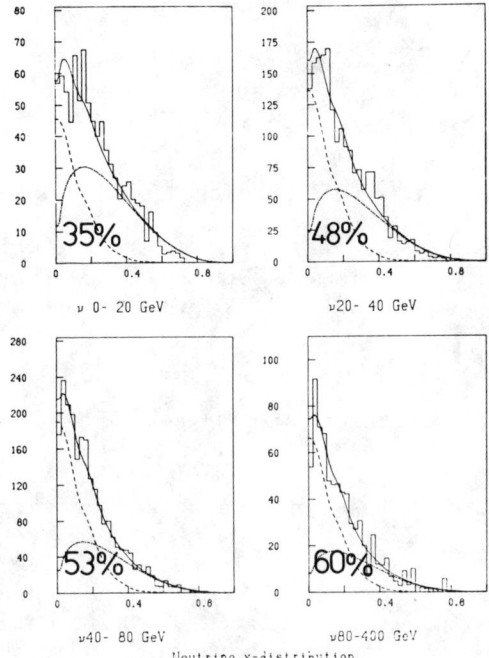

Figure 13: x-distribution of ν-induced dimuon events (CDHS-collaboration) for four bins in $\nu = E_H + P_{\mu+}$. The lines indicate the interpretation of the distributions as a sum of strange sea quarks (----) and valence quarks (-·-·-). The relative contribution from the sea is also given in %.

This ration shows a very strong rise from .35 at ν = 15 GeV to .60 at ν = 100 GeV indicating a large amount of scaling violations. This is the combined effect of "QCD-type" scaling violations of sea and valence distributions and of kinematic threshold effects due to the production of the heavy charmed quark.

A combined analysis of all structure function measurements will eventually allow to isolate the $(x-Q^2)$ behaviour of the strange sea and to separate threshold effects from "dynamic" scaling violations.

V. INTERPRETATION OF SCALING VIOLATIONS : DO WE SEE QCD-EFFECTS?

Two regimes give independent information about QCD-effects
- in the high x region we observe a definite shrinkage of $F_i(x,Q^2)$ which could be due to gluon bremsstrahlung
- at small x we see a strong rise of the sea both in F2 and $\bar{q}^\nu$ which could be due to gluon-quark pair production.

In fact QCD is able to explain both effects quantitatively as can be seen in figure 14 which shows the slopes of F2 together with a leading order QCD fit. Of course we know other sources of scaling

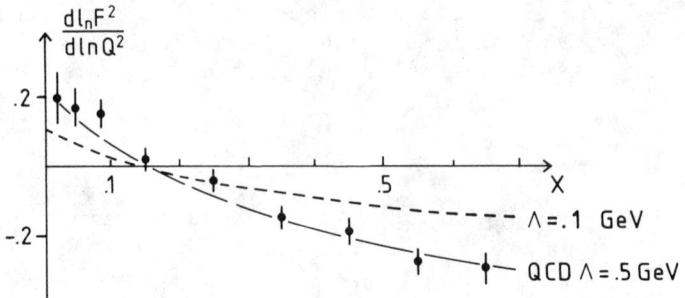

Figure 14: Slope of F2 as measured by the CDHS-collaboration.

violations in addition to perturbative QCD effects. At large x we have to separate them from higher twist effects. At small x we have contributions due to threshold effects, diffractive process and others.

QCD-fits

A rigorous comparison of QCD-predictions with experiments requires
- the inclusion of next to leading order corrections
- kinematic corrections due to target mass and quark mass effects
- the inclusion of higher twist terms.

The first two points are not a problem in principle but are quite ugly in practice. The inclusion of higher twist terms on other hand leads to a principle uncertainty since the magnitude of their contribution cannot be predicted. We know however that their contributions fall like powers of $1/Q^2$, also they should be larger at large x than at small x. The important point about higher twist terms is not that they may be present but that they have to be present to account for the observation of bound quark states in the hadronic final state.

Up to now no group has analyzed their new data including all corrections. Therefore we can draw only preliminary conclusions.

As a first step we may ask if the new data is compatible with the predictions of perturbative QCD alone. The GGM-SPS-collaboration[13] has performed second order QCD-fits to their data using the method of Bialas and Buras[23]. A combined fit to their data on xF3 (all x) and F2 (x>.4) with $Q^2>5$ GeV^2/c^2 leads to the result $\Lambda_{\overline{MS}} = .28 \pm ^{.09}_{.10}$ in the $\overline{MS}$-scheme. The fit is satisfactory as can be seen in fig. 7. It should be noted that no target mass corrections have been applied. The CDHS-collaboration has performed leading order QCD-fits to their data including target mass corrections by numerical integration of the Altarelli-Parisi equation. These fits also give reasonably good description of the observed scaling violations as can be seen in figures 8 a)b). A fit to the non-singlet structure function xF3 and F2 (x>.4) with $Q^2>10$ GeV^2/c^2 leads to a value $\Lambda_{L.O.} = .27 \pm .08(\pm.1 \text{ syst.})$ including target mass corrections.

To summarize we may conclude that also for the new data, QCD alone is able to explain the observed scaling violations rather well.

How to separate higher twist terms?

We have to separate $1/\ln Q^2$ terms from $(1/Q^2)^n$-terms. It has to be said a priori that a separation is not possible without some theoretical input. An arbitrary series in $(1/Q^2)$ is able to mimic even a $1/\ln Q^2$ dependence. The standard theoretical prejudice is that $1/Q^2$-(twist 4) and $1/Q^4$-(twist 6) terms dominate for Q^2-values $\gtrsim 2$ GeV^2/c^2. In addition a plausible functional form in x (which corresponds to a n-dependence for moments) is assumed. A first analysis along these lines had been performed by Abbot and Barnett[24] for the SLAC-MIT data. Their conclusion was that the data in this Q^2-range was not able to distinguish higher twist effects from QCD. Practically all scaling violations could be explained by twist 4 and twist 6 contributions with a mass-scale of ≈ 1 GeV[*]) and a very small value of Λ. This conclusion is no longer valid if the published neutrino data of the CDHS-collaboration[19]) is also included. A moment analysis of all published non-singlet structure functions by Pennington and Ross[25]) leads to the conclusion that the available Q^2-range is sufficient to rule out higher twist contributions as the only source of scaling violations. The scale of twist 4 effects in addition to perturbative QCD was found consistent with zero and less than 500 MeV. Though their method allows to test perturbative QCD and the presence of higher twist terms in a model independent way their conclusions suffer from the uncertainties which are unvoidable when moments at fixed Q^2 are calculated from the data.

The best (qualitative) evidence for large perturbative QCD effects in the present data comes probably from the observed rise of the sea at small x. In QCD this rise involves the strength of the strong coupling constant $\alpha_s (Q^2,\Lambda)$ and the unknown gluon distribution. However for the second moment of the sea distribution $<\bar{q}>_2$ the observed rise with Q^2 does not depend on the gluon distribution:

$$d<\bar{q}>_2/d\ln Q^2 = \alpha_s [a <\bar{q}>_2 + b<1 - F2>_2]$$

where $<1-F2>_2$ is equal to the second gluon moment. If one adds the theoretical prejudice that higher twist contributions are small at small x one comes to the conclusion that Λ cannot be small. These arguments have recently been stressed by Glück and Reya[27]).

What does the new data tell us about higher twist contributions? The GGM-collaboration has performed QCD-fits to their data including a twist 4 contribution of the form

$$F(x,Q^2) = F(x,Q^2)^{QCD} (1 + \frac{a}{(1-x)Q^2})$$

---

[*] Note: a mass scale of 1 GeV does no longer seem unresonable since higher twist effects are probably established in the Drell-Yan process[26]) with exactly this mass scale

which had been used before by Abbot and Barnett[24]. The result of these fits are shown in figure 15 which gives the 90% C.L. contours of $\Lambda_{\overline{MS}}$ versus a for two cuts in Q^2.

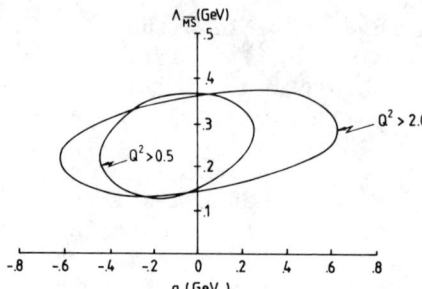

Figure 15: Allowed range of fit parameters (90% C.L.) for QCD-fits to the GGM-SPS-data including a twist 4 term $\sim \dfrac{a}{(1-x)Q^2}$

Obviously the Q^2-range of this experiment is sufficient to allow a separation of $1/\ln Q^2$ term from twist 4 terms of this specific form and the twist 4 term alone is no longer able to describe the observed scaling violations.

The CDHS-collaboration has performed similar studies by including twist 4 and twist 6 terms of various functional x-dependences in their leading order QCD-fits. Figure 16 shows the result of a pure QCD fit and a pure higher twist fit (twist 4 + twist 6) to their non-singlet data with $x > .4$ and $Q^2 > 2$ GeV$^2/c^2$.

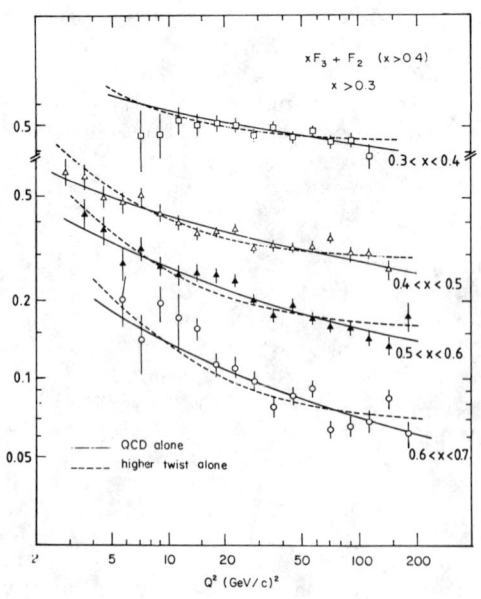

Figure 16: non-singlet structure function as measured by the CDHS-collaboration. (———) leading order QCD fit (---) "higher twist" fit to the functional from
$$F(x,Q^2) = F(x)\left(1 + \dfrac{ax}{(1-x)Q^2} + \dfrac{bx}{(1-x)^2 Q^4}\right)$$

The figure illustrates the main features of the data which allow to separate $1/\ln Q^2$ terms from $(1/Q^2)^n$ terms. The essential point is the large shrinkage observed for high values of Q^2. This is naturally explained by a $1/\ln Q^2$-term with $\Lambda_{L.O.} \approx .3$ GeV which gives at the same time about the measured rise towards low Q^2. In a pure higher twist fit on the other hand, the drop at high Q^2 requires a very large $(1/Q^2)$-term with a mass scale of about 4 GeV which leads to a rise towards low Q^2 which is by far too strong. Of course this rise can be compensated by large negative contributions $\sim 1/Q^4$, $1/Q^6$ etc. which doesn't seem to be a very physical solution however. The sensitivity to separate $1/\ln Q^2$ terms from $(1/Q^2)^n$-terms has been strongly increased by using the SLAC e-d-data $(x>.4)$, $Q^2>2$ GeV^2) in addition to the CDHS data. A preliminary analysis leads to the following conclusions:
a) twist 4 + twist 6-terms alone give a bad fit to the data ($\chi^2/DF = 117/87$)
b) a leading order QCD fit works reasonably well ($\chi^2/DF = 105/87$)
c) the inclusion of a twist 4 term in addition to QCD of the form $F(x,Q^2) = F^{QCD}(1+\frac{ax}{(1-x)Q^2})$ gives a substantial improvement ($\chi^2/DF = 84/86$) and favours a rather large higher twist contribution with values of $\Lambda_{L.O.} \approx .1$ and a ≈ 1 GeV.

It's most likely however that the inclusion of second order corrections will modify this result and reduce the prefered twist 4 contribution.

The present situation can be summarized as follows:
- We need some theoretical input to separate higher twist effects from QCD
- the available data has the statistical power and Q^2-range to separate $1/\ln Q^2$-terms from twist 4 and twist 6 contributions
- twist 4 and twist 6 contributions alone are not able to explain the data
- most likely we do see perturbative QCD-effects in the present Q^2-range (Λ cannot be very small) and QCD alone is able to describe the data reasonably well
- at present we cannot exclude the presence of rather large twist 4 contributions which could explain up to 50% of the observed scaling violations.

We should be aware that any analysis of this sort exploits the very detailed structure of the data and will be seriously affected by systematic errors. Nevertheless I am convinced that further careful analysis of the available data will finally allow to establish the mass scale of higher twist contributions.

REFERENCES:

1. S.J. Barish et al., ANL-HEP_PR-78-30 (78)
2. N.P. Samios, Proc. Int. Symp. on Lepton and photon int.,Stanford (75) P.527
3. W. Lerche et al., Nucl. Phys. B142, 65 (78)
4. BEBC-TST neutrino collaboration: Bari-Birmingham-Brussels-Ecole Polytechnique-Rutherford-Saclay-U.C.London. W. Venus, communication to this conference
5. IIT Chicago-Maryland-Stony Brook-Tufts-Tohoku-Collaboration G.A.Snow, communication to this conference
6. Inst. f. High Energy Physics, Serpukhov. V.V. Makeev, communication to this conference
7. E. Fernandez et al., Phys. Rev. Lett. 43, 1975 (79)
8. E.M. Riordan et al., SLAC-PUB-1634 (75)
9. P. Allen et al. Aachen-Bonn-CERN-Munich-Oxford-Collaboration. Paper contributed to this conference
10. A. Bodek et al., Phys. Rev. D20, 1471 (79)
11. R.P. Feynman, Photon hadron interactions (W.A. Benjamin Inc. 1972) P.150
12. G.R. Farrar,and D.R. Jackson, Phys. Rev. Lett. 35, 1416 (75)
13. GGM-SPS-Collaboration, Aachen-Brussels-Milano-Strassbourg-U.C. London, J.G. Morfin, communication to this conference
14. M. Jonker et al. CHARM-Collaboration, paper contributed to this conference
15. A.J. Buras and K.F. Gaemers, Nucl. Phys. B 132, 249 (78)
16. D. Theriot: Recent experimental measurements of neutrino charged current cross sections, Proc. Int. Symp. on Lepton and Photon int., Batavia (79), P. 337
17. F.J. Sciulli, plenary talk, Int. Conf. on HEP, Madison (80)
18. Havard-Pennsylvania-Wisconsin-Fermilab-Collaboration, A.K. Mann, communication to this conference
19. J.G.H. de Groot et al., Z. Physik C1, 143 (79)
20. R.C. Ball et al. MSU-CSL-80
21. J.J. Aubert et al., European Muon Collaboration, contribution to the Int. Conf. on HEP, Madison (80)
22. A. Para, Proc. Int. Symp. on Lepton and Photon Int., Batavia (79) P.343
23. A. Bialas and A.J. Buras, FERMILAB-PUB-79/73-THY
24. L.F. Abbot and R.M. Barnett, SLAC-PUB-2325 (79)
25. M.P. Pennington and G.G. Ross, OUTP 80-19, Oxford preprint
26. E.L. Berger, SLAC-PUB-2362 (79)
27. M. Glück and E. Reya,DO-TH-80/7, preprint Universität Dortmund

DISCUSSION (Chairman D. Cundy)

H. FAISSNER, Aachen: Do you consider $R'=(F_2-2x\,F_1)/F_2$ as significantly different from zero?

F. EISELE, Dortmund: Not yet.

H. FAISSNER, Aachen: If it was established different from 0, it would mean that this fraction of the total scattering is contributed by mesons, and thus R' should depend on Q^2.

D.A. PERKINS, Oxford: 1) Your statement about distinction of high twist effects from QCD effects depends on the form of the x-dependence assumed as $\frac{1}{q^2(1-x)^n}$.

2) De Glück-Reya analysis depends on assuming no high twist corrections to the energy momentum sum-rule.

A. BODEK, Rochester: Looking at the published CDHS data, I find the amount of $\bar{s}/d$ extracted from ν dimuon events is different by a factor of 2 from that extracted from $\bar{\nu}$ dimuon events (assuming the same semileptonic B.R.). Could you comment on this?

A.K. MANN, Philadelphia: It might be mentioned that the strange sea component appears to increase with energy by roughly a factor of two in going from below to above about 80 GeV. This might be part of the effect mentioned by the previous questioner.

WEAK NEUTRAL CURRENT-SEMILEPTONIC INTERACTIONS

Bruno Borgia

Istituto Nazionale di Fisica Nucleare

Roma, Italy

ABSTRACT

This review will discuss progress in measurements of semileptonic neutral current interactions without assuming the "standard" model from the beginning. The role of inclusive neutrino scattering in the determination of the Weinberg angle will be analysed in some detail. Also the most recent determinations of the lepton-quark couplings will be reviewed.

INTRODUCTION

Most often the emphasis on interpretation of neutral current experiments has been based on the Weinberg-Salam model, while they constitute a per se phenomena in the field of weak interactions, the study of which should be as free as possible from model dependent assumptions. In this light, it is customary to recall the Sakurai [1] tetrahedron which describes in graphic way the 13 coupling constants entering the observable interactions. In this review I will assume e-μ universality, and ignore the weak interactions between the electron and muon, therefore the tetrahedron is reduced to a triangle.

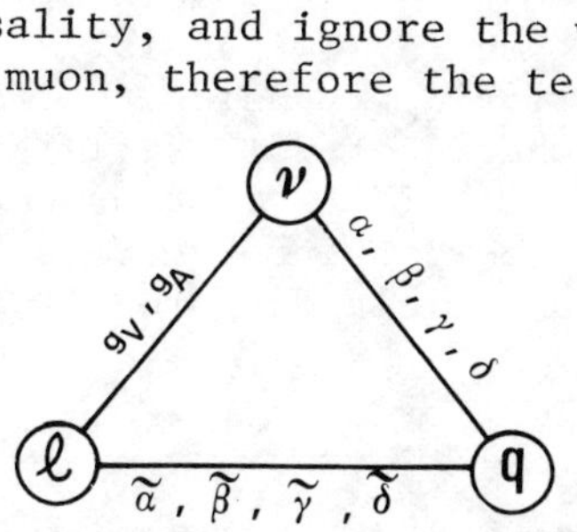

Fig. 1. The Sakurai "triangle"

An alternative definition of the ν-quark couplings has been given by Sehgal [2] in terms of chiral and isospin structure. Table 1 shows the relations with the Sakurai couplings.

Table 1. Neutrino-quark coupling constants

Sakurai couplings	Sehgal couplings	Space	Isospin
α =	$(u_L - d_L) + (u_R - d_R)$	Vector	Isovector
β =	$(u_L - d_L) - (u_R - d_R)$	Axial vector	Isovector
γ =	$(u_L + d_L) + (u_R + d_R)$	Vector	Isoscalar
δ =	$(u_L + d_L) - (u_R + d_R)$	Axial vector	Isoscalar

u_L, d_L : Left-handed (V-A) "up", "down" quark coupling

u_R, d_R : Right-handed (V+A) "up", "down" quark coupling

Actually, the Lorentz structure of the neutral current is already assumed to be only V, A. Evidences in favour of this hypothesis will be briefly recalled.

Once the coupling constants are indipendently derived from measurements, thus fixing the vector and isospin character of the interaction, we can be more specific and ask if the experimental data can be interpreted by a model of the interaction mediated by a single Z neutral boson.

This hypothesis leads to three factorization relations [3]

$$\tilde{\gamma}/\tilde{\alpha} = \gamma/\alpha \quad ; \quad \tilde{\delta}/\tilde{\beta} = \delta/\beta \quad ; \quad g_V/g_A = \alpha\tilde{\beta}/\beta\tilde{\alpha}$$

thus reducing the coupling constants to 7.

Since last year, it is now possible to derive [4] the 10 coupling constants without implicitly using the factorization relations. Finally under the hypothesis of SU(2)XU(1) gauge models, it is of particular interest to test if a right-handed doublet exists, and in the framework of Glashow, Weinberg, Salam[5] theory, it remains to test the Higgs mechanism, leading to the relation

$$\rho = \frac{M_W^2}{M_Z^2 \cos^2\theta_W} = 1$$

SEMILEPTONIC INTERACTIONS

and to determine the weak angle θ_w.

The experimental data included in this review that are relevant to the aims specified above, are listed below:
  i) ν elastic reactions
  ii) semi-inclusive reactions
  iii) inclusive ν reactions on nucleons
  iv) inclusive ν reactions on isoscalar targets
  v) parity violation in e.m.-weak interference

Experiments i) to iv) are all equally important in the determination of the space-time and weak isospin structure of the neutral current [6], while the information on the righthanded e^+ doublet derives from v). The determination of $\sin^2\theta_w$ relies essentially on data obtained from iv) and v).

LORENTZ STRUCTURE OF THE NEUTRAL CURRENT INTERACTION

Since the ratio of NC total cross section between neutrino and antineutrino is different from 1, the interaction is a mixture of at least two type of amplitudes among V, A, S, P and T. It is generally assumed that only V and A partecipate to the interaction, basing this hypothesis on the observation that in the polarized electron scattering experiment the parity violation is maximal. V and A amplitudes are helicity conserving as the e.m. current, while S, P and T flip the helicity, thus are unable to interfere with the e.m. current.

Also an indirect support is given by the y distribution in the inclusive NC scattering, as a different shape is predicted for V, A combinations and for S, P, T terms. In particular V, A terms cannot show a rising distributions. The opposite is however not true as stated by the "confusion theorem".

A direct proof of the presence of only V, A amplitudes would be given by an experiment where it is measured the presence of helicity flip terms in the scattering of polarized muons on nuclei.

This process is discussed in a paper by C.M.Wu[7] where an extimate is obtained of the helicity flipping cross section (S, P, T amplitudes) over the helicity conserving one:

$$\frac{\sigma(\mu_{LR})}{\sigma(\mu_{LL})} \sim 10^{-2}$$

at $q^2 = 100$ GeV$^2/c^2$ and assuming the helicity flip amplitude 10%

of the usual V, A.

Since the μ polarization is measured by the forward-backward asymmetry of the electron decay, the net effect is so much reduced as to doubt of the experimental feasibility.

The chiral (V-A or V+A) structure is however deducible from ν and $\bar{\nu}$ comparison, while the isospin character is obtained from specific reactions where either different targets or the emerging hadrons can be compared. The discussion on the following measurements will try to show their relative merits in understanding the neutral current structure and the recent progress made in this field.

NEUTRINO NUCLEON ELASTIC SCATTERING

In general the cross section for the process can be expressed in terms of the neutral current form factors. Their Q^2 dependence is assumed to be equal to the charged current ones and their normalization at $Q^2=0$ is

$$G_E(0) = \tfrac{1}{2}(\alpha + 3\gamma)$$
$$G_M(0) = \frac{4.7}{2}(\alpha + 0.56\gamma) \qquad (1)$$
$$G_A(0) = \frac{1.25}{2}(\beta + 0.6\delta)$$

The data are obtained at finite value of Q^2, since is experimentally difficult to observe the process when the proton momentum is low, and also because the neutron background is dominant at low energy. In the interpretation of the data, dipole fit to the form factors is assumed, with the same masses as in the charged current counterparts.

The Brookhaven-Harvard-Pennsylvania group has now published his result[8] where, in clean experimental conditions, is given the integrated cross sections for neutrino and antineutrino scattering, and also its distribution with Q^2.

The experiment of the Aachen-Padova group,[9,10] suffers from a notable neutron background, nevertheless they are able to subtract it after a detailed study of its spatial and energy distributions. In their experiment a further complication arises from the target since it is constitued by the Al nuclei of the spark chambers. However, making good use of it, they claim to know with good precision the efficiency for the detection of the neutron recoil. Therefore the measurement of the ratio R, neutral current over charged current events, is the sum of a term originated from the proton plus a fraction given by the neutron target. In formula $R_N = R_p + 0.31\ R_n$.

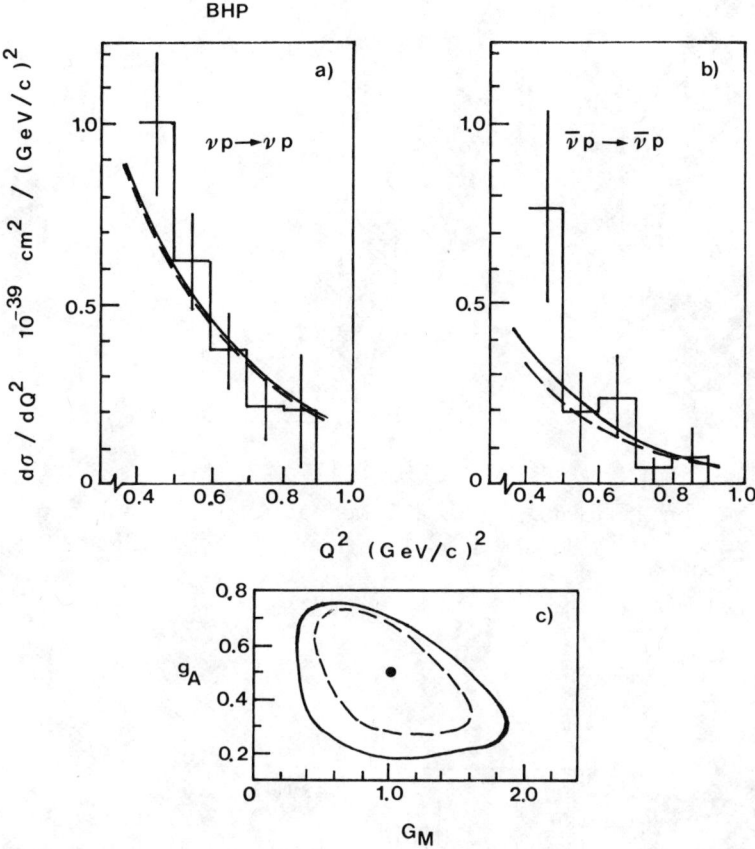

Fig. 2. The Q^2 distributions obtained by BHP collaboration for neutrino (a) and antineutrino (b) elastic scattering. In (c) it is shown the contour in the g_A, G_M plane at 1 and 2 standard deviations (ref. 8).

In table II the experimental results of the two experiments are summarized and the form factors are compared with the prediction of the "GWS" model, assuming $\sin^2\theta_W = 0.23$.

From the equations (1) it is clear the importance of this process for the determination of the coupling constants, being one of the crucial measurement in resolving the so called vector-axialvector and isoscalar-isovector ambiguities. The interpretation of the data, however, suffers from the undetermination of the mass M_A entering the dipole fit of the axial form factor.

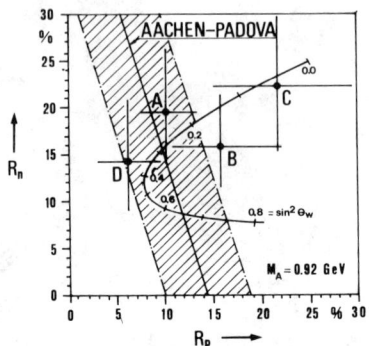

Fig. 3 Region in plane R_p-R_n allowed at 1σ by measurement of R_N in neutrino scattering (Ref. 9).

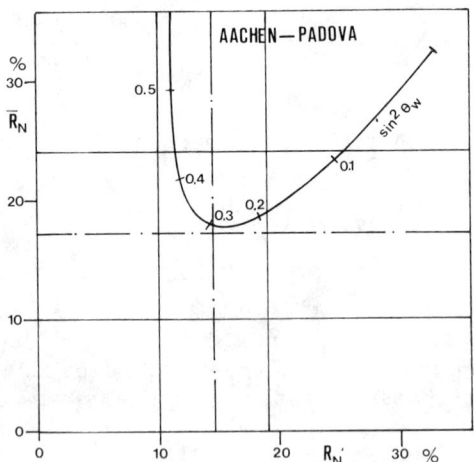

Fig. 4 Regions allowed at 1σ by measurements of R_N and $\bar{R}_N$ in neutrino and antineutrino elastic scattering (Ref. 10).

SEMILEPTONIC INTERACTIONS

Table II. Neutrino elastic scattering experiments

Experiment	R_p	$\bar{R}_p$	R_n	$\bar{R}_n$	$Q^2 (GeV^2/c^2)$
AACHEN – PD[9-10]	.10±.03	.12±.05	.15±.06	.17±.08	0.2–1.0
BNL-HARV.-PENN.[8]	.11±.02	.19±.05			0.4–0.9

Experiment	$G_E(0)$	$G_M(0)$	$G_A(0)$
BNL-HARV.-PENN.[8]	$.05^{+0.25}_{-0.5}$	$1.0^{+0.35}_{-0.40}$	$0.5^{+0.2}_{-0.15}$
GWS ($\sin^2\theta=0.23$)	0.04	1.07	0.63

SEMI-INCLUSIVE REACTIONS

As long ago as 1972 Feynman[11] proposed to use the parton model for the semi-inclusive lepton scattering in the "current fragmentation" region.

According to this assumption, the hadronic composition of the final state is directly related to the couplings of the weak current with the quark "target".

For the neutral current, the chiral couplings are:

neutrino   : u/d $(u_L^2 + 1/3 u_R^2)/(d_L^2 + 1/3 d_R^2)$

antineutrino: u/d $(u_R^2 + 1/3 u_L^2)/(d_R^2 + 1/3 d_L^2)$

If the fragmentation of the emerging quark "factorizes", i.e. is indipendent from the production, as visualized in fig. 5, then

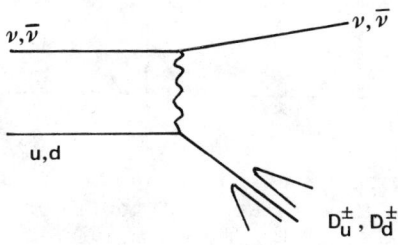

Fig.5. Fragmentation of quarks in neutral current interactions.

the charge ratio of the hadrons in the final state can be related to the coupling constants as follows:

$$\frac{h^+}{h^-} = \frac{(u_L^2 + \frac{1}{3} u_R^2) D_u^+ + (d_L^2 + \frac{1}{3} d_R^2) D_d^+}{(u_L^2 + \frac{1}{3} u_R^2) D_u^- + (d_L^2 + \frac{1}{3} d_R^2) D_d^-}$$

The fragmentation functions $D_q^{\pm}$ can be derived by measurements in charged current interactions, and the validity of the factorization assumption can be tested comparing the observed $h^{\pm}$ distributions in the scaling variable z with the Field and Feynman prediction. This comparison has been performed by the ABCMO collaboration in the experiment[12] where they measure the charge ratio of hadrons in neutrino interactions on hydrogen.

The sensitivity[13] of this kind of experiments is best represented in Fig. 6, where it is drawn, in the plane d_L, u_L for neutrinos, and in the plane d_R, u_R for antineutrinos the region allowed by a hypothetical π^+/π^- measurement with 10% error. In the same plot, also the existing limits imposed by inclusive reactions on isoscalar target are shown. The full point with error bars represents the result of a global fit of J.E.Kim et al.[35] to all data. I shall discuss this fit later.

A final sample of 360 neutral current events has been collected by the ABCMO collaboration in an exposure of BEBC to the neutrino Wide Band Beam. As it is shown in Fig. 7, the ratio h^+/h^-, corrected for the proton contamination, is almost indipendent from the cut

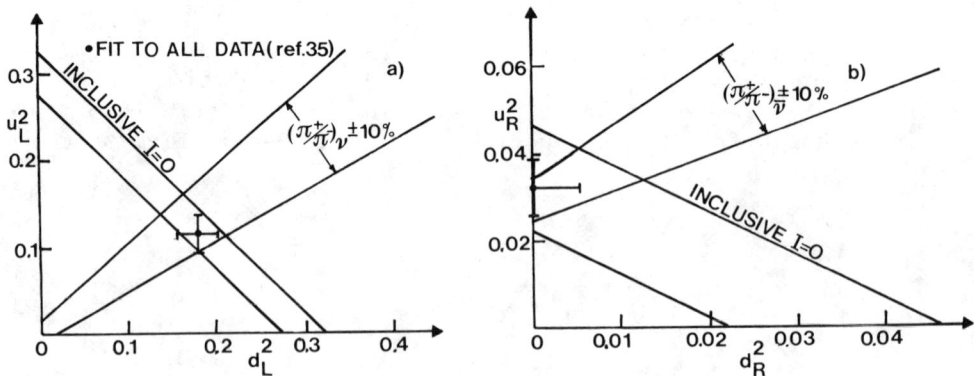

Fig. 6. Limits imposed on chiral coupling by a 10% measurement of the charge ratio in NC (a) neutrino and (b) antineutrino reactions (Ref. 13). The point is the result of a fit[35] to all data.

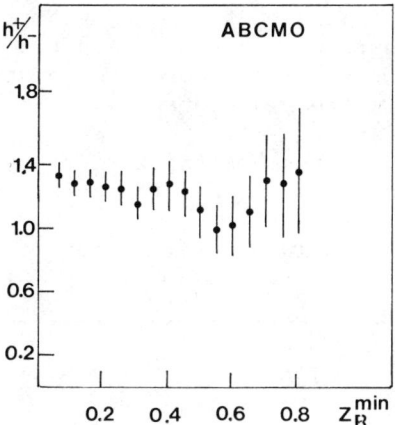

Fig. 7. Charge ratio for tracks with $z_R > z_R^{min}$ in NC events. (Ref. 12)

on z_R above 0.25. (z_R is the hadron energy divided by the visible energy of the hadronic system). The contribution from target fragmentation is believed to be negligible for tracks above this value.

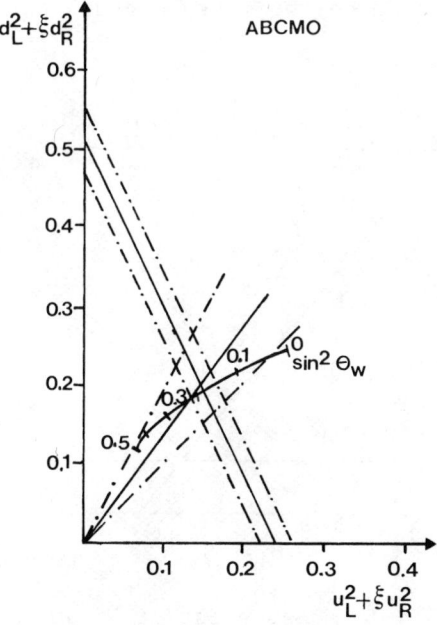

Fig. 8. Limits on chiral couplings impose by the charge ratio data combined with the inclusive result obtained in hydrogen (Ref. 12, 17).

Fig. 8 shows in the plane $d_L^2+\xi d_R^2$, $u_L^2+\xi u_R^2$ the region determined by the charge ratio measurement together with the inclusive NC result obtained in the same experiment. Here the constant ξ is less than 1/3 because of cuts in the y acceptance.

At high energies the "factorization" hypothesis should be more reliable. Recent data in this region are summarized in Table III.

Table III. Charge ratios in neutral current interactions.

Experiment	(h^+/h^-)	(h^+/h^-)
FIIM[14]		1.59+0.27
ABCDLOS[15]	1.07+0.17	1.54+0.45
ABCMO[12]	1.25+0.11	

This kind of experiments, with the further assumption $D_u^+ = D_d^-$, $D_u^- = D_d^+$, can give directly the ratio $(u_L^2+\xi u_R^2)/(d_L^2+\xi d_R^2)$, as it was done by the ABCMO collaboration with the result 0.72±0.23.

INCLUSIVE INTERACTIONS ON PROTONS AND NEUTRONS

The process $\nu+p \rightarrow \nu+X$ and the corresponding one $\nu+n \rightarrow \nu+X$, even though is affected by nuclear effects of deuterium, are on invaluable source of information on the isospin character of the current,

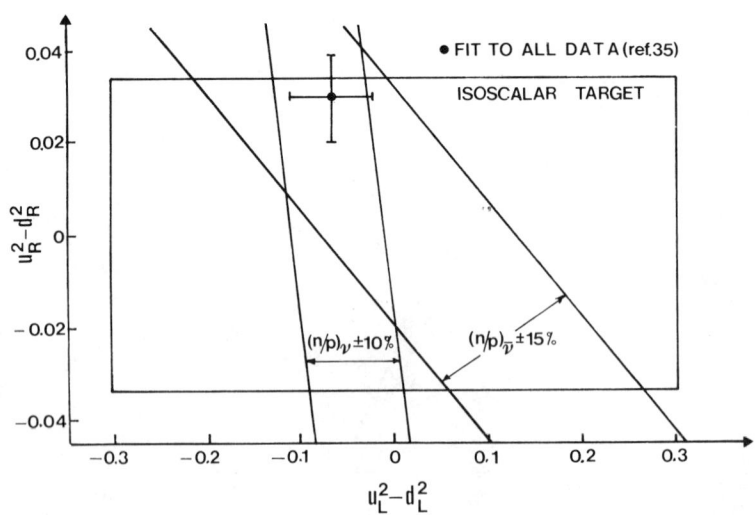

Fig. 9. Limits imposed by a measurement of NC inclusive reaction on p/n target with 10% accuracy for neutrinos and 15% for antineutrinos. The frame delimits the region allowed by inclusive data on isoscalar target. The point is the result of fit to all data by J.E.Kim et al.[35]

due to the different quark content of the targets. In fact the ratio R_p, neutral current over charged current events, is proportional to $2u_L^2+d_L^2+$right-handed couplings, while in the scattering off an isoscalar target, the corresponding ratio is related to $u_L^2+d_L^2$, or $u_R^2+d_R^2$ for antineutrino. Alternatively, the cross sections for NC interactions on proton and neutron can be combined to give the following relations with the chiral couplings:

$$\frac{\sigma(\nu p)-\sigma(\nu n)}{\sigma(\nu p)+\sigma(\nu n)} = u_L^2-d_L^2+\frac{1}{3}(u_R^2-d_R^2)$$

$$\frac{\sigma(\bar\nu p)-\sigma(\bar\nu n)}{\sigma(\bar\nu p)+\sigma(\bar\nu n)} = 1.9\{u_R^2-d_R^2+\frac{1}{3}(u_L^2-d_L^2)\}$$

A measurement of these quantities with 10% (15%) accuracy for neutrinos (antineutrinos) will produce the limits depicted in Fig. 9.

The result of a new experiment[16] in the 15' bubble chamber of FNAL filled with deuterium has been presented to this Conference by the Illinois-Maryland-Stony Brook-Tohoku-Tufts collaboration (IMSTT). The hidden CC interactions are subtracted by using the ob-

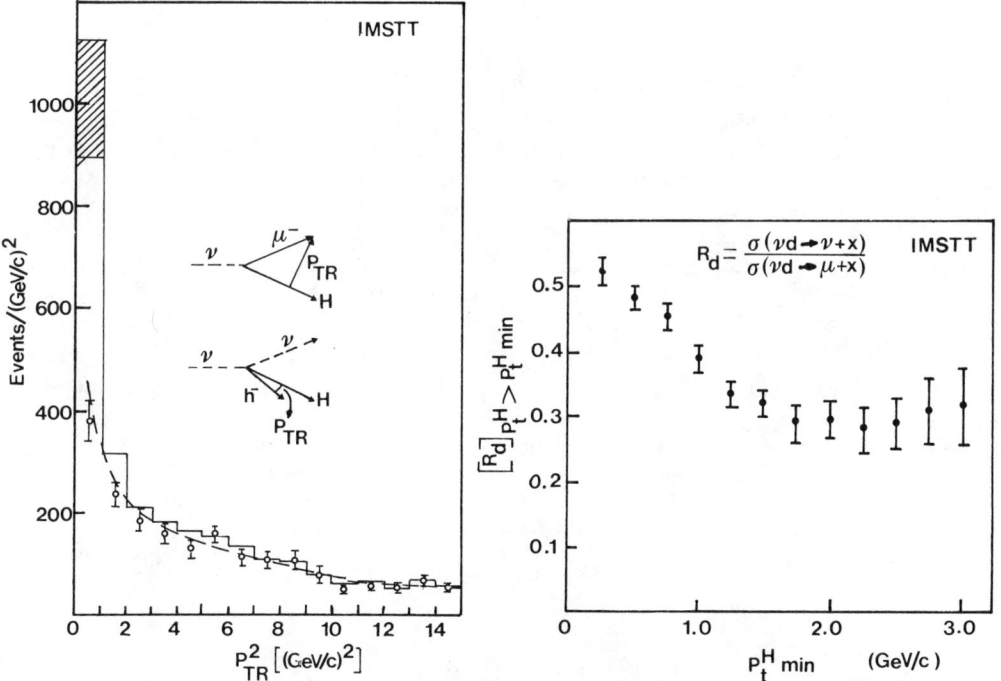

Fig. 10. Distributions in kinematic variables used to subtract
(a) hidden CC events, (b) hadronic induced events (Ref.16).

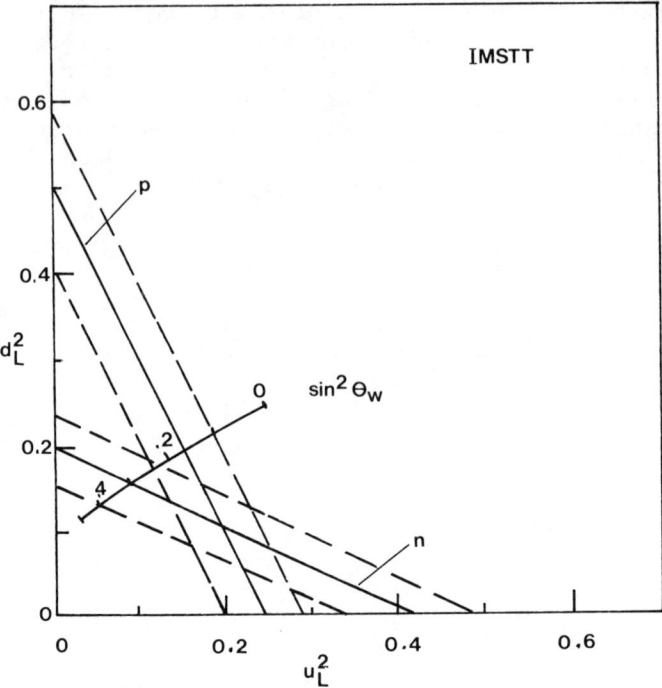

Fig. 11. Limits on chiral coupling imposed by measurement of R_p and R_n by the IMSTT collaboration (Ref. 16).

servation that the unidentified muon has a large component p_{TR} of its momentum perpendicular to momentum sum of the remaining hadrons. (Fig. 10a). The elimination of the hadronic induced events is based on a cut on the total longitudinal momentum, and on the transverse component of the total hadron momentum (Fig. 10b). The measured quantities R_p and R_n are then expressed in terms of the chiral couplings and the allowed regions are represented by two crossing bands in the plane u_L, d_L (Fig. 11).

The ABCMO collaboration last year has published[17] the data obtained from an exposure of BEBC filled with hydrogen. In order

Table IV. Inclusive scattering experiments on nucleon

Experiment	R_p	R_n	$\nu n/\nu p$	$\bar\nu n/\bar\nu p$
FIIM[18]				1.06+0.20
ABCMO[17]	0.51+0.04			
IMSTT[16]	0.50+0.08	0.21+0.03	1.2+0.19	
G W S ($\sin^2\theta_w = 0.23$)			1.1	1.0

SEMILEPTONIC INTERACTIONS

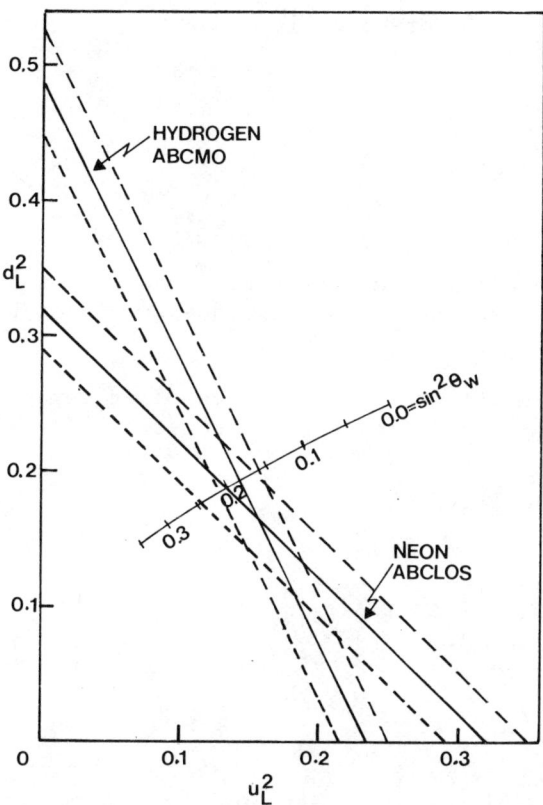

Fig. 12. Limits on chiral couplings determined by measurements on proton and isoscalar targets (Ref. 17).

to define a limited region in the plane u_L, d_L, they use the result[19] of a measurement done in neon (I=0). The comparison between the two results (Fig. 11 and 12) shows that the data from the neutron target allow in principle a greater sensitivity. Table IV summarizes the experimental data at high energy where the contribution of the elastic scattering is negligible[20].

INCLUSIVE REACTIONS ON ISOSCALAR TARGET

The effort sustained in this kind of experiment is justified on theoretical grounds and by the practical consideration on the possibility to build large mass targets properly instrumented. Marble and iron have in fact equal or almost equal number of protons and neutrons. Likewise theoretical arguments indicate that the derivation of the coupling constants from interactions off the isoscalar target is the least affected by details of the QPM.

The measured ratio R of neutral over charged current events

(R̄ for antineutrino interactions) can be related to the chiral couplings in three possible ways.

Let's recall the relation between the NC and CC cross sections:

$$\sigma_{NC}^{\nu,\bar{\nu}} = (u_L^2+d_L^2)\sigma_{CC}^{\nu,\bar{\nu}} + (u_R^2+d_R^2)\sigma_{CC}^{\bar{\nu},\nu} +0(s,\bar{s})$$

Above all one should notice that the accuracy of these measurements is such that the presence of strange quarks cannot be neglected anymore.

The first way to determine the coupling parameters uses the equation written above, from which one gets:

$$u_L^2+d_L^2 = \frac{R-r^2\bar{R}}{1-r^2} + 0\,(s,\bar{s})$$ (2)

$$u_R^2+d_R^2 = \frac{\bar{R}-R}{r^{-1}-r} + 0\,(s,\bar{s})$$

where $\quad r = \dfrac{\sigma_{CC}^{\bar{\nu}}}{\sigma_{CC}^{\nu}}$

The second alternative makes use of the explicit expressions of the cross sections in terms of the coupling constants and of the quark parton distribution.[35] These distributions, measured in CC interactions, are conveniently parametrized, taking into account QCD effects (scaling violations). A calculable expression is then obtained

$$R = \frac{\iiint \phi(E)\,\frac{d^2\sigma^{NC}}{dydx}\,dydxdE}{\iiint \phi(E)\,\frac{d^2\sigma^{CC}}{dydx}\,dydxdE} =$$

$$= a_1 u_L^2 + a_2 d_L^2 + a_3 u_R^2 + a_4 d_R^2 = b_1+b_2\sin^2\theta_w+b_3\sin^4\theta_w$$

and similarly for antineutrino interactions.
The value of $\sin^2\theta_w$ is therefore directly computed. To derive the com
bined coupling parameters, one must use a modified version of the equations (2), having on the right hand side a small correction term
still depending from the couplings.

The third possibility relies on the Paschos Wolfenstein[21] equa tion that I shall discuss in some detail. The appeal of this formula rests on the generality of its derivation. It can be shown that, if the hadronic part of the current is only isovector, under strong isospin invariance the following relation holds:

$$D \equiv \frac{\frac{d\sigma}{dxdy}(\nu \to \nu) - \frac{d\sigma}{dxdy}(\bar{\nu} \to \bar{\nu})}{\frac{d\sigma}{dxdy}(\nu \to \mu^-) - \frac{d\sigma}{dxdy}(\bar{\nu} \to \mu^+)} = \tfrac{1}{2}\alpha\beta \qquad (3)$$

in the limit of $\cos\theta_c = 1$.
The same relation is still valid if the hadronic isoscalar current is pure vector. In the Weinberg Salam model the equation (3) reads:

$$D = \tfrac{1}{2} - \sin^2\theta_w$$

A number of corrections can be computed, invoking the QPM, and as we shall see their size is at present smaller than the experimental errors. I shall follow mainly the work of Paschos[22] who investigated this problem. Firstly in case of presence of an isoscalar neutral current, the QPM enable us to derive[23]:

$$D = \tfrac{1}{2}(\alpha\beta+\gamma\delta) = (u_L^2+d_L^2) - (u_R^2+d_R^2)$$

Higher order QCD terms can generate such a current and give a correction $\Delta\sin^2\theta_w = 0.0008$. Secondly, relaxing the approximation $\cos\theta_c=1$, and introducing the full description of the weak mixing angles between six quarks as in the representation of the Kobayashi-Maskawa matrix, the maximum correction arising from this source could be $\Delta\sin^2\theta_w = +0.005$, based on present limits of the mixing angles. Thirdly, since the equation (3) is experimentally obtained integrating the kinematical domain allowed by the spectrum of the neutrino and antineutrino fluxes, one has to take into account scaling violations that modify the linearity of the cross sections with energy. This effect is extimated in the narrow band beam of CERN to be $\Delta\sin^2\theta_w = 0.001$. Finally it should not be forgotten the corrections[24] arising from higher order weak and electromagnetic diagrams. Their contribution is extimated to be of the order of percent.

In conclusion, the Paschos Wolfenstein equation gives a determination of the weak angle substantially free from the QPM, and only the corrective terms need it to be calculable. However their size is small enough not to impair the validity of this relation.

Therefore, if one is convinced that the theory of Glashow, Weinberg and Salam gives the correct description of the weak interactions, the inclusive neutrino scattering on isoscalar target seems to give the least model dependent derivation of the fundamental parameter $\sin^2\theta_w$.

Table V. Background and corrections in NC analysis of CHARM experiment

Corrections to NC	Neutrino	Antineutrino
Raw Events	2361	1126
WBB background	-4.0%	-12.8%
$K_{e3} \to \nu_e$ background	-6.9%	- 2.3%
Hidden CC events	-4.1%	- 2.0%
π/K decay	+2.3%	+ 1.3%
Corrected events	2059+54	948+43

At present the experimental progress in this field is represented by a new detailed analysis of a measurement[25] performed by the CHARM collaboration at the narrow band beam of Cern. The experiment makes use of the fine grain of the calorimeter to separate NC and CC induced events. On event by event basis, the charged current interaction is identified by the presence of a non interacting track with a range corresponding to a muon momentum of at least 1 GeV/c. The background subtractions to the NC events are summarized in Table V.

The CDHS collaboration is refining the analysis of the data presented[26] last year, but their result is not yet available. I recall that in this experiment the NC separation is statistically done on the basis of the event length, as the muon penetrates in the calorimeter more than the pure hadronic shower.

Table VI shows the ratios R and $\bar{R}$ for neutrino and antineutrino scattering of the two experiments compared with the average values

Table VI. Ratios of neutral current over charged current events for ν and $\bar{\nu}$

Experiment		R	$\bar{R}$
Previous average (Tokyo 1978)[37]		0.29+0.01	0.35+0.025
CDHS[26]	(E_H >10 GeV)	0.307+0.003+0.008	0.373+0.014+0.0
CHARM[25]	(E_H >2 GeV)	0.320+0.009+0.003	0.377+0.020+0.0

The first error is statistical while the second is systematic given by C.Baltay[37] at Tokyo. The $\sin^2\theta_W$ evaluated from the two experiments, are given in Table VII. I shall call "Model dependent"

Table VII. Derivation of $\sin^2\theta_w$ from recent experiments

Experiment	Model dep. analysis	Paschos-Wolfenstein
CDHS[26]		0.228±0.018
CHARM[25]	0.220±0.013±0.005	0.230±0.022

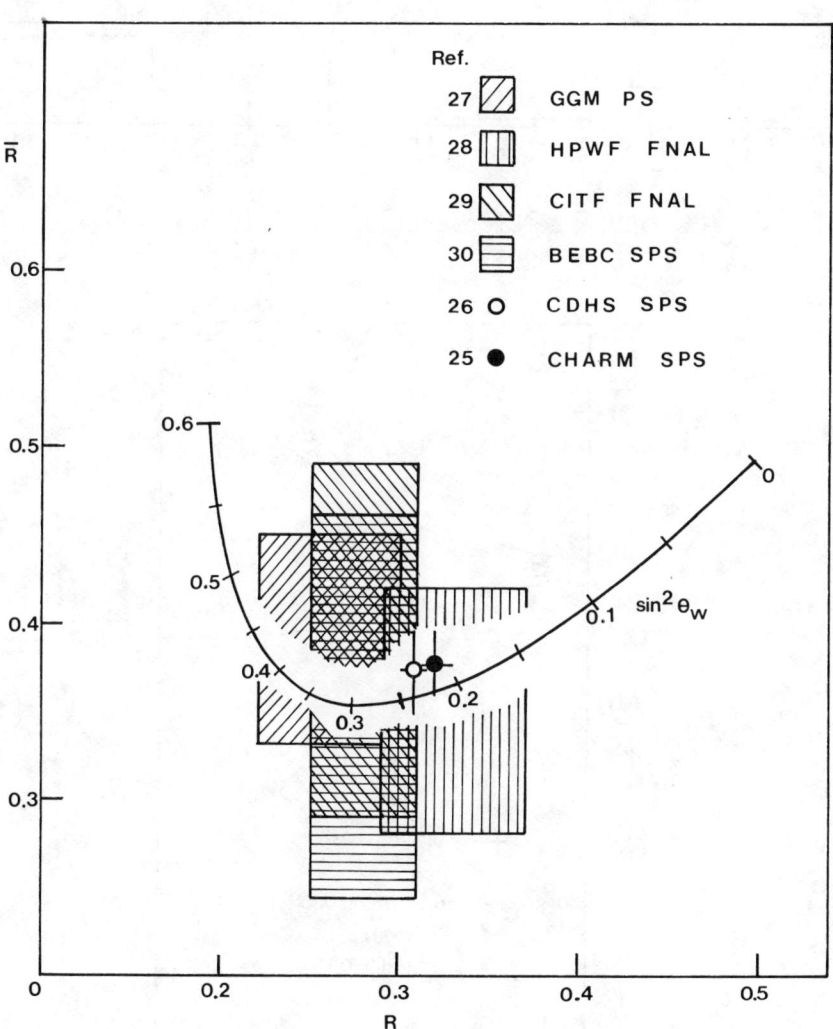

Fig. 13 Determination of $\sin^2\theta_w$ in inclusive neutrino scattering on isoscalar target.

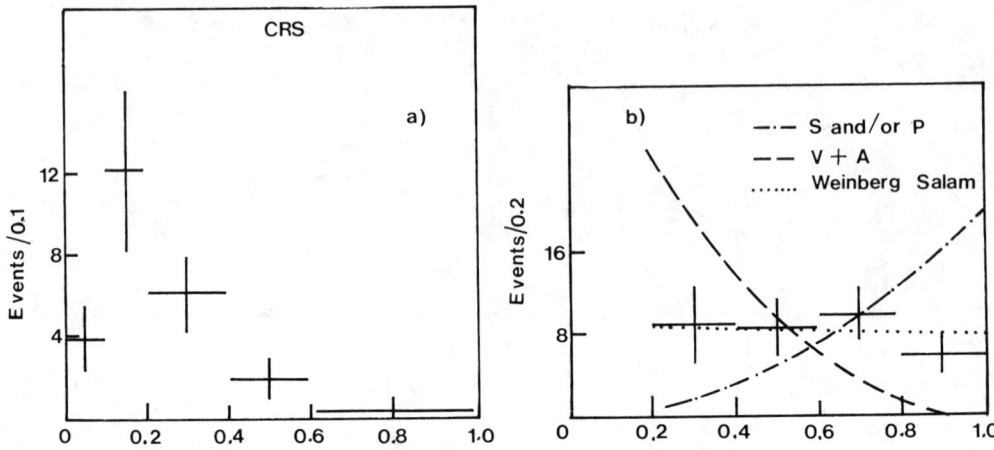

Fig. 14 Scaling-variable distributions for neutral current events: (a) x distributions, (b) y distributions (Ref. 31).

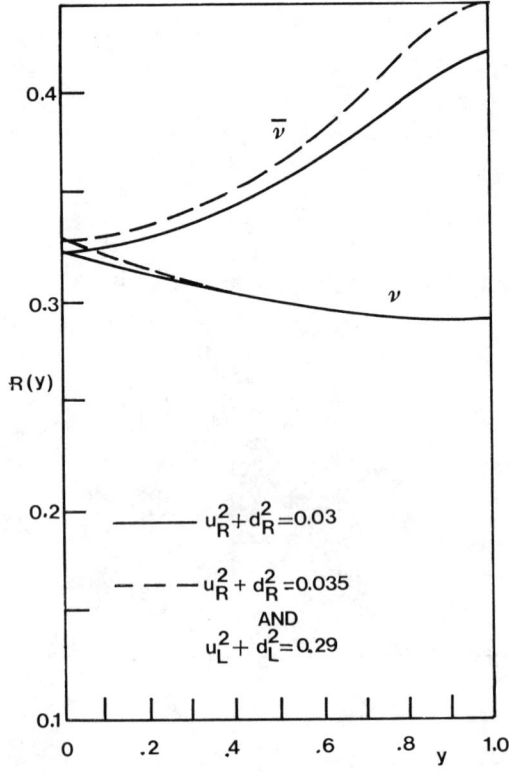

Fig. 15 Computed sensitivity to the ratio R = NC/CC to variations of the right-handed coupling (Ref. 13).

the derivation of $\sin^2\theta_w$ with one of the methods described before the Paschos Wolfenstein equation. One can immediately notice that in the CHARM experiment there is ample space to improve the statistical accuracy since the error is almost three times the systematic uncertainty. From Fig. 13 it is apparent the net improvement of the mentioned experiments compared to the previous results[27-30]. This fact will have consequences on the fit of the coupling constants to all available data on neutral current interactions.

Neutral current distributions in the scaling variables

The only new piece of information on this chapter, comes from the definitive results[31] of Columbia, Rutgers, Stevens collaboration obtained in the 7' bubble chamber exposed to the narrow band beam of BNL. The applied corrections for unseen hadrons are tested in the charged current induced events. Within the large errors, they find the NC distributions in the scaling variables x and y very similar to those of CC interactions, and in agreement with the predictions of the GWS theory.

The computed sensitivity of the y distribution to variations of the righthanded couplings is shown graphically in Fig. 15, but the measurements so far don't reach the necessary accuracy.

PARITY VIOLATION IN ELECTROMAGNETIC -WEAK INTERFERENCE

In the lepton-quark sector, the knowledge of the weak coupling constants derives from measurements of parity violation effects in the interference between the electromagnetic and the weak amplitudes. The observation of parity violation in atomic transitions is based on the rotation of the polarization plane of a laser beam or on a difference between the absorption of right and left circularly polarized laser light. The effect on heavy atoms is proportional to the weak charge:

$$Q_w = -\{\tilde{\alpha}(Z-N) + 3\tilde{\gamma}(Z+N)\}$$

Presently it exists some conflict between the experiments, but as the new data[32] have a trend to approach the results of the Novosibirsk and Berkeley experiments, I will quote the last two, and use them to check the factorization hypothesis:

$$\text{Novosibirsk}^{32} \quad \text{Bi} \quad Q_w = -140 \pm 40$$

$$\text{Berkeley}^{33} \quad \text{Tl} \quad Q_w = -280 \pm 140$$

The parity violation effect can also be measured in the scattering of polarized electrons on deuterium target. If weak and electromagnetic interference exists, then the asymmetry in cross section for scattering of electrons with right-handed or left-handed helicity will be different from zero. It can be shown[38] that the asymmetry is

$$A \equiv (\sigma_R - \sigma_L)/(\sigma_R + \sigma_L) = \left[a_1(x) + a_2(x)\left(\frac{1-(1-y)^2}{1+(1-y)^2}\right)\right] Q^2$$

with the constants a_1 and a_2 related to the coupling constants:

$$a_1 = \frac{G}{\sqrt{2}e^2}(9\tilde{\alpha} + 3\tilde{\gamma})/5 \quad ; \quad a_2 = \frac{G}{\sqrt{2}e^2}(9\tilde{\beta} + 3\tilde{\delta})/5$$

independent from x in deuterium.

A "unique" experiment was performed at SLAC,[34] and is too well known to describe it. Owing to the measurement of the y dependence, both coefficients could be determined:

$$a_1 = (-9.7 \pm 2.6) \cdot 10^{-5} (\text{GeV}/c)^{-2}; \quad a_2 = (4.9 \pm 8.1) \cdot 10^{-5} (\text{GeV}/c)^{-2}$$

Restricting myself only to the lepton-quark sector, we see immediately from Fig. 16 that in the plane $\tilde{\alpha} - \tilde{\gamma}$ the intersection between the bands allowed by the polarized electron scattering and by the

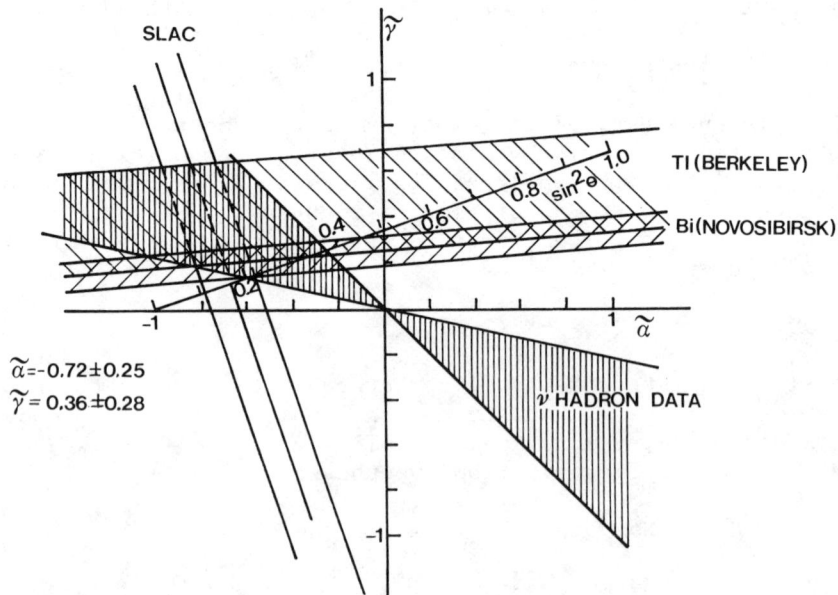

Fig. 16 Determination of the coupling constants $\tilde{\alpha}$ and $\tilde{\gamma}$ and test of the factorization relation $\alpha/\gamma = \tilde{\alpha}/\tilde{\gamma}$. (From. ref. 4).

atomic parity violation experiments falls in the region determined by the factorization relation from the neutrino quark data. Furthermore the measured y dependence is in complete disagreement with the "Hybrid model"[39] with a χ^2 probability of 6×10^{-4}. This result rules out the possibility to have the right handed electron in a doublet with a neutral heavy lepton.

COUPLING CONSTANTS ANALYSIS

The work of Hung and Sakurai[4] is directed to show that the present measurements enable to determine in a completely model independent way most of the 10 coupling constants, not even assuming tacitely that the strength of neutrino-neutrino scattering is $c_\nu^2 = 1$. In their analysis a special emphasis is given to verify with the present knowledge the factorization relations, consequence of the hypothesis of a single Z boson. It is also derived from data the value of the coupling c_ν of the Z to the neutrino with the result $c_\nu^2 = \pm(0.86 \pm 0.26)$. In SU(2) X U(1) models, with $T_{3L}^\nu = 1/2$, the relation $c_\nu^2 = \rho$ holds, thus giving an independent estimate of the parameter measuring the relative strength of NC and CC interactions.

Since last year, J.E.Kim et al.[35], and I.Liede and M.Roos[36] have updated their analyses, as reported previously in the reviews of U.Amaldi[40] and K.Winter[41]. As expected no major changes are however reported. Here it suffices to recall that the first one aimed to determine the coupling constants without recourse to a specific model of the weak interactions. In addition they examine the consequences of the fit on specific models based on groups allowing right-handed fermion doublets. In this context they found limits on right-handed couplings (see Table VIII). Furthermore, within the GWS model, the values for ρ and $\sin^2\theta_w$ are also obtained, and finally it is derived the one parameter fit of $\sin^2\theta$ with $\rho = 1$. The second analysis by Liede and Roos is directly aimed to obtain the best value of $\sin^2\theta_w$ to compare with the prediction of the Grand Unified Theories. In Fig. 17 it is shown for comparison how a single expe-

Table VIII. Determination[35] of parameters in generalized SU(2)xU(1) models

ρ	$\sin^2\theta_w$	T_{3R}^u	T_{3R}^d	T_{3R}^e
1.018	0.249	−0.010	−0.101	0.039
±0.0045	±0.031	±0.040	±0.058	±0.047

riment[25] on inclusive neutrino scattering determines the two parameters ρ and $\sin^2\theta_w$ of the GWS model.

CONCLUSIONS

The neutral current interactions represent a field where after a rapid progress, we assist to an evolution with the characteristic rate dictated by experiments aiming at precise measurements. In this light, the model independent analysis of the coupling constants seems firmly founded, reducing the alternatives to gauges models other than that based on the group SU(2) x U(1). New experiments have been presented to this conference on the p/n and h^+/h^- ratios that will have an impact on the determination of the coupling constants. In fact all the right handed sector needs improved experimental information.

In the context of SU(2) X U(1) theories, the presence of only a doublet in the Higgs mechanism seems to be supported by the determination of the parameter ρ.

Progress in neutrino hadron scattering is represented by the improvement in the systematic error of one experiment. The determination of the weak angle in the Weinberg Salam model depends largely by this kind of measurements, where also an effort is started to controll the assumptions and to extimate the corrections in the derivation of $\sin^2\theta_w$.

Table IX. Determinations of ρ and $\sin^2\theta_w$ in the Glashow Weinberg Salam model

Analysis	ρ	$\sin^2\theta_w$
J.E.Kim et al[35]	0.992+0.017 (+0.011)	0.224+0.015 (+0.012)
ρ = 1	1	0.229+0.009 (+0.005)

In parenthesis: extimated theoretical uncertainties.

I.Liede, M.Roos[36a]	1.04+0.04	0.245+0.026
ρ = 1 [36b]	1	0.238+0.011
CHARM[25] Model dep.	1.027+0.023	0.247+0.038
ρ = 1	1	0.220+0.014
CDHS[26] Paschos-Wolfenstein	1	0.228+0.018

SEMILEPTONIC INTERACTIONS

The difference between the expected value[42] of $\sin^2\theta_W$ from SU(5) and that determined at the present range of energies is beginning to be significative. Therefore better data from the inclusive scattering on isoscalar targets will have a decisive role, due to the least assumptions in the interpretation of the results.

In this panorama of continuous improvements, precise structure functions in neutral current interactions and values of the couplings to strange and charm quarks are still lacking. Presumably the present experiments will be soon in position to answer, at least partially, to these questions before the large e^+e^- storage rings enter into the game.

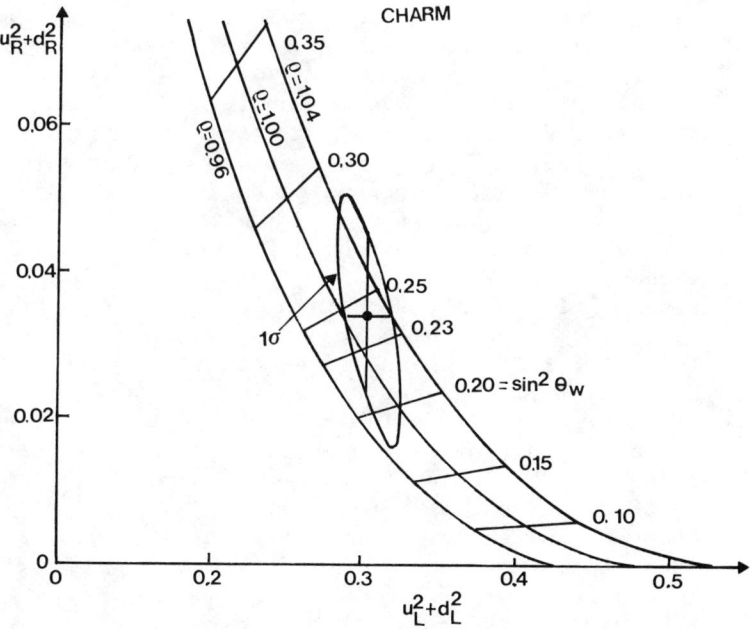

Fig. 17 Determination of the chiral coupling constants and their relation with the parameters ρ and $\sin^2\theta_W$ in the experiment of CHARM collaboration (Ref. 25).

REFERENCES

1. J. J. Sakurai, Proc. Int. Conf. on Neutrino 1977 Baksan Valley vol. 2, pg. 242.
2. L. M. Sehgal, Proc. Int. Conf. on Neutrino 1978 Purdue, pg. 253.
3. P. Q. Hung, J. J. Sakurai, Phys. Lett. 69B:323 (1977).
4. P. Q. Hung, J. J. Sakurai, Phys. Lett. 88B:91 (1979).
5. S. Glashow, Nucl. Phys. 22:579 (1961).
   S. Weinberg, Phys. Rev. Lett. 19:1264 (1967).
   A. Salam, Elementary Particle Theory, ed. N. Svartholm (Almquist and Wiksells, Stockholm 1969).
6. Exclusive reactions as single pion and Δ (1238) production were used to discriminate against a predominant isoscalar solution. The interpretation of the data however depends largely from details of the model. New results on single π production are recently available from M. Derrick et al., Argonne, Carnegie, Kansas, Purdue collaboration Phys. Lett. 92B:363 (1980). Direct information on the coupling constant β is obtained in deuterium fission by $\bar{\nu}_e$, E. Pasierb, H. S. Gurr, J. Lathrop, F. Reines, H. W. Sorbel, Phys. Rev. Lett. 43:96 (1979).
7. C. M. Wu, Phys. Lett. 89B:218 (1980).
8. A. Entenberg et al., Phys. Rev. Lett. 42:1198 (1979).
9. H. Faissner et al., Phys. Rev. D21:555 (1980).
10. H. Faissner et al., contribution to this Conference.
11. R.P.Feynman, Proc. of Int. Conf. on Neutrino 1972, Balaton.
12. ABCMO collaboration, contribution to this Conference BONN-WA21-int-004.
13. L. M. Sehgal, CERN Neutrino Workshop 1978 CERN/SPSC 78-53.
14. J. Bell et al., FIIM collaboration, contribution to Int. Conf. on Neutrino 1979, Bergen, Proceedings pg. 592 vol. 2.
15. ABCDLOS collaboration, contribution to Int. Conf. on Neutrino 1979 Bergen, Proceedings pg. 397 vol. 2.
16. S. Sommars et al., IMSTT collaboration, contribution to this Conference.
17. J. Blietschau et al., ABCMO collaboration, Phys. Lett. 88B:381 (1979).
18. J. Bell et al., FIIM collaboration, contribution to Int. Conf. on Neutrino 1979, Bergen, Proceedings pg. 592 vol. 2.
19. H. Deden et al., ABCLOS collaboration, Nucl. Phys. B149:1 (1979)
20. A discussion on the correction due to unseen elastic scattering is contained in the review by U. Amaldi, Proc. Int. Conf. on Neutrino 1979, Bergen, pg. 367, vol. 1.
21. E. Paschos, L. Wolfenstein, Phys. Rev. D7:91 (1973).
22. E. Paschos, Lecture given at the 15th Recontre de Moriond 1980 DO-TH 80:12.
23. J. J. Sakurai, invited talk at the Topical Conf. on Neutrino Physics at Accelerators 1978 Oxford - UCLA/78/TEP/18 and Invited Lectures to Int. Summer Inst. Desy, Hamburg 1975, CERN TH-2099.
24. QED corrections: J. Kiskis, Phys. Rev. D8:2129 (1973);

A. De Rujula, R. Petronzio, A. Savoy-Navarro, Nucl. Phys. B154:394 (1979).

WS corrections: S. Sakakibara DO-TH 80:01 Dortmund (1980).

25. M. Jonker et al., CHARM collaboration, contribution to this Conference.
26. CDHS collaboration, contribution to Int. Conf. on Neutrino 1979 Bergen, Proceedings pg. 392 vol. 2.
27. J. B. Blietschau et al., Gargamelle collaboration, Nucl. Phys. B118:218 (1977).
28. P. Wanderer et al., HPWF collaboration, Phys. Rev. D17:1679 (1978).
29. F. S. Merrit et al., CITF collaboration, Phys. Rev. D17:2199 (1978).
30. P. C. Bosetti et al., ABCLOS collaboration, Phys. Lett. 76B:505 (1978), see also ref. 19.
31. C. Baltay et al., Phys. Rev. Lett. 44:916 (1980).
32. L. M. Barkov, review talk at this Conference and L. M. Barkov, M. S. Zolotorev, JETP Lett. 27:379 (1978).
33. P. Conti et al. Phys. Rev. Lett. 42:343 (1979).
34. C. Y. Prescott et al., Phys. Lett. 84B:524 (1979).
35. J. E. Kim, P. Langacker, M. Levine, H. H. Williams, UPR-158T (1980) submitted to Rev. Mod. Phys.
36.a M. Roos, I. Liede, Proc. Int. Conf. on Neutrino 1979, Bergen, pg. 309, vol. 1.
   b I. Liede, M. Roos, HU-TFT-79-27.
37. C. Baltay, Proc. 19th Int. Conf. on High Energy Physics, Tokyo 1978, pg. 882.
38. J. D. Bjorken, Phys. Rev. D18:3239 (1978).
    L. Wolfenstein, COO-3066-111 (Aug. 1978).
    H. Fritzsch, CERN TH-2607 (Nov. 1978).
39. R. N. Cahn, F. J. Gilman, Phys. Rev. D17:1313 (1978).
40. U. Amaldi, Proc. Int. Conf. on Neutrino 1979, Bergen, pg. 367, vol. 1.
41. K. Winter, Proc. Int. Symp. Lepton and Photon, Batavia 1979, pg. 258.
42. See for example the review talk of F. Wilczek, Proc. Int. Symposium on Lepton, Photon, Batavia 1979 pg. 437 where it is quoted the value $\sin^2\theta_w = 0.19 \pm 0.02$ in the framework of the SU(5) model.

DISCUSSION (Chairman: H. Faissner)

C. GEWENIGER, Heidelberg: I would like to comment on the CDHS data presented here. (i) The momentum cut-off for the muon is not 10 GeV/c but rather about 3 GeV/c, which is given by an event length cut about two meters of iron. (ii) The systematical errors are preliminary. The analysis is still in progress and the final errors will be smaller.

REVIEW ON PURELY LEPTONIC INTERACTIONS OF WEAK NEUTRAL CURRENTS[*]

Luke W. Mo

Virginia Polytechnic Institute and State University

Blacksburg, Virginia 24061

Abstract

The purely leptonic interactions of weak neutral currents are reviewed in the context of the electroweak theory. Experiments on neutrino-electron scattering are summarized. Experimental work to be done in the future is discussed.

---

[*] Work supported by the National Science Foundation under Grant No. PHY80-00241

It has been well-known for many years that the neutrino-electron scattering reactions provide the best testing ground for the fundamental principles of weak interactions. There are mainly four processes:

$$\nu_\mu + e^- \to \nu_\mu + e^-,$$
$$\bar{\nu}_\mu + e^- \to \bar{\nu}_\mu + e^-,$$
$$\nu_e + e^- \to \nu_e + e^-,$$
$$\bar{\nu}_e + e^- \to \bar{\nu}_e + e^-. \quad (1)$$

Perhaps we should add ν_τ and $\bar{\nu}_\tau$ to this list in the near future. As shown in Fig.1, the first two reactions involve only the neutral currents; while in the last two reactions, the charged and neutral currents both contribute. The main advantages of doing these experiments are because the neutrinos and the electrons are point-like particles and they possess only electromagnetic and weak interactions The interpretation of the experimental results is free from the complications arising from particle structures and strong interactions. The disadvantage is that the cross sections are too small, $\sim 10^{-42}$ cm^2/GeV, which has prevented the accumulation of good statistics.

Since there are no new results contributed to this Conference, the elementary physics involved on this subject will be reviewed. The existing experiments will be briefly summarized, and the topics that remain to be done in the future will then be discussed.

I. REVIEW OF PHYSICS

(a) The Charged Current Interactions:

The weak charged currents are now well established to be of $(V - A)$.[1] It follows directly from the γ_5-invariance.[2] The two conditions

LEPTONIC INTERACTIONS

$$O\gamma_5 = 0,$$
$$[O, \gamma_5]_+ = 0, \tag{2}$$

can uniquely determine that the operator O should be V - A. The effective Lagrangian is given by:

$$\mathscr{L} = \frac{G_F}{\sqrt{2}} J_\mu^\dagger J_\mu, \tag{3}$$

where $J_\mu = \bar{e} \gamma_\mu (1 + \gamma_5) \nu_e$ for the case of $\nu_e e^-$ elastic scattering, and G_F is the Fermi coupling constant. The cross sections for $\overset{(-)}{\nu}_e e^-$ elastic scattering are given respectively by:[3]

$$\frac{d\sigma_\nu}{dy} = \frac{2 G_F^2 m_e E_\nu}{\pi},$$

$$\frac{d\sigma_{\bar{\nu}}}{dy} = \frac{2 G_F^2 m_e E_{\bar{\nu}}}{\pi} (1 - y)^2, \tag{4}$$

where $y \equiv E_e/E_\nu$. The y-dependence of the cross sections can be explained by the helicity conservation of neutrinos which are illustrated in Fig. 2. For y = 1, the $\bar{\nu}_e e^-$ scattering is forbidden because the angular momentum is not conserved. Historically, these cross sections serve as the classical examples illustrating the inadequacy of the phenomenological theory of weak interactions. As the energy of the neutrino increases to ~370 GeV in the C. M. system, the s-wave unitarity limit will be violated. Therefore, the intermediate vector bosons are needed to damp the linear rise of the cross sections with energy.[4] Such a theory has to be renormalizable to allow for the calculations of higher-order effects. This quest has led to the advent of the electroweak theory by Glashow, Salam, and Weinberg.[5]

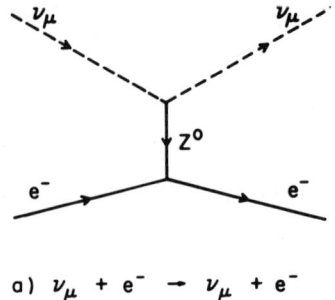

a) $\nu_\mu + e^- \rightarrow \nu_\mu + e^-$

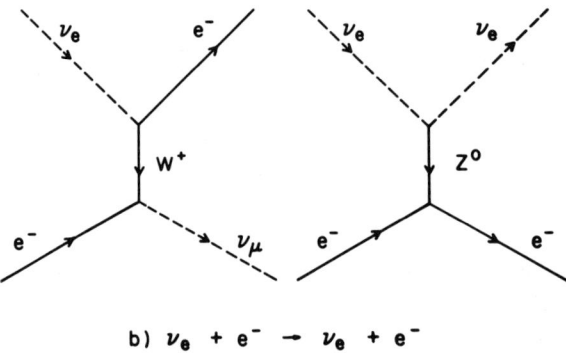

b) $\nu_e + e^- \rightarrow \nu_e + e^-$

Figure 1   The Feynman diagrams of the neutrino-electron elastic scattering processes.

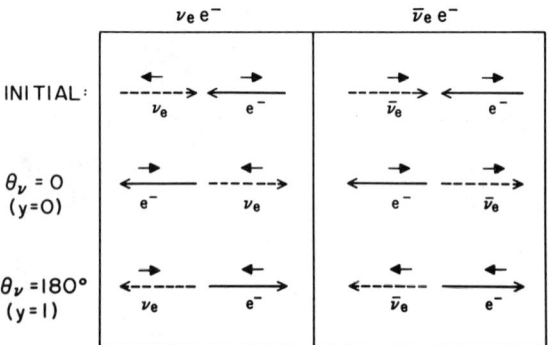

Figure 2   Forward and backward scattering of ν_e and $\bar{\nu}_e$ off electron target. For $y = 1$, the $\bar{\nu}_e e^-$ scattering cross section vanishes because of the angular momentum conservation.

(b) The Neutral Current Interactions

For neutral current interactions, the γ_5-invariance gives only one condition, $[0,\gamma_5]_+ = 0$. Therefore, it allows an arbitrary admixture of both (V - A) and V + A). The neutral current of leptons can be written as:

$$J_\mu^{NC} = \bar{\nu}_e \gamma_\mu (1 + \gamma_5) \nu_e + \bar{\nu}_\mu \gamma_\mu (1 + \gamma_5) \nu_\mu + \bar{e} \gamma_\mu (g_V + g_A \gamma_5) e, \quad (5)$$

and the effective Lagrangian for $\nu_\mu e^-$ elastic scattering can be written as

$$\mathscr{L}^{NC} = \frac{G_F}{\sqrt{2}} [\bar{\nu}\gamma_\mu (1 + \gamma_5)\nu] [\bar{e}\gamma_\mu (g_V + g_A \gamma_5) e]. \quad (6)$$

The cross sections for $\nu_\mu e^-$ and $\bar{\nu}_\mu e^-$ elastic scattering are then given by:[6]

$$\frac{d\sigma^{\nu,\bar{\nu}}}{dy} = \frac{G_F^2 m_e E_\nu^{(-)}}{2\pi} \left[(g_V \pm g_A)^2 + (g_V \mp g_A)^2 (1 - y)^2 - \frac{m_e y}{E_\nu^{(-)}} (g_V^2 - g_A^2) \right], \quad (7)$$

where the upper sign inside the bracket is for ν_μ, and the lower sign is for $\bar{\nu}_\mu$; also, the last term is unimportant except for experiments at low energies. Since the expression for the cross section is quadratic in g_V and g_A, their loci in the g_V-g_A plane are ellipses. Experiments with $\nu_\mu e^-$ and $\bar{\nu}_\mu e^-$ elastic scattering will yield four pairs of solutions. The removal of the 4-fold ambiguity depends on the other two experiments with ν_e and $\bar{\nu}_e$.

(c) The Charged and The Neutral Current Interactions:

For $\nu_e e^-$ elastic scattering, both the charged and the neutral current contribute. The effective Lagrangian for the charged current is given by

$$\mathscr{L}^{CC} = \frac{G_F}{\sqrt{2}} [\bar{\nu}_e \gamma_\mu (1 + \gamma_5) e] [\bar{e} \gamma_\mu (1 + \gamma_5) \nu_e]. \tag{8}$$

By making a Fierz transformation to the charge-retention form

$$\mathscr{L}^{CC} = \frac{G_F}{\sqrt{2}} [\bar{\nu}_e \gamma_\mu (1 + \gamma_5) \nu_e] [\bar{e} \gamma_\mu (1 + \gamma_5) e], \tag{9}$$

and adding it to the neutral current part,

$$\mathscr{L}^{NC} = \frac{G_F}{\sqrt{2}} [\bar{\nu}_e \gamma_\mu (1 + \gamma_5) \nu_e] [\bar{e} \gamma_\mu (g_V + g_A \gamma_5) e], \tag{10}$$

it can be observed immediately that the cross section formula, Eq. (7), can still be applied to this case provided the following simple replacements are made:

$$g_V \to 1 + g_V,$$
$$g_A \to 1 + g_A. \tag{11}$$

The $\nu_e e^-$ and $\bar{\nu}_e e^-$ elastic scattering are also the best reactions to study the interference between charged and neutral current in weak interactions.[7]

(d) **The Glashow-Salam-Weinberg Theory**

The most natural explanation for the (V + A) contribution to the neutral current interactions is given by the standard $SU(2)_L \times U(1)$ model due to Glashow, Salam, and Weinberg,[5] which basically consists of the Yang-Mills field theory[8] and the Higgs mechanism.[9] As illustrated in Fig. 3 by the case of nucleon beta decay, the Yang-Mills quanta have I = 1 and I = 0 components, and they coupled to the leptons with strength g and g' respectively. The left-handed fermions are assigned to the iso-doublets, and the right-handed charged fermions are assigned to the iso-singlets. After breaking the symmetry spontaneously with a single Higgs doublet, it arrives at the following important result:

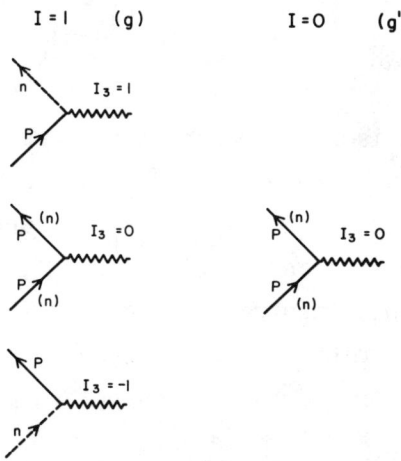

Figure 3 The isospin emission in nucleon beta decay. The intermediate vector bosons are coupled to leptons with strength g and g' for I = 1 and I = 0 respectively.

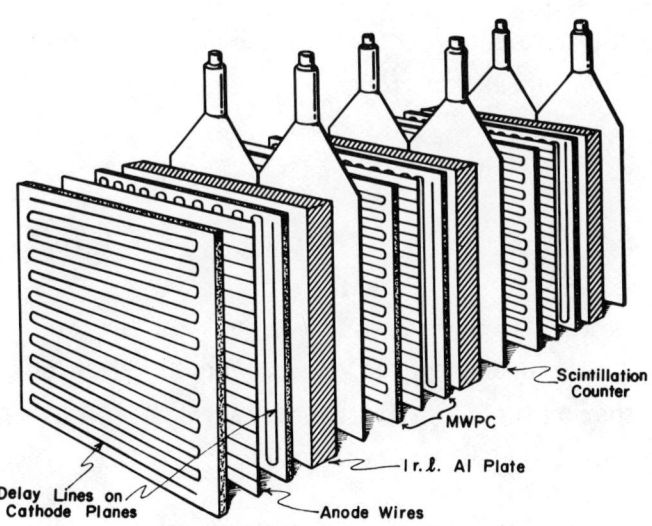

Figure 4 Schematic of the E-253 apparatus at Fermilab which measures the reaction $\nu_\mu + e^- \to \nu_\mu + e^-$.

$$J^{NC}_{L,R} = \rho \left[I_3 - \sin^2\theta_W J^{EM} \right]_{L,R} \qquad (12)$$

where I_3 is the third isospin component of weak current, J^{EM} is the electromagnetic current, the parameter θ_W is defined by $\tan\theta_W = g'/g$, and ρ is the ratio of the neutral to charged current coupling, G^{NC}/G_F. Since I_3 is a pure (V - A) and J^{EM} is a pure V, their mixing naturally gives rise to the weak neutral current both the (V - A) and the (V + A) component according to a definite prescription which applies to leptons as well as to quarks. The specific predictions of this so-called standard $SU(2)_L \times U(1)$ model are:

$$\rho = \frac{M_W^2}{M_Z^2 \cos^2\theta_W} = 1,$$

$$g_A = -1/2, \qquad (13)$$

$$g_V = 2\sin^2\theta_W - 1/2.$$

The condition that $\rho = 1$ may not be true if the Higgs are not in the iso-doublet. In that case, the leptonic coupling constants are modified by the following expressions:

$$g_A = \rho \left[-1/2 - I_3(e_R^-) \right],$$
$$g_V = \rho \left[2\sin^2\theta_W - 1/2 + I_3(e_R^-) \right], \qquad (14)$$

where $I_3(e_R^-)$ is the third isospin component of the right-handed electron. In the standard model, $I_3(e_R^-)$ equal to zero. But it can assume some other value in a different representation. Also, the radiative corrections can make the value of ρ differ from unity.

II. REVIEW OF EXPERIMENTS

The main goal of the neutrino-electron scattering experiments is to determine the model independent leptonic coupling constants, g_V and g_A, or equivalently to determine the parameters ρ and $\sin^2\theta_W$. Then we can compare the value of $\sin^2\theta_W$ with various theoretical models which strive to calculate the value of $\sin^2\theta_W$ rather than leaving it as an experimental parameter. For example, the predictions of $\rho = 1$ and $g_A = -1/2$ of the standard model can be tested critically. The grand unification theory, SU(5), predicts that $\sin^2\theta_W$ should be approximately 0.19 - 0.207 but not much bigger.[10] There are also various predictions from the left-right symmetric model,[11] SU(2/1),[12] and the [SU(2N)][4] model,[13] etc.

Table 1

Summary of $\bar{\nu}_\mu e^-$ Scattering Results

Experiment	Event Number	σ/E_ν $(10^{-42}$ cm^2/GeV)
GGM[14]	~0.0	<2.7
BEBC[15]	0.5	<3.4
15' Fermilab[16]	~0.0	<2.1
GGM[17]	2.6	$1.0^{+0.9}_{-0.6}$
Aachen-Padova[18]	9.6	2.2 ±1.0
CHARM[19]	41.0	1.8 ±0.4
$<\sin^2\theta_W> \simeq 0.29^{+0.04}_{-0.05}$	Mean	1.69 ±0.33

Thus far, the neutrino-electron scattering experiments were done either with bubble chambers or with counters. The rate is small in the bubble chamber and "large" in the counter, but so far it has not been achieved. Energy resolution is poor in the bubble chamber and better in the counter. The angular resolution is somewhat comparable, maybe it is a little bit better with the bubble chamber. The background for the counter experiments purely depends upon the angular resolution. Table 1 and Table 2 summarize the existing results on $\bar{\nu}_\mu e^-$ and $\nu_\mu e^-$ scattering.

Table 2

Summary of $\nu_\mu e^-$ Scattering Results

Experiment	Event Number	σ/E_ν (10^{-42} cm^2/GeV)
GGM[20]	≤0.7	<3.0
Serpukhov[21]	≤0.8	<3.8
Aachen-Padova[18]	11.5	1.1 ±0.6
BNL-Columbia[22]	11.0	1.8 ±0.8
GGM[23]	8.6	$2.4^{+1.2}_{-0.9}$
CHARM[24]	6.5	2.6 ±1.6
E-253 Fermilab[25]	34.0	1.4 ±0.3
$<\sin^2\theta_W> \simeq 0.24^{+0.05}_{-0.03}$	Mean	1.49 ±0.24

LEPTONIC INTERACTIONS

The weighted average values listed in these two tables are the averages of those experiments which gave definite results. They should be compared with the results from neutrino-nucleon interactions[26] and the polarized eD scattering at SLAC[27] which give the most precise value of $\sin^2\theta_w$, 0.238 ± 0.011.[28] Naturally, this kind of averaging process favors only those results of higher statistics. For $\bar{\nu}_\mu e^-$ elastic scattering, it is dominated by the CHARM result,[19] which was announced in the 1979 Proceedings of the Fermilab Symposium without detailed explanation. For $\nu_\mu e^-$ elastic scattering, the average value is dominated by the E-253 result.[25] Some of the pertinent technical details of that experiment will be briefly explained here.

The apparatus of E-253 is shown schematically in Fig. 4, which is quite different from all other experiments. The (x,y) coordinates of the centroids of the electron shower were measured at every radiation length by the cathode-plane delay-lines as time intervals.[29] The chambers were filled with a gas mixture of 80% argon, 19.7% CO_2, and 0.3% Freon-B1-13. This gas mixture helped in quenching the very low energy photons, thus limiting the transverse size of the electromagnetic shower. Also, as shown in Fig. 5, we only used the chambers in the initial 5 r.l. to determine the shower angle. After the shower maximum, there is a large fluctuation in shower direction because of the uneven absorption of the low energy components. Therefore, it is not prudent to include that part of the shower in determining its direction. These practices led to the angular resolution shown in Fig. 6. The fine transverse spacing of the delay-lines on the cathode-planes, 1.5 mm, also helped.

Fig. 7 shows the scatter plot in the $\theta_x - \theta_y$ plane of the "single" electromagnetic showers observed in E-253. The concentration of events within the angular range of 0 - 10 mr can be clearly seen. The background subtraction had to be performed with the "density" of this angular bin and that of the neighboring bin of 10 - 20 mr. Fig.

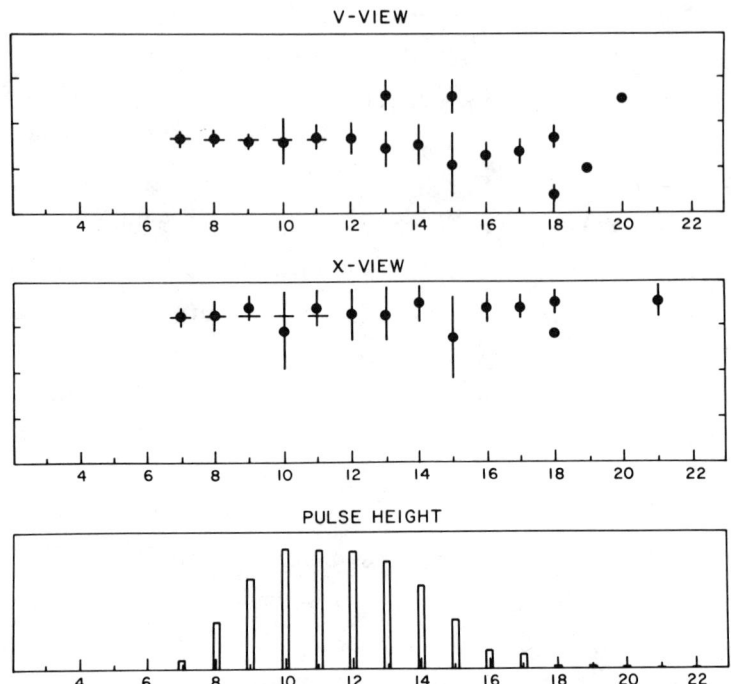

Figure 5 A typical electron shower induced by a neutrino. Numbers on the horizontal scale label the detector modules. The two pictures at the top show the y- and x- projection of the shower. The bottom picture shows the pulse height of each layer of plastic scintillators in linear scale.

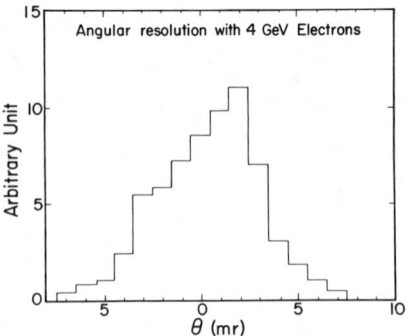

Figure 6 Angular resolution measured with 5 chambers and 4 GeV electrons. The radiator between adjacent chambers was 1 r.ℓ. thick each.

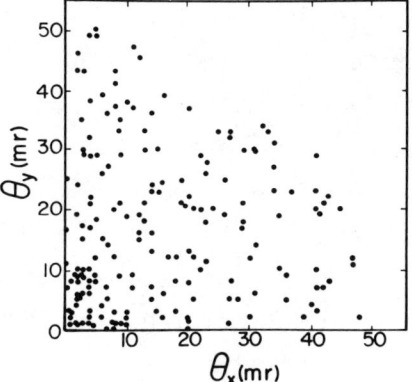

Figure 7   Scatter plot of electromagnetic showers with projected angles equal or less than 50 mr.

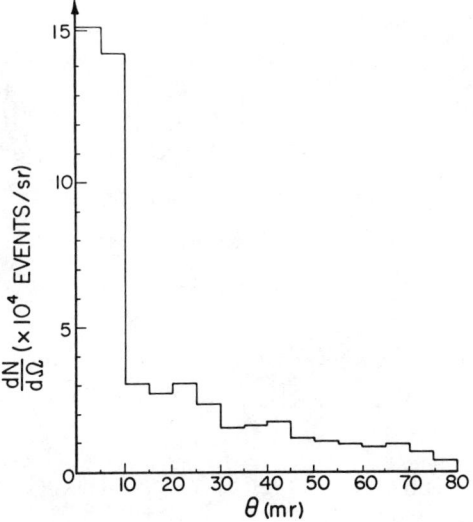

Figure 8   Angular distribution of the $\nu_\mu e^-$ elastic scattering candidates. Within the forward peak of 0 to 10 mr, there are 46 events.

8 shows the angular distribution of the "density" $dN/d\Omega$ vs θ. By a straightforward extrapolation, the 46 events in the angular range of 0 - 10 mr can be easily separated into 34 $\nu_\mu e^-$ events and 12 background events. Alternatively, one can adopt a more detailed subtraction procedure with the $E\theta^2$ distributions. Fig. 9 shows the scatter plot in the $E - \theta^2$ plane of those events having $\theta \leqslant 10$ mr. Fig. 10 shows the $E\theta^2$ distribution of the same events. Because of the finiteness of the angular resolution of the apparatus (± 5 mr), the range of $E\theta^2$ extended to 6 MeV, beyond the theoretical limit of $2 m_e(1 - y)$ $\leqslant 1$ MeV. Fig. 11 shows the $E\theta^2$ distribution of the events in the angular bin of 10 - 20 mr. Their average value is 12 events in the $E\theta^2$ range of 0 - 6 MeV, which is regarded as background to those events shown in Fig. 10. In this manner, the same result was arrived, i.e., 34 $\nu_\mu e^-$ events and 12 background events. This purely experimental approach does not depend on calculations. As a matter of fact, the calculation of the background due to $\nu_e + n \rightarrow p + e^-$ is trivial. Their contribution is less than 6 events. The calculation of the single π^0 production is harder because of the event selection criteria used in the analysis (only the single visible electromagnetic showers were accepted). The $\nu_\mu e^-$ elastic scattering cross section was determined to be $\sigma/E_\nu = (1.40 \pm 0.30) \times 10^{-42}$ cm^2/GeV. Using the condition $g_A = -1/2$ of the standard model, we arrived at $\sin^2\theta_W = 0.25 ^{+0.07}_{-0.05}$. The systematic error in cross section was estimated to be $\sim\pm 0.2 \times 10^{-42}$ cm^2/GeV, which was mainly due to the uncertainty in our knowledge about the neutrino flux.

The latest result of E-253 is shown in Fig. 12. Within the angular range of 0 - 10 mr, there are now 52 events, separating into 40 $\nu_\mu e^-$ events and 12 background events. Figs. 13 and 14 show their P_T- and $E\theta^2$ -distribution respectively. The slightly better statistics yields the result: $\sin^2\theta_W = 0.25 ^{+0.05}_{-0.03}$.

There is no measurement yet on the reaction $\nu_e + e^- \rightarrow \nu_e + e^-$.

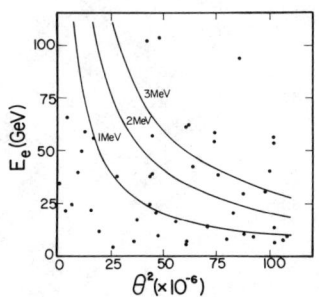

Figure 9 Distribution of the observed $\nu_\mu e^-$ elastic scattering candidates with $\theta \leq 10$ mr in the E-θ^2 plane.

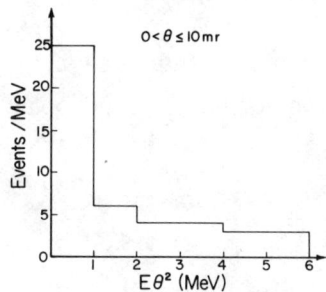

Figure 10 $E\theta^2$ distribution of observed events in the angular range of $\theta \leq 10$ mr.

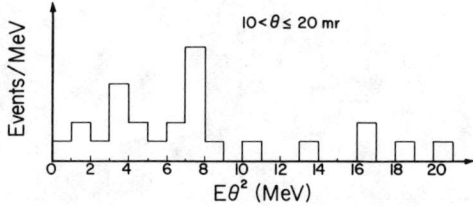

Figure 11 $E\theta^2$ distribution of observed events in the angular range of $10 < \theta \leq 20$ mr.

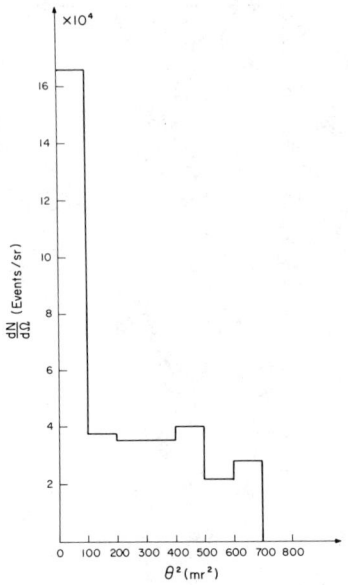

Figure 12 dN/dΩ vs θ^2 plot of the latest E-253 result. There are 52 events within the forward peak of 0 to 10 mr.

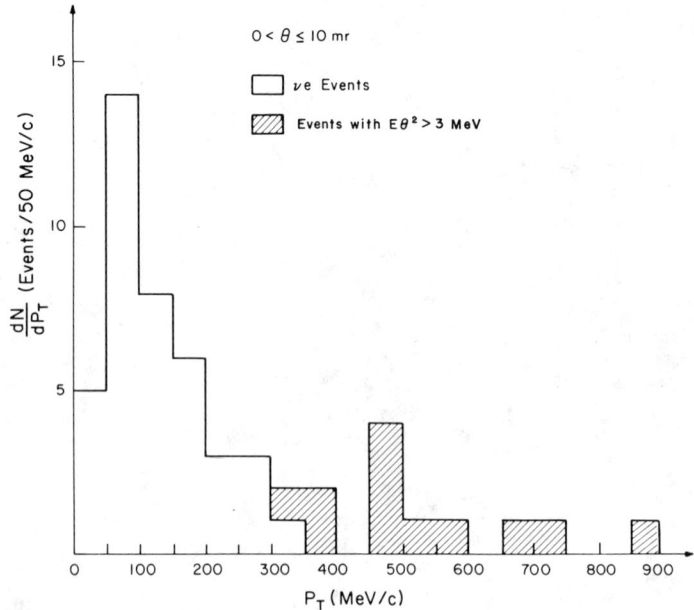

Figure 13 P_T- distribution of 52 events with $\theta \leq 10$ mr.

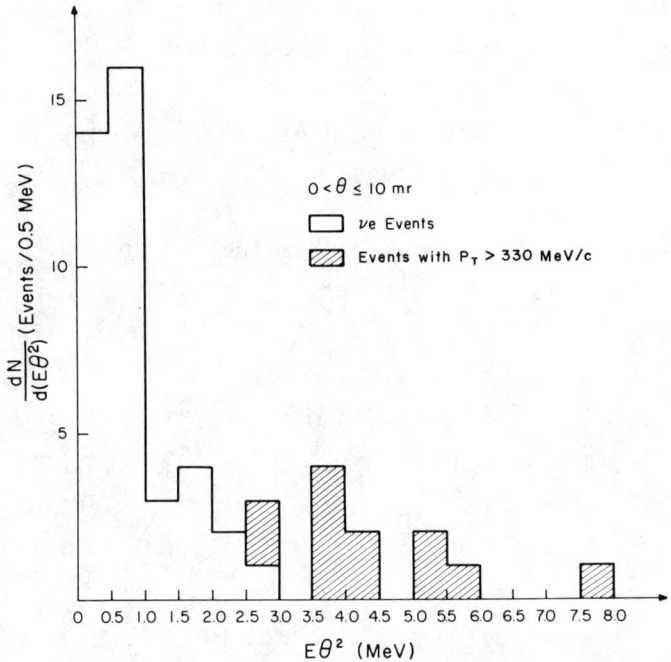

Figure 14 $E\theta^2$- distribution of 52 events with $\theta \leqslant 10$ mr.

There existed only one experiment on $\bar{\nu}_e + e^- \to \bar{\nu}_e + e^-$, which was done at the Savannah River Reactor by Reines et al.[30] The background subtraction of this experiment was quite severe, ~85%. It left with ~7.1 events/day; and ~64 operation days yielded ~460 $\bar{\nu}_e e^-$ elastic scattering events. The cross sections were given in terms of the (V - A) cross section σ_{V-A}:

$$\sigma = (0.85 \pm 0.25)\sigma_{V-A}, \quad E_{\bar{\nu}} = 1.5 - 3.0 \text{ MeV},$$
$$\sigma = (1.70 \pm 0.44)\sigma_{V-A}, \quad E_{\bar{\nu}} = 3.0 - 4.5 \text{ MeV}. \quad (15)$$

This result is in consistence with either V - A or the standard model. If the latter is favored, it gives $\sin^2\theta_W = 0.29 \pm 0.05$.

Fig. 15 summarizes the present situation on neutrino-electron scattering. Because there are only three kinds of experiments, we still have two possibilities. The A-dominating solution agrees with the WSG model with $\sin^2\theta_W \simeq 1/4$ and $I_3(e_R^-) \simeq 0$. The V-dominating solution also agrees accidentally with an alternative model in which the right-handed electron being put into a doublet. If $\sin^2\theta_W = 1/4$, the V-dominating solution will agree with this right-handed doublet model which has $I_3(e_R^-) = -1/2$. This solution is eliminated by the factorization relation proposed by Sakurai,[31] which is illustrated in Fig. 16. The νe scattering is characterized by two coupling constants, g_V and g_A. For the νq (where q represents quarks) interactions, four coupling constants are required to describe it because it involves V, A, and I = 1, 0; similarly is the case for eq scattering. Altogether ten independent parameters are required to describe the four processes shown in Fig. 16. After factoring out the $\nu\nu$ elastic scattering, the ratio of g_V/g_A can be expressed as:

$$\frac{g_V}{g_A} = \left[\frac{\tilde{\beta} + \tilde{\delta}/3}{\tilde{\alpha} + \tilde{\gamma}/3}\right]\left[\frac{\alpha + \gamma/3}{\beta + \delta/3}\right]. \quad (16)$$

LEPTONIC INTERACTIONS

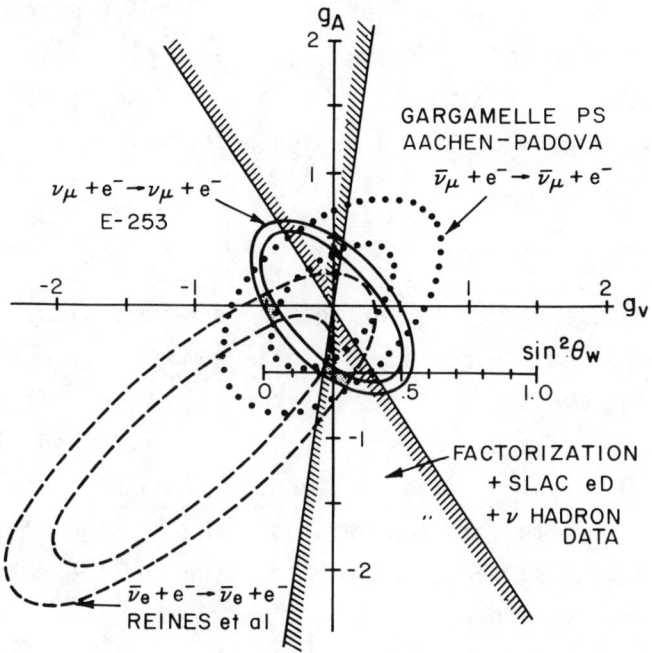

Figure 15 Summary of the neutrino-electron scattering experiments. The factorization relationship together with the $\stackrel{(-)}{\nu}_\mu N$ and polarized eD data rules out the V-dominating solution.

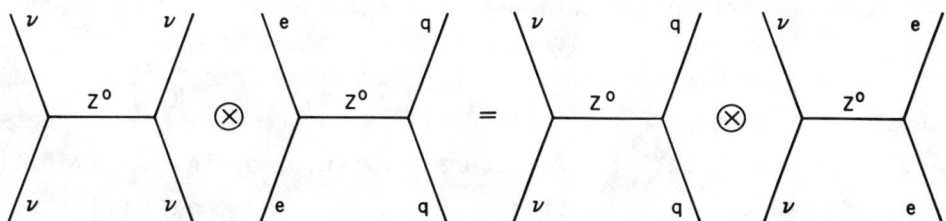

Figure 16 The factorization relationship. Ten independent parameters are required to describe these four processes: g_V and g_A for νe scattering; α, β, γ, and δ for νq scattering; $\tilde{\alpha}$, $\tilde{\beta}$, $\tilde{\gamma}$, and $\tilde{\delta}$ for eq scattering.

The first factor was measured by the polarized eD scattering at SLAC,[27] and the second factor was measured by $\overset{(-)}{\nu}_\mu N$ interactions.[26] The allowed phase space in the g_V-g_A plane of νe scattering is shown by the shaded region of Fig. 15 which accepts the A-dominating solution only.

III. THE FUTURE OUTLOOK

The progress on νe scattering has been very slow. In spite of all the experimental efforts, the total number of $\nu_\mu e^-$ or $\bar{\nu}_\mu e^-$ elastic scattering events is still less than 10^2 each. It would be much more meaningful if these experiments can be performed with statistics as high as 10^3 or even up to 10^6 events. To achieve that goal only the electronics detectors can be used. The tonnage of the detector has to be increased, the angular resolution and the data taking speed have to be improved upon.

In the immediate future, new experiments on $\overset{(-)}{\nu}_\mu e^-$ scattering at Fermilab[32] and BNL[33] are going to operate. At Los Alamos,[34] $\nu_e e^-$ and $\bar{\nu}_e e^-$ scattering will be done at the Meson Factory with neutrinos of energy up to 53 MeV. New experiments on $\overset{(-)}{\nu}_e e^-$ elastic scattering are proposed at Serpukhov[35] and Fermilab.[36] The $\overset{(-)}{\nu}_e$ sources are either from $K_L^0 \to \pi e \overset{(-)}{\nu}_e$ or from the decays of D and $\bar{D}$ which are produced in a high energy proton beam dump. Since the production mechanism of $D\bar{D}$ in the beam dump is not well understood, the neutrino flux has to be determined experimentally which is not a light undertaking.

In case we can do a 1,000 events experiment on $\nu_\mu e^-$ and a 200 events experiment on $\bar{\nu}_\mu e^-$, Fig. 17 will show how well one can determine g_V, g_A, and $\sin^2\theta_W$. The accuracy of ~0.01 on the determination of $\sin^2\theta_W$ can be reached. This accuracy of $\sin^2\theta_W$ would make the experiment most decisive. The determination of ρ is equally important. The radiative corrections[37] will make the ρ value differ from unity. The most interesting process of the radiative corrections is the one-loop correction which has been calculated by Veltman,[38] also by

Chanowitz et al.[39] For the charged current interactions the vacuum polarization loop is made up of a neutrino and a lepton; but for the neutral current case it can be a pair of quarks or leptons as shown in Fig. 18. The value of ρ is related to the fermion mass by the following expression:

$$\rho = 1 + \frac{G_F}{\sqrt{2}} \frac{1}{8\pi^2} m_f^2 + \text{logarithmic terms}$$

$$\simeq 1 + 9.1 \times 10^{-8} m_f^2/m_p^2 + \cdots \qquad (17)$$

Since only the heaviest mass can contribute to the last expression, a precise measurement of ρ will allow the determination of the mass of the heaviest lepton or quark in the range of a few hundred GeV.

Next we will discuss the interesting case of the conjectured neutrino oscillations.[40] If they exist, then the data analysis for neutrino-electron scattering would be very complicated. This has been pointed out by Rosen and Kayser[41] very recently. We can understand this easily by considering a two-level system when the neutrino beam is made of only ν_e and ν_μ. At any given time the neutrino beam is a superposition of ν_μ and ν_e. Then the cross section measured at any given time is the sum of the probability multiplied by the cross section. Since the $\nu_e e^-$ cross section is approximately 7 times bigger than that of $\nu_\mu e^-$, therefore if the initial beam is ν_μ, then the measured cross section is always bigger than the $\nu_\mu e^-$ cross section; for the initially ν_e beam, the measured cross section will always be less than that of the $\nu_e e^-$ cross section. We have to make corrections before making comparisons with the Weinberg-Salam-Glashow model. There are experimental limits on $\nu_\mu \not\rightleftarrows \nu_e$ oscillation. It was measured by Fiorini et al., at GGM[42] and later by Baltay et al., at Fermilab.[43] The ratio of the neutrino intensities, $I(\nu_e)/I(\nu_\mu)$, was determined to be less than 3×10^{-3}. This limit could introduce a maximum error in $\sin^2\theta_W$ of approximately 0.004.

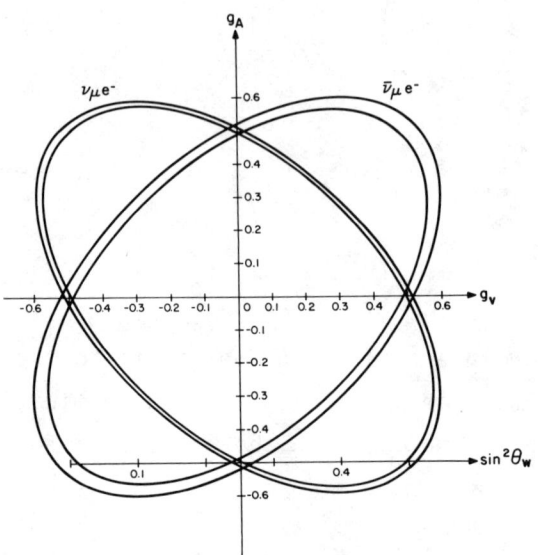

Figure 17 Determination of g_V, g_A, and $\sin^2\theta_W$ with a possible measurement of 1,000 $\nu_\mu e^-$ and 200 $\bar{\nu}_\mu e^-$ elastic scattering events.

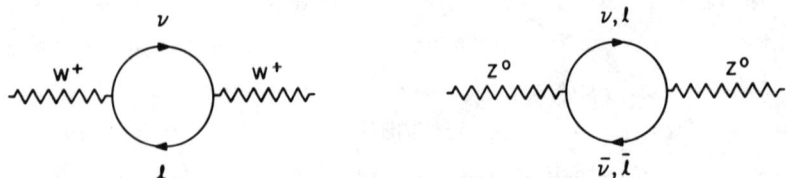

Figure 18 One-loop correction to charged and neutral current interactions.

Recently, a very interesting model on neutrino oscillations was proposed by Wolfenstein[44] and Zee,[45] in which the Higgs sector of the standard SU(2) x U(1) model was enlarged by incorporating a Higgs doublet which couples only to the leptons, and a Higgs singlet which changes the flavor as illustrated in Fig. 19. The mass eigen-values are given by Wolfenstein[44] as

$$m_1 = m_0 \sin 2\alpha,$$
$$m_2 = m_0 (1 + \frac{\sigma}{2} \sin 2\alpha),$$
$$m_3 = m_0 (1 - \frac{\sigma}{2} \sin 2\alpha), \qquad (18)$$

where m_0 is a mass scale; α, the neutrino mixing angle; σ, a small parameter, ~ 0.01. The differences in mass squared are given by

$$\delta m_{12}^2 \simeq \delta m_{13}^2 \simeq m_0^2,$$
$$\delta m_{23}^2 = 2 m_0^2 \sigma \sin 2\alpha. \qquad (19)$$

The first expression is responsible for the short wavelength oscillation $\nu_\mu \rightleftarrows \nu_e$, and the second expression is responsible for the long wavelength oscillation ν_e or $\nu_\mu \rightleftarrows \nu_\tau$. Taking $m_0 \simeq 20$ eV, $\sin 2\alpha \leqslant 0.06$, and a source-to-detector distance of $\ell \simeq 1,000$ m, the probability of finding ν_e in an initially pure ν_μ beam due to $\nu_\mu \rightleftarrows \nu_e$ oscillation is given by

$$P(\nu_e) = \sin^2 (2\alpha) \sin^2 \left[\frac{\delta m_{12}^2 \ell}{4E_\nu} \right]$$
$$\simeq 0.0036 \sin^2 (\frac{500}{E_\nu}). \qquad (20)$$

If the neutrino oscillation indeed exists and we neglect to take it into account, then $\sin^2 \theta_W$ will be an oscillating function of the neu-

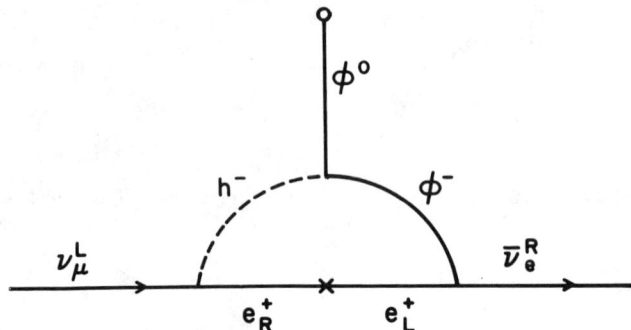

Figure 19 The Wolfenstein-Zee model.

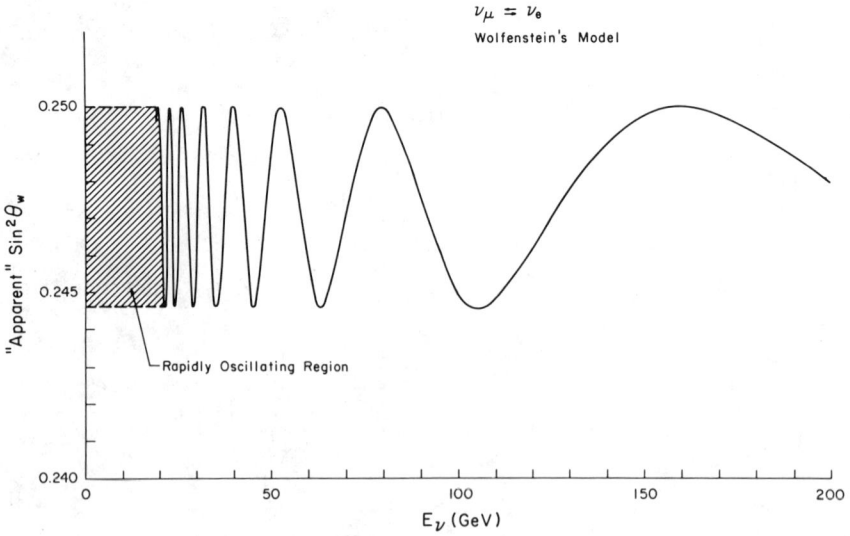

Figure 20 Example illustrating the effect of $\nu_\mu \rightleftarrows \nu_e$ oscillation on the value of $\sin^2\theta_W$ if it is not corrected. The Wolfenstein model and $\sin^2\theta_W = 1/4$ are assumed.

trino energy, as shown in Fig. 20. One can learn from this example that the data analysis will be very complicated if all modes of the neutrino oscillations exist.

Acknowledgements

I wish to express my sincere thanks to Professors A. Abashian, J. D. Bjorken, L.N. Chang, R. E. Marshak, Riazuddin, M. Veltman, and L. Wolfenstein for helpful discussions.

References

1. E. C. G. Sudarshan and R. E. Marshak, Proc. of the Padua-Venice Conf. on "Mesons and Newly Discovered Particles" (1957); R. P. Feynman and M. Gell-Mann, Phys. Rev. 109, 193 (1958); J. J. Sakurai, Nuovo Cimento 7, 649 (1958).
2. R. E. Marshak, Riazuddin and C. P. Ryan, Theory of Weak Interactions in Particle Physics (Wiley, New York, 1968), p. 86.
3. T. D. Lee and C. S. Wu, "Weak Interaction" in Annual Rev. of Nucl. Sci. Vol. 15 (1965), p. 420; see also Ref. 2, p. 214.
4. See Ref. 3, p. 397.
5. S. Weinberg, Phys. Rev. Lett. 19, 1264 (1967); A. Salam, in "Elementary Particle Theory", ed. by N. Svartholm, (Almquist and Wiksells, Stockholm, 1968), p. 367; S. L. Glashow, Nucl. Phys. 22, 579 (1961).
6. G. 't Hooft, Phys. Lett. 37B, 195 (1971).
7. B. Kayser et al., Phys. Rev. 20D, 87 (1979).
8. C. N. Yang and R. Mills, Phys. Rev. 96, 191 (1954).
9. P. W. Higgs, Phys. Rev. Lett. 13, 508 (1964); T. W. B. Kible, Phys. Rev. 155, 1554 (1967).
10. H. Georgi, H. R. Quinn and S. Weinberg, Phys. Rev. Lett. 33, 451 (1974); J. Ellis, CERN Report Ref.TH.2723-CERN (1979).
11. R. E. Marshak and R. N. Mohapatra Festschrift for Maurice Goldhaber, New York Academy of Sciences (to be published, 1980); Riazuddin and Fayyazuddin, Trieste preprint IC/79/154 (1979).

12. Y. Ne'eman, Phys. Lett. 81B, 190 (1979); J. G. Taylor, Phys. Lett. 83B, 331 (1979).
13. V. Elias, J. C. Pati and A. Salam, Phys. Rev. Lett. 40, 920 (1978); V. Elias and S. Rajpoot, Phys. Rev. 20D, 2445 (1979).
14. A. Pullia et al., Proc of Neutrino 79, Vol. 2, p. 230, Bergen, Norway (1979), ed. by A. Haatuft and C. Jarlskog.
15. N. Armenise et al., Phys. Lett. 81B, 385 (1979).
16. J. P. Berge et al., paper contributed to Neutrino 79, Bergen, Norway (1979).
17. F. J. Hasert et al., Phys. Lett. 46B, 121 (1973); J. Blietschau et al., Nucl. Phys. B114, 189 (1976).
18. H. Faissner et al., Phys. Rev. Lett. 41, 213 (1978).
19. K. Winter, Proc. of 1979 Int. Symposium on Lepton and Photon Interactions at High Energies, Fermilab, Batavia, Ill. (1979). p. 258, ed. by T. B. W. Kirk and H. D. I. Abarbanel.
20. J. Blietschau et al., Phys. Lett. 73B, 232 (1978).
21. A. E. Asratyan et al., Proc. of Neutrino 79, Vol. 2, p. 246, Bergen, Norway (1979), ed. by A. Haatuft and C. Jarlskog.
22. A. M. Cnops et al., Phys. Rev. Lett. 41, 357 (1978).
23. P. Alibran et al., Phys. Lett. 74B, 422 (1978).
24. M. Jonker et al., Proc of Neutrino 79, Vol. 2, p. 219, Bergen, Norway (1979), ed. by A. Haatuft and C. Jarlskog.
25. R. H. Heisterberg et al., Phys. Rev. Lett. 44, 635 (1980).
26. M. Holder et al., Phys. Lett. 72B, 254 (1977); P. Wanderer et al., Phys. Rev. 17D, 1679 (1978).
27. C. Prescott et al., Phys. Lett. 77B, 347 (1978); 84B, 524 (1979)
28. M. Roos, paper contributed to this Conference.
29. T. A. Nunamaker et al., preprint VPI-HEP-80/3, to be published in Nucl. Instr. and Meth., 1980.
30. F. Reines et al., Phys. Rev. Lett. 37, 315 (1976).
31. P. Q. Hung and J. J. Sakurai, UCLA Report UCLA 79/TEP/9 (unpublished); Proc. of Neutrino 79, Vol. 1, p. 267 Bergen, Norway

(1979), ed. by A. Haatuft and C. Jarlskog.
32. J. Walker et al., a Fermilab-Michigan State-MIT-Northern Illinois collaboration.
33. H. White et al., a BNL-Brown-Pennsylvania collaboration.
34. H. Chen et al., a Los Alamos-UC/Irvin collaboration.
35. T. Riemann, H. E. Ryseck and M. Walter, paper contributed to this Conference.
36. C. Baltay et al., Fermilab Proposal.
37. N. Byers et al., Physica $\underline{96A}$, 163 (1979).
38. M. Veltman, Nucl. Phys. $\underline{B123}$, 89 (1977); Proc. of 1979 Int. Symposium on Lepton and Photon Interactions at High Energies, Fermilab, Batavia, Illinois (1979), p. 529, ed. by T. B. W. Kirk and H. D. I. Abarbanel.
39. M. S. Chanowitz, M. A. Furman and I. Hinchliffe, Phys. Lett. $\underline{78B}$, 285 (1978).
40. B. Pontecorvo, Zh. Eksp. Teor. Fiz. $\underline{33}$, 549 (1957); $\underline{34}$, 247 (1958); $\underline{53}$, 1717 (1967); S. M. Bilenky and B. Pontecorvo, Phys. Report, Vol. 41C, No. 4 (1978); Z. Maki et al., Prog. of Theo. Phys. Vol. 28, No. 5, 870 (1962); M. Nakagawa et al., Prog. of Theo. Phys. Vol. 30, No. 5, 727 (1963); S. Eliezer and A. Swift, Nucl. Phys. $\underline{B105}$, 45 (1976); A. K. Mann and H. Primakoff, Phys. Rev. $\underline{15D}$, 655 (1977); L. Wolfenstein, Phys. Rev. $\underline{17D}$, 2369 (1978); A. De Rújula et al., CERN Report Ref.TH.2788-CERN(1979); L. Maiani, CERN Report Ref. TH. 2846-CERN (1980); V. Barger et al., Wisconsin preprint COO-881-135 (1980).
41. P. Rosen and B. Kayser, Purdue preprint (1980).
42. E. Bellotti et al., Nuovo Cim. Lett. $\underline{17}$, 553 (1976).
43. A. M. Cnops et al., Phys. Rev. Lett. $\underline{40}$, 144 (1978).
44. L. Wolfenstein, Carnegie-Mellon preprint COO-3066-149 (1980).
45. A. Zee, University of Pennsylvania preprint UPR-0150T (1980).

DISCUSSION (Chairman: H. Faissner)

A. GARFINKEL, Lafayette: Please describe the sources of background events and explain why they are dominantly at high energy.

A. ABASHIAN, Washington: In response to Garfinkel's question, there are two reasons why background processes might give rise to large electron energies. The first is ν_e elastic scattering by a nucleon where the outgoing electron has an energy equal to that of the incoming neutrino. In the case of $\nu_\mu e$ elastic scattering, the average energy of the outgoing electron is one half that of the incoming neutrino. The second background process would be single π^0 production which would more likely be emitted within a cone of 10 mr for higher neutrino energies because of the larger Lorentz factor.

T. BOWLES, Los Alamos: The ν_e-e elastic scattering experiment at LAMPF, which is a collaboration between University of California at Irvine and Los Alamos, will begin data taking next February. We expect to measure the cross-section to 10% accuracy with signal to background of 10 to 1. The interference between charged and neutral currents is predicted to be negative and a 25% effect and we expect to have a result sensitive to this interference in about $2\frac{1}{2}$ years.

L.W. MO, Blacksburg: That's very nice to hear. We are all eager to see the result.

H. FAISSNER, Aachen: Is there anybody from the CHARM collaboration which can comment on their results?

P. STÄHELIN, Hamburg: Our preliminary CHARM results (1979) has been correctly reported by the rapporteur. CHARM has a larger effective sampling steps and, threfore poorer resolution. Data taking is still in progress. We hope that better statistics will compensate for the poorer resolution.

PARITY VIOLATION IN NUCLEI

R.G.H. Robertson

Cyclotron Laboratory
Michigan State University
East Lansing, MI, U.S.A.*
and
Physics Division
Argonne National Laboratory
Argonne, IL, U.S.A.

INTRODUCTION

Nuclear parity violation[1] provides a window on a part of the hadronic weak interaction which is otherwise invisible. The idea of a universal weak current which acts in hadrons as well as leptons is of course central to all modern theories of the electroweak force but it is one which does not readily admit testing. We have some experimental knowledge of $\Delta S=1$ and $\Delta C=1$ non-leptonic weak processes, and there is little cause for satisfaction in our understanding of those processes. The only immediate prospect for probing the $\Delta S=0$ interaction is nuclear parity violation, and in the early days it was hoped that much of a fundamental nature might be learned. This optimism soon yielded to gloom when the cluttered nature of the nuclear workshop became apparent. Now, following substantial efforts both by theorists and experimentalists there is a renewed, more conservative optimism that nuclear parity violation (PV) can be understood at a level which tests our ability to calculate hadronic interactions, although it is not likely to influence the development of the underlying theories of the weak interaction.

THE TWO-NUCLEON SYSTEM

For obvious reasons, the two-nucleon system is an interesting one and we now have the remarkable luxury of not one but four
-----
*Present address

successful observations of parity violation in the scattering of polarized protons by nucleons. At 15 MeV the Los Alamos collaboration has observed a difference in the cross-section for the scattering of longitudinally-polarized protons from hydrogen for the two helicities of the beam. After lengthy and careful analysis (still in progress) of possible systematic effects, the result of Potter et al.[2] is

$$A_{pp} = \frac{\sigma_+ - \sigma_-}{\sigma_+ + \sigma_-} = (-1.2 \pm 0.6) \times 10^{-7}.$$

The second measurement of this quantity has been made at the Sweizeriches Institut für Nuklearforschung using a 45 MeV polarized proton beam from the SIN injector cyclotron. The experimenters find[3]

$$A_{pp} = (-2.3 \pm 0.8) \times 10^{-7}.$$

The third measurement is not strictly a measurement of the quantity A_{pp}. Lockyer et al.,[4] using the 5 GeV polarized proton beam from the Argonne ZGS, measured the helicity-dependence of the transmission through a water target. They find an extremely large effect,

$$\bar{A}_{pN} = (+2.6 \pm 0.6) \times 10^{-6}.$$

In their new apparatus, a low-dispersion spectrometer removes hyperon decay products which, in earlier versions of the experiment, produced large asymmetries. The main systematic effect now (about $+1.8 \times 10^{-6}$) results from passage of the beam through air and some monitoring equipment while its polarization is still transverse.

The fourth result is also a transmission measurement and is now in progress at LAMPF by the same group that carried out the 15 MeV measurement. At 800 MeV they find [5]

$$\bar{A}_{pN} = (+3.0 \pm 1.0) \times 10^{-6},$$

also a much larger effect than expected. Both the 5 GeV and 800 MeV results are to be considered very preliminary.

Figure 1 shows the 4 measurements in a log-log plot to illustrate that the high energy results, while large, are not totally at variance with the trend of the low energy data. Of course, there is a sign difference, and the quantities being measured are not strictly the same. In the more detailed theoretical analysis of Henley and Krejs,[6] the higher energy data are at least an order of magnitude larger than expected. The solid curve in the figure has the energy dependence given by Brown, Henley and Krejs[7] and Henley and Krejs,[6] and is normalized in a manner I will describe later.

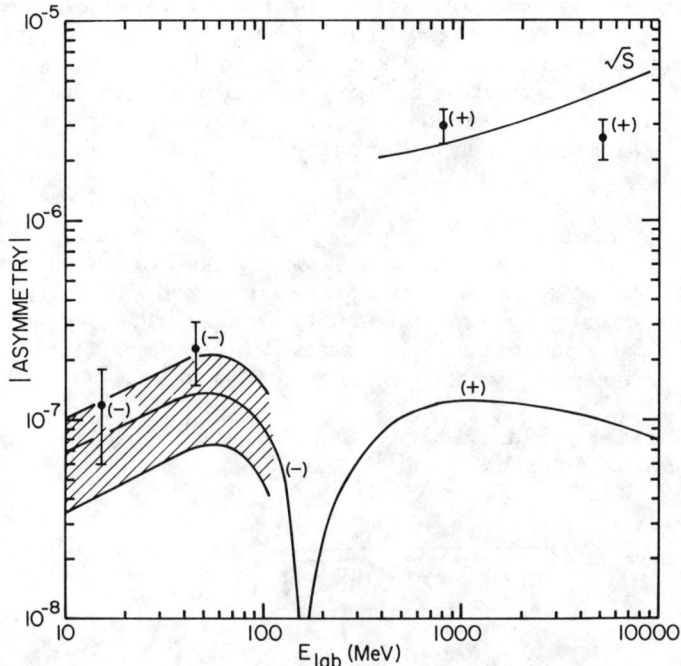

Fig.1 Plot of parity-violating asymmetries observed in polarized proton scattering. The signs of the measurements and of the theoretical curve are indicated.

There is real significance in the fact that the two lower energy points lie on a line of slope 1/2, because the asymmetry at low energies depends simply on the amplitude for mixing from S-to P-states, which in turn is proportional to the momentum. The importance of this agreement cannot be exaggerated because it implies that two separate, very difficult experiments performed on different accelerators by different groups of physicists are measuring the same quantity.

At higher energies, as shown by Simonius,[8] the theoretical curve turns over and eventually passes through zero at about 150 MeV because of a cancellation in the strong interaction phase shifts. Above 300 MeV the analysis becomes vastly more difficult, at least in the inclusive type of experiment, because of the opening of inelastic channels. Below 300 MeV the magnitude of the asymmetry A_{pp} depends on all 3 isospin components of the weak force, $\Delta T=0$, 1 and 2, although it is not influenced by the long-range pion exchange allowed by neutral currents, since (by Barton's theorem[9]) only charged pions contribute in a CP-conserving interaction. At higher energies the asymmetry might be expected to scale as the square root of the invariant c.m. squared energy, s, if there were no damping of the incident waves through absorption; theory suggests substantial absorption.

Radiative capture in the n-p system also provides information of a rather basic nature. As was shown by Danilov,[10] a measurement of the circular polarization of γ rays emitted in thermal neutron capture by protons is sensitive only to the $\Delta T=0,2$ parts of the force, while the directional asymmetry of γ rays from capture of polarized neutrons is sensitive to the $\Delta T=1$ part.

The circular polarization has been measured in a celebrated experiment by Lobashov's group[11] (Fig. 2). As is well known, they

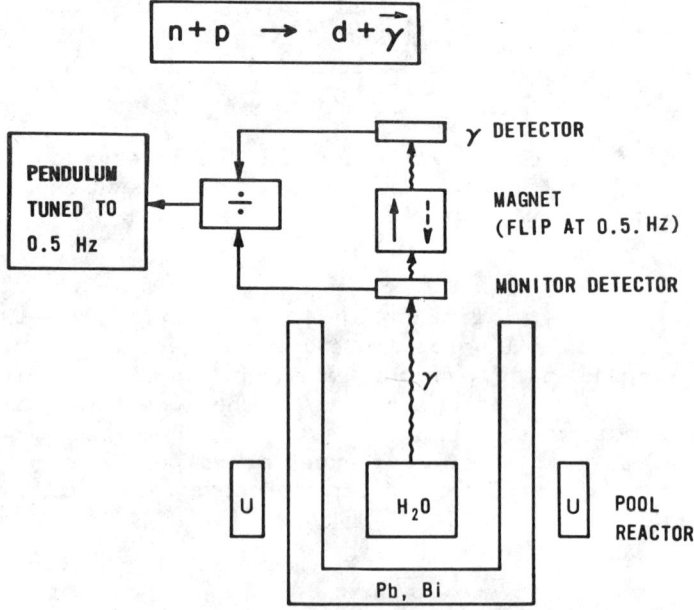

Fig. 2 Schematic diagram of experiment of Lobashov et al. (Ref. 11).

find a result $P_\gamma = (-1.3\pm0.5)\times10^{-6}$, a value which, despite intensive theoretical effort, has not been satisfactorily explained. As McKellar[12] has pointed out, there is in this case a spectacular cancellation between the parity violating effects in the initial and final states, a cancellation which is not sensitive to the potentials chosen to describe the states. Thus all calculations which proceed from first principles or from parameterization of other data fail by a factor of 100 or so to give the observed effect.

In view of the importance of this result, there is considerable interest in confirming it. Two experiments are in preparation. At Chalk River, McDonald, Earle and Knowles[13] are setting up the inverse experiment, (Fig. 3) in which the deuteron is photodisintegrated with circularly polarized bremsstrahlung. Lee[14] has shown that, even for photons well above threshold, this reaction measures the same thing as the Lobashov experiment. Circularly polarized bremsstrahlung are produced from longitudinally polarized electrons. A SLAC-type GaAs source[15] and 4 MV linac are used.

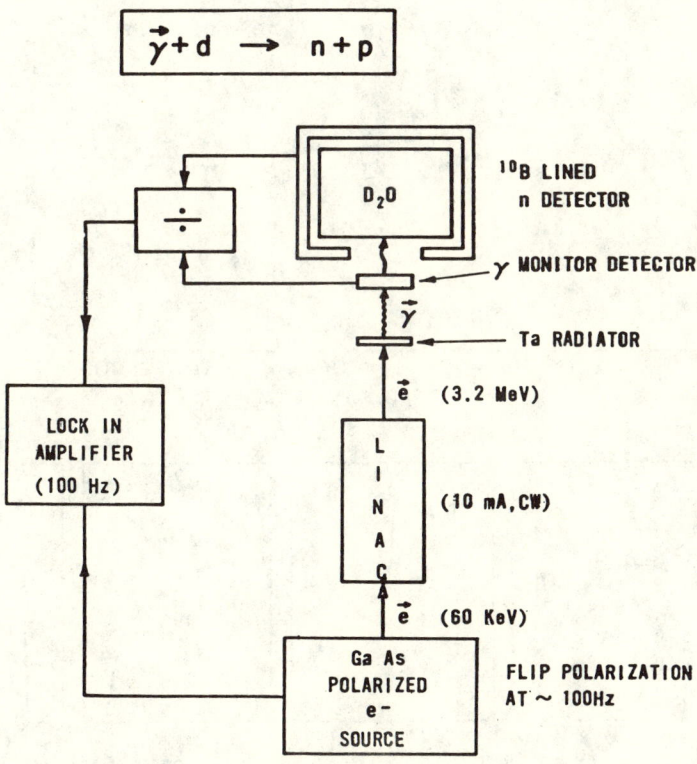

Fig. 3   Schematic diagram of deuteron photodisintegration experiment of McDonald et al. (Ref. 13).

Lisowski and Bowles[16] are investigating the possibility of using the intense neutron pulses available at the LAMPF Neutron Research facility. The unique time structure would permit detailed study of the circular polarization of the signal and the background (Fig. 4). It will be recalled that circularly polarized bremsstrahlung from β activities was one of the major concerns of the Lobashov group in their reactor-based experiment.

For the time being theorists are pragmatically omitting the Lobashov result from their analyses because, if it is right, theory needs such radical revision that current analyses are incorrect, and, if it is wrong, there is no point in including it.

ΔT=1 PARITY VIOLATION

Let us turn now to one of the most interesting aspects of

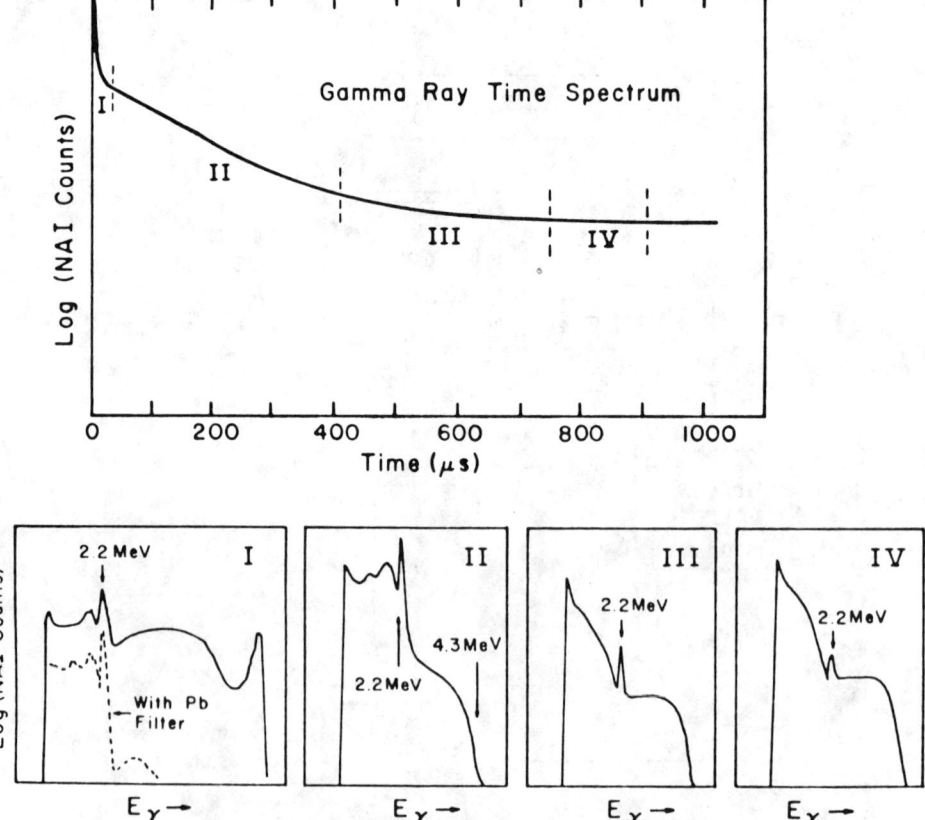

Fig. 4 Time structure of γ-rays following beam burst, and corresponding γ-spectra (Ref. 16).

nuclear PV studies, the question of whether there is a parity-violating hadronic neutral current interaction. In the absence of neutral currents (n.c.) the $\Delta T=1$ components in the Hamiltonian arise only from strangeness-changing currents and are Cabibbo-suppressed by the factor $\sin^2\theta_c$, about 0.052. If, in addition, a n.c. acts, then isovector parity violation may not be so suppressed (the term "enhanced" is sometimes used to describe this situation!).[17] However, the uncertainties in the n.c. part of the effective weak Hamiltonian are so large that one can only say that the presence of an isovector enhancement definitely indicates parity-violating hadronic neutral currents, while its absence is merely inconclusive. It is convenient to define an enhancement factor F which is the ratio of actual isovector parity-irregular amplitude in a nuclear wavefunction to that expected from charged currents alone.

Nature has been parsimonious in providing examples where $\Delta T=1$ PV might be observed in isolation. In fact only 4 cases satisfy the twin constraints of experimental and theoretical tractability.

$\vec{n}+p \rightarrow d+\gamma$

The two-nucleon system is the most amenable to theoretical treatment, and for $\Delta T=1$ does not suffer from the cancellations that render the $\Delta T=0,2$ parts so small and uncertain. Even so, the expected effect[18] is exceedingly small, a γ-ray asymmetry A_γ of $7F \times 10^{-9}$. Cavaignac, Vignon and Wilson[19] have performed an experiment at the Institut Laue-Langevin using a liquid D_2 moderator, Fe-Co mirror neutron polarizer, a liquid para-H_2 target and scintillators to detect the γ rays. Their result,

$$A_\gamma = (0.6\pm2.1) \times 10^{-7},$$

sets an upper limit of about 40 on the enhancement factor. A new version of the experiment with improved neutron polarization and intensity, and better geometry and light collection from the scintillators, is planned. An accuracy of 2 to 3×10^{-8} could be attained.[20]

$^6Li(3.56) \rightarrow \alpha + d$

The alpha decay of the 0^+, T=1 state of ^6Li is energetically allowed but forbidden by both parity and isospin conservation. Many attempts have been made to observe this process (or, more commonly, the corresponding $\alpha+d$ capture through the 3.56 MeV state), with the experiment of Bellotti et al.[21] achieving the highest sensitivity. Using a gas cell containing D_2, a ^4He beam, and a Ge(Li) detector to search for capture γ rays through the 0^+, T=1 state, they were able to set an upper limit of 8×10^{-4} eV on the parity forbidden alpha width $\Gamma_{\alpha d}$.

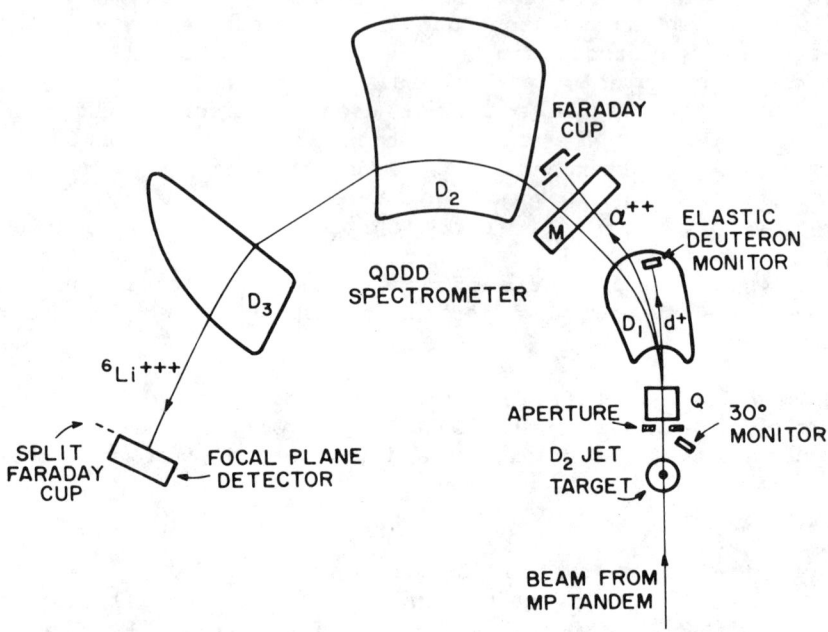

Fig. 5 Apparatus used for direct detection of ^{6}Li recoils from the ^{2}H(α,γ) reaction.

A Michigan State-Argonne-Chalk River collaboration[22] is engaged in a new experiment to improve on this limit. Rather than detecting γ rays, we use a windowless D$_2$ jet target and detect ^{6}Li^{+++} ions directly on the focal plane of a large magnetic spectrograph (QDDD type) (Fig. 5). In this way ^{6}Li produced by direct (non-resonant) capture has been observed for the first time at the 20nb level, virtually free of background. Figure 6 shows the a_0 term (the isotropic component of the angular distribution) plotted as a function of beam energy. The solid curve is a Gaussian peak of predetermined width fitted for amplitude and position on a constant background. From these data we conclude that $\Gamma_{\alpha d}$=(0.6±0.8)x10^{-6} eV, and $\Gamma_{\alpha d} \leq$ 2x10^{-6} eV (90% C.L.).

No detailed theoretical treatment of this case has been presented in the literature. It is unusual in a nuclear PV context

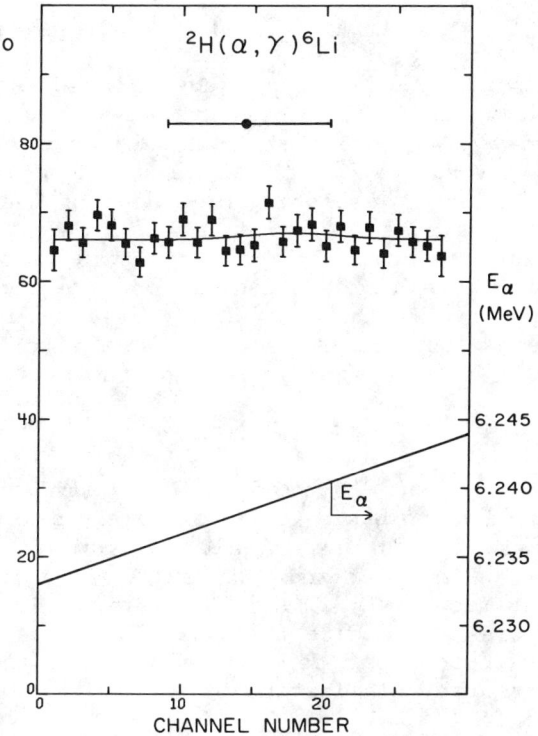

Fig. 6 Excitation function of isotropic component of $^2H(\alpha,\gamma)^6Li$ reaction. The expected resonance position is indicated by the horizontal bar.

because the parity impurity occurs in an isolated state embedded in the continuum rather than as an admixture between two close-lying levels. Michel[23] presented a simple theorem which lends itself to this problem. If the full effective Hamiltonian is of the form

$$H = H_o + G\ \vec{\sigma}\cdot\vec{p}$$

where H_o is the unperturbed Hamiltonian and $G\ \vec{\sigma}\cdot\vec{p}$ a one-body parity violating effective potential, then, to second order,

$$H = e^{-iS} H_o e^{iS}$$

where $\hbar S = m_N G\ \vec{\sigma}\cdot\vec{r}$. The perturbed wavefunctions may thus be obtained directly by the transformation

$$\psi_i = e^{-iS}\psi_i^{(o)} = [1 - im_N \hbar^{-1} G \vec{\sigma} \cdot \vec{r}] \psi_i^{(o)},$$

in an obvious notation. One may thus proceed directly from unperturbed to parity-irregular wavefunctions without introducing any dubious unknown 0^-, T=0 levels at high excitation. Robertson and Riska,[24] using a Vergados[25] shell-model wavefunction for ^{6}Li, and treating also the parity violating component of the deuteron wavefunction, have calculated $\Gamma_{\alpha d} = 3.4 \ F^2 \times 10^{-10}$ eV. Thus the limit on F from experiment is about 80. However, this is a very schematic calculation which uses harmonic oscillator wavefunctions, neglects the parity impurity in the alpha particle, treats only weak π exchange, neglects core excitations, etc., so a more refined calculation could well give a significantly different result.

^{6}Li + $\alpha \rightarrow$ ^{10}B(5.164)

A pair of levels in ^{10}B, the 5.110 MeV 2^-,T=0 and 5.164 MeV 2^+, T=1 states, provides the basis for an extremely sensitive search for neutral currents.[26] An experiment is now in preparation at Argonne[27] in which a polarized ^{6}Li target is formed on a heated oxygenated W surface (Fig. 7). This target is bombarded with 1.2 MeV α particles (to excite the 5.16 MeV state) and parity violation is evidenced by a dependence of the total capture cross section on the vector polarization of the ^{6}Li:

$$\sigma_{\alpha\gamma} = \sigma_o(1 - \alpha RP + \frac{A}{2}),$$

where P is the longitudinal vector polarization (i.e., $m_{+1} - m_{-1}$), A the tensor polarization $(1 - 3m_0)$, α the parity mixing amplitude

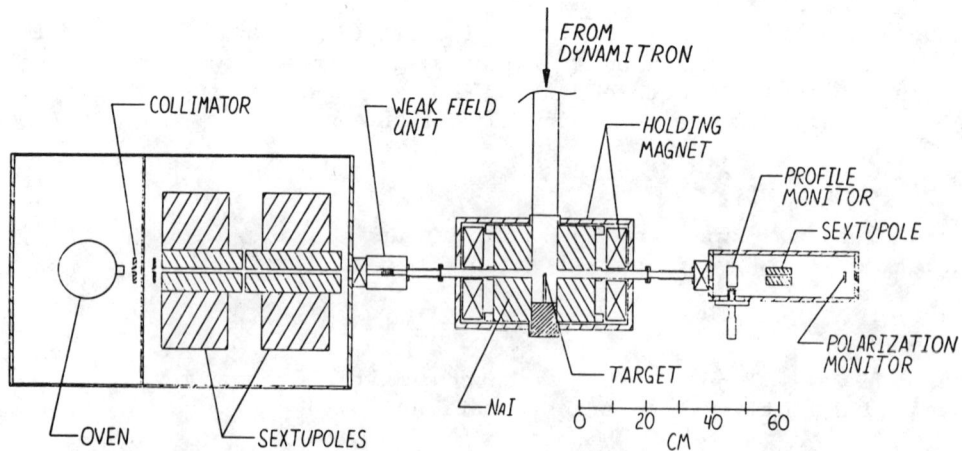

Fig. 7   Atomic beam apparatus for production of polarized ^{6}Li target.

PARITY VIOLATION IN NUCLEI   229

$\langle 2^+|H_{pv}|2^-\rangle / \Delta E$ and R a structure enhancement which depends on the relative alpha widths and penetrabilities for the two states.

$$R = 3\sqrt{\frac{3}{5}} \left[\frac{\Gamma_{\alpha(2^-)}}{\Gamma_{\alpha(2^+)}}\right]^{1/2} \left[\frac{P_\alpha(5.164)_{L=1}}{P_\alpha(5.110)_{L=1}}\right] \approx 105.$$

A preliminary estimate for the mixing matrix element, including only the long-range π exchange part of the ΔT=1 force, has been made by Teeters and Kurath,[28] who find

$$\langle 2^+|H_{PV}|2^-\rangle \approx 0.036 \, F \, eV.$$

It is anticipated that the experiment will actually be able to probe to the level $F \approx 1$.

$^{18}F(1.081) \to {}^{18}F + \vec{\gamma}$

The best direct limit on isovector parity violation yet obtained comes from the doublet of states in ^{18}F, the 0^+,T=1 level at 1.042 MeV and the 0^-,T=0 level at 1.081 MeV. Parity mixing between these close-lying states gives rise to a circular polarization in the γ ray de-exciting the relatively long-lived 1.081 MeV state. Since the first measurement of this quantity by the Caltech-Seattle-Cal State Los Angeles collaboration[29] there have been two new attempts, one by Waffler's group at the Max-Planck-Institut für Chemie, Mainz,[30] and the other by Maurenzig et al. at Florence.[31] Data from the Mainz experiment are shown in Fig. 8. None of the experiments finds a non-zero circular polarization. (It is curious that two[29,30] seem to show positive effects for the 1.02 MeV transition which feeds the 1.081 MeV level from above. This may be simply an artifact associated with summing of 511-keV annihilation quanta at this energy.) The three measurements and their weighted average are summarized in Table 1. As will be discussed later, this limit severely restricts the possible range of neutral current enhancement in nuclei.

Table 1. Measurements of circular polarization of 1.081 MeV γ-rays in ^{18}F.

Reference	Result x 10^3
29 (Barnes et al.)	−0.7 ± 2.0
30 (Wäffler)	+0.8 ± 1.8
31 (Maurenzig et al.)	−1.0 ± 3.0
Average	−0.1 ± 1.2

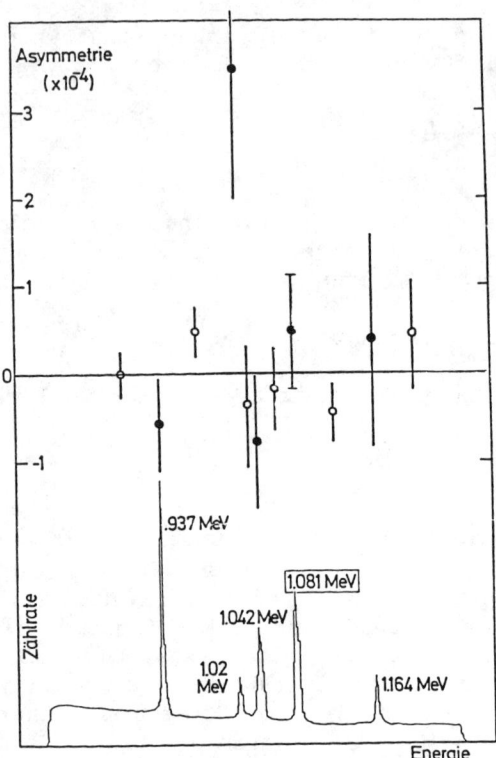

Fig. 8   Gamma spectrum from $^{16}O(^3He,p)^{18}F$ reaction and experimental "asymmetries" for various spectral regions. The circular polarizations are proportional to these asymmetries. (Ref. 30).

We conclude this section with the observation that there is no prima facie case for $\Delta T=1$ nuclear parity violation, but, as we shall describe, recent experimental and theoretical work in mass 18, 19 and 21 lead to a rather well-defined and startling result.

THEORETICAL CONSIDERATIONS

Proceeding from a fundamental weak interaction theory to experiment is now seen to be an extremely complicated task, involving detailed nuclear structure, short range correlations, an effective weak Hamiltonian and symmetry principles. Recently some new approaches have been developed which have been very valuable in pinpointing problem areas in theory and experiment. McKellar and collaborators and Desplanques and Missimer have pioneered techniques which allow a critical assessment of whether the existing experimental data are internally consistent, and whether or not one can, even in principle, hope to understand the data in terms of a fundamental theory. McKellar[12] has started with a general set of spin-isospin effective operators and has assumed a meson exchange form of weak potential in order to parameterize existing data. Desplanques and Missimer[32] have avoided the meson exchange picture by making use of the low-energy weak scattering amplitudes as fundamental parameters. The advantage is a more model-independent synthesis of all the data, but the price paid is the remote connection between weak interaction theory and the scattering amplitudes. Both of these approaches have led to two important conclusions: a) The large result from the Lobashov experiment[11] cannot be accommodated without radical changes in the theory, and, b) Until this year the isovector components were essentially undetermined.

Two theoretical advances of the highest significance have occurred this year. Desplanques, Donohue and Holstein[33] (DDH), recognizing that symmetry principles by themselves are non-predictive, incomplete or inapplicable, have combined the $SU(6)_W$ symmetry for non-leptonic weak decays with a quark model to arrive at a detailed description of the weak nucleon-nucleon meson couplings. They conclude that strong interaction uncertainties make it impossible to assign definite values to the weak NNm couplings but that the strengths are, with high probability, constrained to be within well-defined limits. In the non-relativistic limit, DDH parameterize the NN PV potential in terms of π, ρ and ω exchange, decomposed according to isospin exchange character. The ranges allowed for the coupling constants are listed in Table 2.

The remaining part of the problem has been tackled by Haxton, Gibson and Henley[34] and by Brown, Richter and Godwin[44] in full two-body shell model calculations for three nuclei near mass 20. The $\Delta T=1$ ^{18}F experiments have already been mentioned; the others are a measurement of the γ-ray asymmetry in the decay of polarized ^{19}F in its $1/2^-$, 110 keV state to the ground state (Adelberger et al.,[35] $A_\gamma = -8.5(26) \times 10^{-5}$); and a measurement of the circular polarization of the γ-ray de-exciting the $1/2^-$ 2.79 MeV state in ^{21}Ne to the ground state (Snover et al.,[36] $P_\gamma = +2.3(29) \times 10^{-3}$). The last two

Table 2. Probable ranges for NNm coupling constants (Ref. 33).

Parameter*	Cabibbo	Weinberg-Salam
f_π^a	0 to 1	0 to 30
h_ρ^o	15 to -64	30 to -81
h_ρ^1	0 to -0.7	-1 to 0
h_ρ^2	-58	-20 to -29
h_ω^o	6 to -22	15 to -27
h_ω^1	0 to -2	-5 to -2

*In units of 3.8×10^{-8} ($h = m_n = c = 1$)

$^a F = f_\pi / 3.8 \times 10^{-8}$

measurements, being in T =-1/2 nuclei, are influenced both by ΔT=0 and 1 parts of the PV interaction. However, since ^{19}F is an odd-proton nucleus and ^{21}Ne an odd-neutron nucleus, it might be expected that the ΔT=0 and 1 pieces would have different relative signs if the wave functions were the same. In fact, the states involved are not that simply related, but a separation of the two components indeed occurs. Thus the recent ^{21}Ne experimental result, even though null, is extremely important because it can only vanish by virtue of a cancellation between the ΔT=0 and 1 parts of the interaction.

The three observations can be expressed in terms of PV mixing matrix elements $\langle H_{pv} \rangle$ (all three are accurately two-state mixing cases; $\langle H_{pv} \rangle$ is in eV):

PARITY VIOLATION IN NUCLEI

$$^{18}F \quad P_\gamma = 4.9 \times 10^{-3} \langle H_{PV}^{18}\rangle$$

$$^{19}F \quad A_\gamma = 2.0 \times 10^{-4} \langle H_{PV}^{19}\rangle$$

$$^{21}Ne \quad P_\gamma = 9.5 \times 10^{-2} \langle H_{PV}^{21}\rangle$$

These mixing matrix elements are then expressible in terms of the weak coupling constants of DDH as follows:

$$0.0862\, f_\pi^1 = \langle H_{PV}^{18}\rangle$$

$$0.0546\, f_\pi^1 - 0.0197\, (h_\rho^o + 0.56 h_\omega^o) = \langle H_{PV}^{19}\rangle$$

$$-0.0310\, f_\pi^1 - 0.0124\, (h_\rho^o + 0.56 h_\omega^o) = \langle H_{PV}^{21}\rangle.$$

where $f_\pi^1 = f_\pi - 0.12 h_\rho^1 - 0.17 h_\omega^1$.

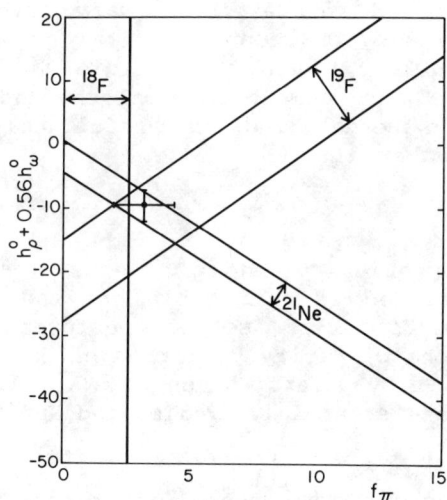

Fig. 9   Dependence of parity violation in ^{18}F, ^{19}F and ^{21}Ne on isoscalar and isovector parts of weak force.

The equations are plotted in Fig. 9. We find

$$h^0_\rho + 0.56 h^0_\omega = -9.4 \pm 2.7$$

and $f^1_\pi = 3.2 \pm 1.1$ with $\chi^2 = 1.2$.

Allowing h^1_ρ and h^1_ω to span the full range given by DDH adds an additional ± 0.3 uncertainty to f_π, and the result is

$$f_\pi = 2.5 \pm 1.1.$$

This small a value for f_π is quite remarkable. Bearing in mind that charged currents alone give values for f_π between 0 and 1, we find little, if any, evidence for hadronic neutral current parity violation.[45]

The isoscalar term is also somewhat smaller than expected. We may test its reliability by referring back to the proton-proton scattering experiments. DDH give for the asymmetry at 15 MeV,

$$A_{pp} = 0.0837 (h^0_\omega + h^1_\omega + h^0_\rho + h^1_\rho + h^2_\rho/\sqrt{6}).$$

Inserting the value obtained for $h^0_\rho + 0.56 h^0_\omega$ and allowing the remaining undetermined parameters to vary (coherently) over the full range specified by DDH, one finds the normalization used for the solid curve in Fig. 1 and the possible variation indicated by the shaded area. The agreement is excellent and further increases one's confidence in the detailed and careful analyses made by Haxton et al. and by Brown et al.

CONCLUSIONS

A summary of parity violating effects in nuclei is given in Table 3. Not included in this table are the novel neutron spin rotation measurements in tin isotopes performed by Forte et al.[37] and the observations of parity violating effects in the fission of ^{233}U, ^{235}U and ^{239}Pu by polarized neutrons.[38] Also omitted are some null results in heavier nuclei. Tadic[1] and Gari[17] give keys to the literature.

Thanks to vigorous experimental and theoretical effort, it now appears that a reasonably well-defined value for the weak isovector π-nucleon coupling constant can be obtained. There is one major uncertainty in the analysis, namely the M2/E1 mixing ratio for the 2.79 MeV transition in ^{21}Ne. This quantity is virtually impossible to calculate reliably and must be measured. If it turns out to be much larger than 1, then a null result in ^{21}Ne is expected no matter

PARITY VIOLATION IN NUCLEI

Table 3. Parity violation in nuclei

Reaction	ΔT	Result	References
$n + p \rightarrow d + \gamma$	0,2	$P_\gamma = -1.3(5) \times 10^{-6}$	11
$\vec{n} + p \rightarrow d + \gamma$	1	$A_\gamma = 0.6(21) \times 10^{-7}$	19
$\vec{p} + p \rightarrow p + p$ (15)	0,(1),2	$A_L = -1.2(6) \times 10^{-7}$	2
$\vec{p} + p \rightarrow p + p$ (45)	0,(1),2	$A_L = -2.3(8) \times 10^{-7}$	3
$\vec{p} + N \rightarrow X$ (800)	0,1,2	$A_L = +3.0(10) \times 10^{-6}$	5
$\vec{p} + N \rightarrow X$ (5000)	0,1,2	$A_L = +2.6(6) \times 10^{-6}$	4
$\vec{n} + d \rightarrow t + \gamma$	0,1,2	$A_\gamma = +8.5(33) \times 10^{-6}$	20
$^6\text{Li}(0^+,1) \rightarrow \alpha + d$	1	$\Gamma_\alpha \leq 2.0 \times 10^{-6}$ eV (90% C.L.)	22
$^{16}\text{O}(2^-,0) \rightarrow \alpha + {}^{12}\text{C}$	0	$\Gamma_\alpha = 1.03(28) \times 10^{-10}$ eV	39
$^{16}\text{O}(0^-,0) \rightarrow \alpha + {}^{12}\text{C}$	0	$\Gamma_\alpha < 5 \times 10^{-4}$ eV (90% C.L.)	40
$^{16}\text{O}(3^+,0) \rightarrow \alpha + {}^{12}\text{C}$	0	$\Gamma_\alpha < 8 \times 10^{-5}$ eV (90% C.L.)	40
$^{18}\text{F}(0^-,0) \rightarrow {}^{18}\text{F} + \gamma$	1	$P_\gamma = -0.1(12) \times 10^{-3}$	29,30,31
$^{19}\text{F}(1/2^-) \rightarrow {}^{19}\text{F} + \gamma$	0,1	$A_\gamma = -8.5(26) \times 10^{-5}$	35
$^{16}\text{O} + \alpha \rightarrow {}^{20}\text{Ne}$ (11.262)	1	$\Gamma_\alpha < 3.3 \times 10^{-5}$ eV (95% C.L.)	41
$^{19}\text{F} + p \rightarrow {}^{20}\text{Ne}$ (13.168)	1	$\Gamma_\alpha < 7 \times 10^{-6}$ eV	42
$^{19}\text{F} + p \rightarrow {}^{20}\text{Ne}$ (13.479)	1	$A_\alpha = 5.1(30) \times 10^{-3}$	30
$^{21}\text{Ne}(1/2^-) \rightarrow {}^{21}\text{Ne} + \gamma$	0,1	$P_\gamma = +2.3(29) \times 10^{-3}$	36
$^{41}\text{K}(7/2^-) \rightarrow {}^{41}\text{K} + \gamma$	0,1,2	$P_\gamma = 2.0(4) \times 10^{-5}$	1
$^{113}\text{Cd}(n,\gamma){}^{114}\text{Cd}$	0,1,2	$A_\gamma = -3.4(7) \times 10^{-4}$	43
$^{117}\text{Sn}(n,\gamma){}^{118}\text{Sn}$	0,1,2	$P_\gamma = -6.0(15) \times 10^{-4}$	43
		$A_\gamma = 6.3(20) \times 10^{-4}$	1
$^{175}\text{Lu}(9/2^-) \rightarrow {}^{175}\text{Lu} + \gamma$	0,1,2	$P_\gamma = 5.5(5) \times 10^{-5}$	1
$^{180}\text{Hf}(8^-) \rightarrow {}^{180}\text{Hf} + \gamma(.501)$	0,1,2	$P_\gamma = -2.5(3) \times 10^{-3}$	1
		$A_\gamma = -0.017(2)$	1
$^{181}\text{Ta}(5/2^+) \rightarrow {}^{181}\text{Ta} + \gamma$	0,1,2	$P_\gamma = -5.2(5) \times 10^{-6}$	1

what the weak interaction, so an experimental determination is urgently needed. The most promising approach is perhaps a measurement of the pair internal conversion coefficient.

Of course, a "direct" measurement of a pure isovector case is highly desirable, and it is to be hoped that the four $\Delta T=1$ experiments will be pushed still further, and that improved calculations will be made for the ^{6}Li case. Nuclear parity violation seems to be rapidly approaching an interesting and useful synthesis.

ACKNOWLEDGMENTS

I am very grateful to all those who sent their data and calculations prior to publication; Drs. E. Adelberger, E. Bellotti, P. Bizzetti, T. Bowles, K. Fifield, W. Haxton, C. Hoffman, B. Holstein, D. Kurath, N. Lockyer, A. McDonald, B. McKellar, N. Ramsey, B. Vignon and H. Waffler. The preparation of this talk has been supported by the U.S. Dept. of Energy and by the Alfred P. Sloan Foundation in the form of a Fellowship.

REFERENCES

1. D. Tadic, Rep. Prog. Phys. 43:67 (1980).
2. J.M. Potter, J.D. Bowman, C.F. Hwang, J.L. McKibben, R.E. Mischke, D.E. Nagle, P.B. Debrunner, H. Frauenfelder, and L.B. Sorensen, Phys. Rev. Lett. 33:1307 (1974); D.E. Nagle, J.B. Bowman, C. Hoffman, J. McKibben, R. Mischke, J.M. Potter, H. Frauenfelder, and L. Sorensen, in "High Energy Physics with Polarized Beams and Polarized Targets" ed. G.H. Thomas, Am. Inst. Phys. New York (1978).
3. R. Balzer, R. Henneck, Ch. Jacquemart, J. Lang, M. Simonius, W. Haeberli, Ch. Weddigen, W. Reichart, and S. Jaccard, Phys. Rev. Lett. 44:699 (1980); M. Simonius, in "Fifth Int. Conf. on Polarization Phenomena" Santa Fe (1980).
4. N. Lockyer, T.A. Romanowski, J.D. Bowman, C.M. Hoffman, R.E. Mischke, D.E. Nagle, J.M. Potter, R.L. Talaga, E.C. Swallow, D. Alde, and D.R. Moffett, Bull. Am. Phys. Soc. 25:525 (1980).
5. C.M. Hoffman, private communication.
6. E.M. Henley and F.R. Krejs, Phys. Rev. D 11:605 (1975).
7. V.R. Brown, E.M. Henley and F.R. Krejs, Phys. Rev. C 9:935 (1974).
8. M. Simonius, Phys. Lett. 41B:415 (1972); M. Simonius, Nucl. Phys. A220:269 (1974).
9. G. Barton, Nuovo Cim. 19:512 (1961).
10. G.S. Danilov, Phys. Lett. 18:40 (1965).
11. V.M. Lobashov, D.M. Kaminker, G.I. Kharkevich, V.A. Kniazkov, N.A. Lozovoy, V.A. Nazarenko, L.F. Sayenko, L.M. Smotritsky, and A.I. Yegorov, Nucl. Phys. A197:241 (1972).

12. B.H.J. McKellar in "Int. Conf. on Frontiers of Physics", Singapore (1978).
13. A.B. McDonald, E.D. Earle and J.W. Knowles, contrib. to this conference.
14. H.C. Lee, Phys. Rev. Lett. 41:843 (1978).
15. C.Y. Prescott, et al., Phys. Lett. B77:347 (1978).
16. T.J. Bowles, private communication.
17. M. Gari in "Interaction Studies in Nuclei" ed. H. Jochim and B. Ziegler, North Holland, Amsterdam (1975).
18. K.R. Lassey and B.H.J. McKellar, Nucl. Phys. A260:413 (1976).
19. J.F. Cavaignac, B. Vignon and R. Wilson, Phys. Lett. 67B:148 (1977).
20. B. Vignon, private communication.
21. E. Bellotti, E. Fiorini, P. Negri, A. Pullia, L. Zanotti, and I. Filosofo, Nuovo Cim. 29A:106 (1975).
22. R.G.H. Robertson, R.A. Warner, P. Dyer, R.C. Melin, T.J. Bowles, A.B. McDonald, W.G. Davies, G.C. Ball and E.D. Earle, Progress Report, Michigan State University (1980) (unpublished).
23. F.C. Michel, Phys. Rev. 133:B329 (1964).
24. R.G.H. Robertson and D.O. Riska, unpublished.
25. J.D. Vergados, Nucl. Phys. A220:259 (1974).
26. P.G. Bizzetti and A. Perego, Phys. Lett. 64B:298 (1976).
27. C.A. Gagliardi, A.R. Davis, G.T. Garvey, R.D. McKeown, B. Myslek-Laurikainen, R.G.H. Robertson, S.J. Freedman, and T.J. Bowles, contrib. to "Fifth Int. Conf. on Polarization Phenomena" Santa Fe (1980).
28. W. Teeters and D. Kurath (unpublished).
29. C.A. Barnes, M.M. Lowry, J.M. Davidson, R.E. Marrs, F.B. Morinigo, B. Chang, E.G. Adelberger and H.E. Swanson, Phys. Rev. Lett. 40:840 (1978).
30. H. Waffler, private communication.
31. P.R. Maurenzig, M. Bini, P.G. Bizzetti, T.F. Fazzini, A. Perego, G. Poggi, P. Sona and N. Taccetti, in "Neutrinos 79", Bergen, p. 179 (1979); also P.R. Maurenzig, private communication.
32. B. Desplanques and J. Missimer, Nucl. Phys. A300:286 (1978).
33. B. Desplanques, J.F. Donoghue and B.R. Holstein, Ann. Phys. 124:449 (1980).
34. W.C. Haxton, B.F. Gibson and E.M. Henley, to be published.
35. E.G. Adelberger, H.E. Swanson, M.D. Cooper, J.W. Tape, and T.A. Trainor, Phys. Rev. Lett. 34:402 (1975); also E.G. Adelberger, private communication.
36. K.A. Snover, R. Von Lintig, E.G. Adelberger, H.E. Swanson, T.A. Trainor, A.B. McDonald, E.D. Earle, and C.A. Barnes, Phys. Rev. Lett. 41:145 (1978); also A.B. McDonald, private communication.

37. M. Forte, B. Heckel, N. Ramsey, K. Green, G. Greene, M. Pendlebury, W. Sumner, P.D. Miller and W. Dress, Bull. Am. Phys. Soc. 25:526 (1980).
38. See V.V. Flambaum and O.P. Sushkov, Phys. Lett. B (to be published).
39. K. Neubeck, H. Schober, and H. Waffler, Phys. Rev. C10:320 (1974).
40. E. Bellotti, E. Fiorini, C. Liquori, P. Negri, and L. Zanotti, in "Neutrinos-79" Bergen, p. 175 (1980).
41. L.K. Fifield, private communication.
42. N. Krimmelbein, H. Schober and H. Waffler, in "Int. Conf. on Nuclear Structure and Spectroscopy" Amsterdam, Vol. 1 p. 149 (1974).
43. Y.G. Abov and P.A. Krupchiskii, Sov. Phys. Usp. 19:75 (1976).
44. B.A. Brown, W.A. Richter and N.S. Godwin, to be published.
45. There is now some confusion over the sign of the ^{18}F result from Mainz. A sign change would allow a somewhat larger n.c. enhancement.

DISCUSSION (Chairman: L. Wolfenstein)

H. FAISSNER, Aachen: Also the Grenoble experiment, by Vignon et al., on $\vec{n}$ capture by d will probably be repeated under better systematic conditions. The alternative would be to go to $\vec{n}$ on p right away. Which one would be more reasonable?

R.G.H. ROBERTSON, Argonne: The $\vec{n}$p experiment is more fundamental

F. BOEHM, Pasadena: I understand that Lobashov is currently repeating his experiment. Does anyone know of his preliminary results?

R.G.H. ROBERTSON, Argonne: Yes, he has some data which seems to confirm his earlier large value.

N.W. REAY, Columbus: If one applies a four standard deviation criterion to these parity violation experiments, the experiment of ANL, LASL, Ohio State (Lockyer et al.) stands out as a six sigma effect. Can you comment whether there is any theoretical reason a parity violating effect would increase so much at high energy?

R.G.H. ROBERTSON, Argonne: The high energy experiment is difficult to analyze because of the number of channels open, and the number of phase shifts which apply.

LOW ENERGY NEUTRINO INTERACTIONS

Henry W. Sobel, Frederick Reines, Elaine Pasierb
(Presented by Henry W. Sobel)

Department of Physics
University of California
Irvine, California 92717

This paper was intended to be a review of low energy neutrino interactions. However, in view of our recently reported deuteron results, we have been requested to concentrate on them. To this end, we comment briefly on pion factory neutrino experiments and then discuss reactor neutrino work and, at greater length, the deuteron experiment.

PION FACTORY EXPERIMENTS

At pion factories, a proton beam is made to interact in a beam dump. The π^+'s which are produced stop and decay, the π^-'s and μ^-'s are absorbed. The π^+ decay at rest produces a monoenergetic ν_μ flux and in addition, μ^+'s which decay producing ν_e's and $\bar{\nu}_\mu$'s with a 53 MeV end point (Fig. 1).

There has been only one neutrino experiment completed at a pion factory and that was LAMPF 31.[1] This experiment looked at the reaction

$$\nu_e + d \rightarrow p + p + e^-$$

and measured the cross section to $\pm 35\%$. F. Boehm will treat this work in more detail in connection with his discussion of lepton conservation tests.

Another experiment at LAMPF, (LAMPF 225),[2] is just being mounted. It has three main thrusts:

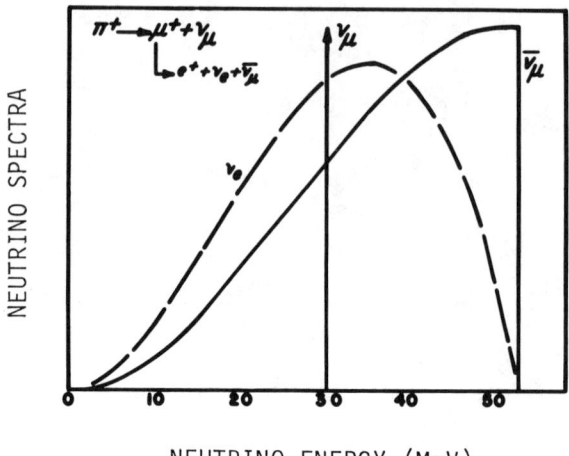

Fig. 1. Neutrino spectra from a pion factory beam dump.

1. The primary interest is in measuring the elastic scattering process

$$\nu_e + e^- \rightarrow \nu_e + e^-$$

with a sandwich array of plastic scintillators and plastic flash chambers (Fig. 2). The intention is to measure the cross section to $\pm 10\%$. The rate as a function of $\sin^2 2\theta_\omega$ for each neutrino component is shown in Fig. 3.

2. As in LAMPF 31, this experiment can also be used as a test of the multiplicative law or of neutrino oscillations. Both tests are based on the normal presence of $\bar{\nu}_\mu$'s but no $\bar{\nu}_e$'s from the beam dump. Because of backgrounds a limit of 1 to 2% on the process

$$\mu^+ \rightarrow e^+ + \bar{\nu}_e + \nu_\mu$$

is expected. This would also imply a limit on Δ of about 0.1 eV^2.

3. A measurement of the inverse beta cross section on ^{12}C

$$\nu_e + {}^{12}C \rightarrow e^- + {}^{"12}N"$$

will be made. Here, "N" represents all final states. About 75% of the events will be tagged with an 11ms delayed coincidence

$$^{12}N \rightarrow {}^{12}C + \beta^+ + \nu_e$$

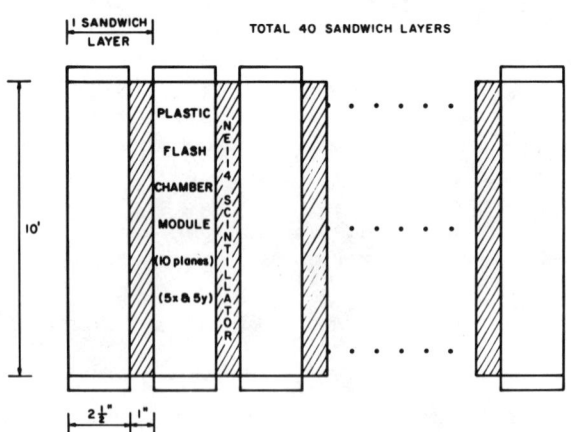

Fig. 2. Sandwich array of plastic scintillator and plastic flash chambers.

LOW ENERGY NEUTRINO INTERACTIONS

REACTOR NEUTRINO EXPERIMENTS

The Reactor as a Neutrino Source

A power reactor produces neutrinos from the fission of ^{235}U, ^{238}U and ^{239}Pu. The neutrino spectrum of each component is different (Fig. 4). At the Savannah River Project (SRP) reactor where we work, ^{239}Pu produced less than 8% of the fissions during our experiment, and ^{238}U produced less than 4%. At the Grenoble reactor all the fissions come from ^{235}Pu.

Neutron activation of the reactor surroundings can also produce $\bar{\nu}_e$'s but they are less than ~1.3 MeV.[3] Neutrino production in this mode is also possible with $\nu_e/\bar{\nu}_e \simeq .0005$ and $E_{\nu_{max}} = 0.8$ MeV.[4] Experimentally, the ratio $\nu_e/\bar{\nu}_e$ has been determined to be < 0.02.[5]

The antineutrino spectrum itself is calculated by adding up the beta decay spectrum of all the fission products. Unfortunately, approximately 30% of the fission products involve unknown decay schemes, and as a consequence, the neutrinos from these decays require extrapolation and modeling. The two newest theoretical predictions for the antineutrino spectrum are shown in Fig. 5. They disagree to a level of about 30%. depending on $\bar{\nu}_e$ energy.

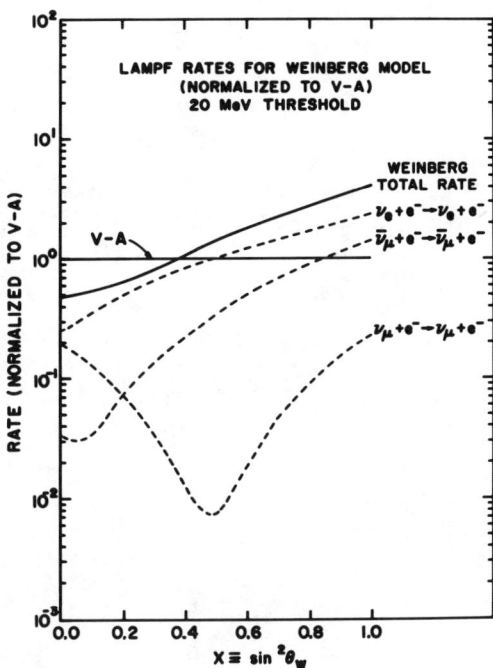

Fig. 3. Reaction rates for each neutrino component as a function of $\sin^2 2\theta_\omega$

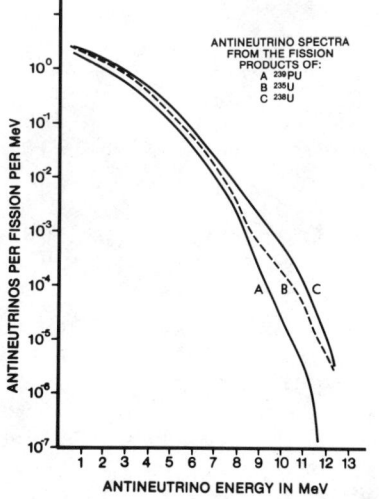

Fig. 4. Antineutrino spectra from fission products.

$\bar{\nu}_e + p \rightarrow n + e^+$ Experiments

The inverse beta decay reaction has been studied in several different experiments and new experiments are being built. The data so obtained can be used in several ways.

1. Since the e^+ takes essentially all of the $\bar{\nu}_e$ energy in the reaction, a measurement of the e^+ spectrum determines the $\bar{\nu}_e$ spectrum. This neutrino spectrum can be used as input in other experiments.

2. The process can be used to study neutrino oscillations.

   a. Experimental results can be compared with theoretical prediction: This technique suffers from the uncertainty in the predicted neutrino spectrum in that any observed discrepancies can be attributed to that source.

   b. We can compare the results of different detectors at various source to detector distances: These data are just beginning to be available in the required precision and analysis is incomplete.[6]

   c. We can compare the results of the same detector taken at different distances: This clearly is the experiment of choice and two groups are actively pursuing this technique.

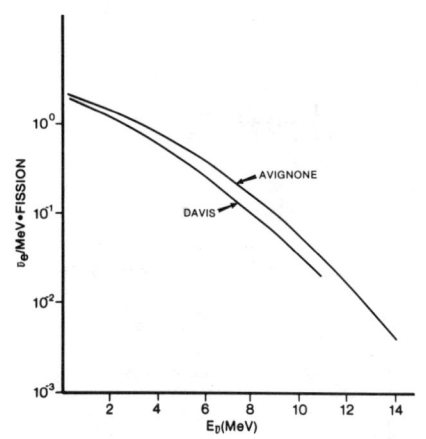

Fig. 5. Predicted $\bar{\nu}_e$ spectrum.

There are three data sets available at this time. They are the 6.5 meter data of Nezrick and Reines,[7] the 11.2 meter data of Reines, Gurr, and Sobel,[8] and the 8.7 meter data of the C.I.T.-Grenoble-Munich group.[9] Within one year, additional data are expected from a 38 meter point with the C.I.T.-Munich detector at a 2700 MW reactor near Zurich, a mobile detector at the SRP reactor which will span a 12 to 35 meter distance,[10] and a single point at 15.4 meters from the SRP reactor by a G.I.T.-U.S.C. group.[11]

Results from the 11.2 meter data are given in Fig. 6 and compared with the predicted positron spectrum of Avignone et al and Davis et al. We note that the observed data agree with the Davis prediction at β^+ energies $\gtrsim$ 2 MeV but diverge below the prediction at higher energies.

LOW ENERGY NEUTRINO INTERACTIONS

All three experiments are compared in Table I. We list the ratio of the observed rates to those predicted by the Avignone and Davis spectra. In order to interpret these results in terms of neutrino oscillations we assume a simple two neutrino case.

Two Neutrino Approach

In this case a physical neutrino can be written as a superposition of two pure states i.e.:

$$\bar{\nu}_e = \nu_1 \cos\Theta + \nu_2 \sin\Theta$$

and $$\bar{\nu}_{\mu/\tau} = -\nu_1 \sin\Theta + \nu_2 \cos\Theta$$

where Θ is the mixing angle. Then, the probability of finding a $\bar{\nu}_\mu$ (or $\bar{\nu}_\tau$) at time t, given that we started with a $\bar{\nu}_e$ is:

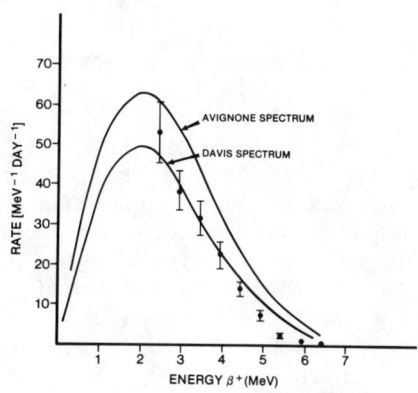

Fig. 6. 11.2 meter data from $\bar{\nu}_e + p \to n + e^+$.

$$\text{Prob.} = |a(t)|^2 = \tfrac{1}{2}\sin^2 2\Theta \left[1 - \cos\frac{(\Delta)(t)}{2E_\nu \tau \eta}\right]$$

where $\Delta = |m_1^2 - m_2^2| c^4$, and the oscillation length

$$\lambda(m) = \frac{2.5 E_\nu (\text{MeV})}{\Delta (\text{eV}^2)}.$$

Table I. Comparison of inverse beta decay results. The ratio of observed/predicted rates are tabulated.

Distance from Core Center (Meters)	Reaction	Neutrino Detection Threshold (MeV)	Ratio Avignone Spectrum	Ratio Davis Spectrum
11.2	ccp	4.0	.68 ± .12	.88 ± .15
11.2	ccp	6.0	.42 ± .09	.58 ± .12
6	ccp	1.8	.65 ± .09	.84 ± .12
6	ccp	4.0	.81 ± .11	1.02 ± .15
8.7	ccp	3.0	.68 ± .15	.87 ± .14

For $\bar{\nu}_e$ sensitive experiments, the ratio of counting rates with and without oscillations is given by:

$$R(d) = 1 - \tfrac{1}{2}\sin^2 2\theta \left[1 - \frac{\int N(E_\nu)\sigma(E_\nu)\cos\left(\frac{\Delta\, d}{2c\hbar E_\nu}\right) dE_\nu}{\int N(E_\nu)\sigma(E_\nu) dE_\nu} \right]$$

where d = distance from source to detector

$N(E_\nu)$ = Neutrino flux

and, $\sigma(E_\nu)$ = Reaction cross section.

We can plot R(d) as a function of $\Delta \cdot d$ using $\sin^2 2\theta$ as a parameter for a particular process and energy interval. The result is a family of curves which can be used to imply Δ for any distance, d. In Fig. 7 for example, we plot such a family for the $\bar{\nu}_e$ + p, charged current proton (ccp), reaction with a neutrino energy greater than 6.0 MeV. If we now use the Table I value 0.42 $\pm$ .09 (11.2 meters, $E_\nu > 6.0$ MeV, Avignone spectrum), Fig. 7 gives, for $\sin^2 2\theta = 1$, Δ = .42 $\pm$ .04 eV^2 and Δ = 1.1 $\pm$ .05 eV^2. We can continue this procedure and generate a plot of Δ vs. $\sin^2 2\theta$, showing a region allowed by this measurement at the one standard deviation level (Fig. 8). Using the same data, if we compare instead to the Davis spectrum we get an allowed region shown in Fig. 9.

From this we can see that the conclusions we reach with this technique depend very strongly on the spectrum we use for comparison.

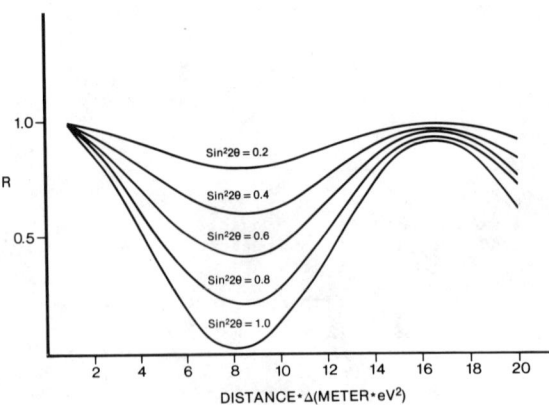

Fig. 7. R as a function of distance times Δ for ccp and $E_\nu > 6.0$ MeV.

THE DEUTERON EXPERIMENT

As part of our reactor program, we have been studying the neutral current reaction,[12]

$$\bar{\nu}_e + d \to n + p + \bar{\nu}_e \text{ (ncd)}$$

and its charged current counterpart

$$\bar{\nu}_e + d \to n + n + e^+ \text{ (ccd)}$$

We have recently realized that this experiment could be used as a neutrino oscillation test. The neutral current branch is independent of ν type, while the charged current branch will only occur for incident $\bar{\nu}_e$'s. In addition, while the predicted rates of the individual branches are sensitive to the predicted neutrino spectrum (Fig. 11), the ratio of the predicted rates is not (Table II).

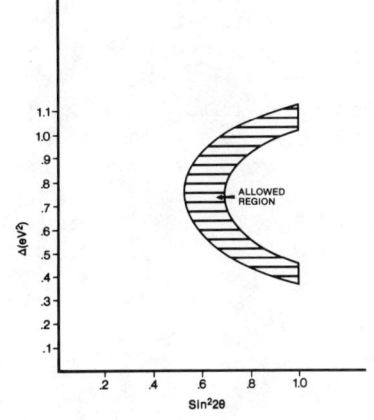

Fig. 8. $\Delta(eV^2)$ vs. $\sin^2 2\theta$ allowed region for 11.2 m ccp data; $E_\nu \geq 6.0$ MeV; Avignone spectrum.

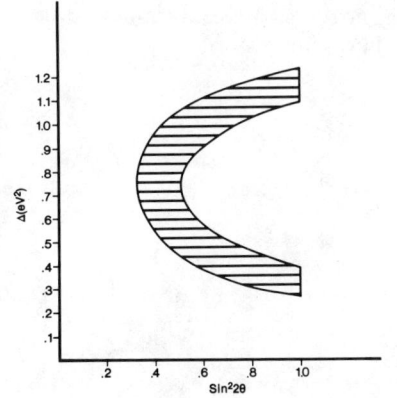

Fig. 9. $\Delta(eV^2)$ vs. $\sin^2 2\theta$ allowed region for 11.2 m ccp data; $E_\nu \geq 6.0$ MeV; Davis spectrum.

We choose to define the quantity:

$$R = \frac{\left(\frac{ccd}{ncd}\right)_{experiment}}{\left(\frac{ccd}{ncd}\right)_{predicted}}.$$

The denominator of this quantity has the following features:

1. It is independent of the reactor neutrino absolute normalization.

2. It is insensitive to the precise shape of the reactor neutrino spectrum.

3. The ncd process is independent of neutrino type.

4. The ccd process only occurs with $\bar{\nu}_e$'s.

5. Assuming the standard model, the ratio of the coupling constants is known to $\sim 5\%$.[13]

The quantity, R, is expected to be unity. A value of R below unity would signal the instability of $\bar{\nu}_e$ as it traversed the distance (centered in this deuteron experiment at 11.2 meters) from its origin to the detector.

Experimental Approach

We have constructed a shielded

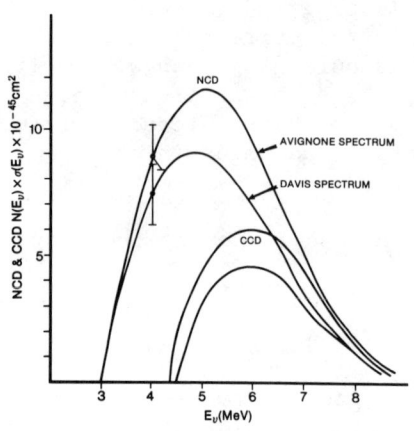

Fig. 10. Flux times cross section for ncd and ccd process as predicted from Avignone and Davis spectra.

volume at the reactor which has a greatly reduced neutron background. This shield made it feasible to search for the n.c. reaction by looking only at the product neutron, so avoiding the proton background problem of the earlier approaches.[14] The c.c. reaction is identified by detecting both product neutrons in a time window of 2 ms (the neutron capture time in this detector is 300 μs). Those cases in which only one of the two neutrons from the c.c. reaction was detected represented background for the n.c. reaction.

Detector Description

The target consists of 268 kg. of D_2O. Immersed in the D_2O are 10 cylindrical, helium-3 filled, neutron proportional counters. A side view of the detector is shown in Fig. 11. The target is enclosed in 10.2 cm of lead, 0.1 cm of cadmium and immersed in a 2200 liter anticoincidence detector. The detector is in turn surrounded by massive lead, concrete and water shielding. Events arise from a neutron capture in the ^{3}He counters,

$$^3He + n \rightarrow p + ^3H + 773 \text{ keV}$$

which meet the trigger requirements. All signals from the ^{3}He counters and the anticoincidence system within 2 ms before and after an event are recorded.

Table II. Predicted cross sections for charged current and neutral current reactions, and the ratio of these predictions.

Cross section (cm^2/fission)	Davis Spectrum	Avignone Spectrum
ncd	2.87×10^{-44}	3.73×10^{-44}
ccd	1.21×10^{-44}	1.64×10^{-44}
Ratio $\frac{ccd}{ncd}$	0.42	0.44

LOW ENERGY NEUTRINO INTERACTIONS

Neutron Detection Efficiency

The neutron detection efficiency was determined using Monte Carlo techniques and a ^{252}Cf source.

1. <u>Monte Carlo</u>. The neutron detection efficiency for our system is dependent on the energy spectrum of the neutrons under consideration. Accordingly, the efficiency is different for neutrons emitted from the ^{252}Cf source and those resulting from the lower energy n.c., c.c. or $\bar{\nu}_e + p \to n + e^+$ (I.B) background reaction.

The Monte Carlo computer code utilized for these calculations is one written by the Savannah River Laboratory. It has an extensive cross section library. Originally written for reactor geometries, it was modified to simulate the geometry of this experiment. The calculation is two dimensional, assuming an infinitely long array, and therefore a correction due to the loss of neutrons from the top and bottom of the detector must be made.

We find that the detection efficiency is relatively insensitive to neutron energy.

2. ^{252}Cf Source

We have checked the results of the Monte Carlo using a ^{252}Cf source positioned at various locations inside the detector.

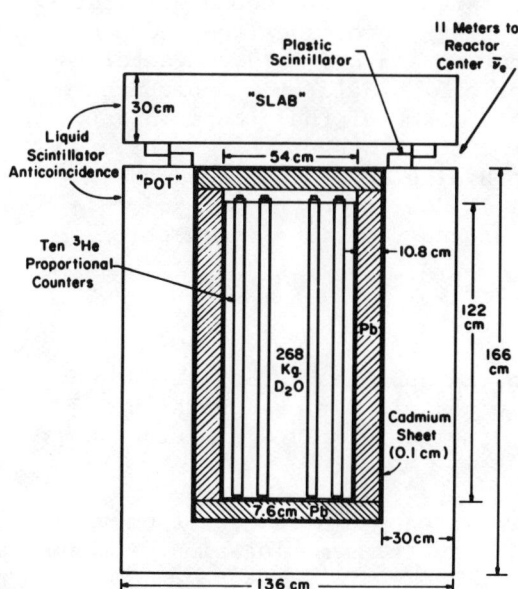

Fig. 11. Schematic diagram of the detector side view.

a. A source calibrated to ± 3% was counted and the results used to establish efficiency as a function of location.

b. The Cf source is a source of multiple neutrons, emitting on the average 3.73 neutrons per fission.

The numbers of single, double, triple and multiple neutron events were recorded for each source position. The neutron detection efficiency can be determined from ratios of these values without a knowledge of the absolute source calibration.

The efficiencies derived from the Monte Carlo, the direct neutron counting, and the neutron multiplicity method are in agreement within the uncertainties given below. The efficiencies for single neutrons produced uniformly throughout the D_2O are:

$$\bar{\eta}_{n.c.} = \bar{\eta}_{c.c.} = .32 \pm .02$$

$$\bar{\eta}_{252_{Cf}} = .28 \pm .02$$

and, $\bar{\eta}_{I.B.} = .36 \pm .02$

The efficiency for detecting two neutrons in the c.c. reaction is $\overline{\eta^2} = 0.112 \pm 0.009$ where the two neutron efficiency is averaged over the D_2O volume.

During single neutron analysis, we use $\vec{\bar{\eta}} = .89\,\bar{\eta}$, the efficiency loss due to a background reduction, cut.

Data

We report the results of two data sets. Each data set consists of a number of reactor on and reactor off sequences alternating in time. Several time groupings were made from these sequences, each consisting of reactor on and off data. Reactor associated (reactor on minus reactor off) single and double neutron rates were obtained for each group. Some of the groups in data set 1 are shown in Table III. From this, we see that we have a reactor associated signal from both single and double neutrons. We now establish that this signal is due to neutrinos, and further, that they are due to the deuteron reactions in question.

Background Tests

1. <u>Neutrons</u>. Our neutron background was established by completely surrounding our detector with an additional neutron shield. The observed change in our signal implied a neutron background of 0.7 ± 0.14 day^{-1}.

2. <u>Gammas</u>. The reactor associated gamma ray spectrum was measured with a 300 kg NaI detector in the same location as the D_2O target.

The background due to the (γ,n) reaction on the deuteron was calculated to be 0.05 day^{-1}.

3. $\bar{\nu}_e$ Background.
   a. Our D_2O is not pure. The ratio of the number of protons to the number of deuterons is .0015. Since the inverse beta process has a relatively large cross section, this small contaminant gives a neutron background of 1.6 ± 0.1 day^{-1}.

   b. The liquid scintillator anticoincidence detector consists of $CH_{1.8}$ and is therefore a large source of proton targets for the I.B. process. We calculate that about 10^4 day^{-1} are occurring in this detector. Most of the neutrons which are produced are thermalized in the scintillator, and captured on hydrogen or our cadmium shield. The probability of a neutron being seen by the ^{3}He detectors was calculated via Monte Carlo and measured with a neutron source immersed in the liquid scintillator. This probability varies from .0018 to .00056 as a function of source location, and implies a background of 34 neutrons per day.

The I.B. process occurs in a live anticoincidence, and as a consequence a large fraction (.77) of the background events are discriminated against by means of the energy deposition of the positron and its annihilation gammas. The residual background is thus 7.9 ± 0.7 day^{-1}.

The total background to the single neutron signal from these sources is therefore 10.2 ± 0.7 day^{-1}.

Deuteron Rates

We can now calculate the c.c.d. rate (R^{ccd}) and the n.c.d. rate (R^{ncd}) from the observed one neutron and 2 neutron signals (S_{1N}, S_{2N}).

Table III. Sample neutron rates in data set 1.

	Group 1	Group 2	Group 3 . . .	Group 7	Weighted Mean
			Single Neutron Data		
Reactor on (day^{-1})	386.90 ±11.83	387.27 ±11.90	406.31 ± 6.72 . . .	439.95 ± 6.98	
Reactor off (day^{-1})	323.73 ± 7.32	320.01 ± 5.09	333.92 ±12.92 . . .	385.61 ± 5.68	
On-Off (day^{-1})	63.17 ±13.49	67.26 ±12.94	72.39 ±14.56	54.34 ± 8.90	68.26 ±4.11
			Double Neutron Data		
Reactor on (day^{-1})	53.51 ±4.40	54.12 ±4.45	58.54 ±2.55 . . .	51.85 ±2.40	
Reactor off (day^{-1})	47.67 ±2.81	51.55 ±2.04	55.87 ±4.96 . . .	46.81 ±1.98	
On-Off (day^{-1})	5.84 ±5.22	2.57 ±4.90	2.67 ±5.58 . . .	5.04 ±3.11	3.66 ±1.52

$$R^{ccd} = \frac{S_{2N}}{\overline{\eta^2}} \quad \text{and} \quad R^{ncd} = \frac{S_{1N}^{ncd}}{\eta_{1N}}$$

where: $S_{1N}^{ncd} = S_{1N} - S_{1N}^{(BKGND)} - S_{1N}^{ccd}$

and, $S_{1N}^{ccd} = 2(\eta_{1N})(1 - \eta_{1N})R^{ccd}$

The experimental ratio of ccd to ncd is therefore:

$$r_{exp} = \frac{R^{ccd}}{R^{ncd}} = \frac{\overline{\eta^2} S_{2N}}{\overline{\eta^2}(S_{1N} - S_{1N}^{BKGND}) - 2(.89)(\overline{\eta} - \overline{\eta^2})S_{2N}}$$

and the error in r_{exp} is calculated from

$$\sigma_{r_{exp}}^2 = (\frac{\partial r}{\partial S_{1N}})^2 \sigma_{S_{1N}}^2 + (\frac{\partial r}{\partial S_{2N}})^2 \sigma_{S_{2N}}^2 + \ldots$$

to be $r_{exp} = 0.167 \pm 0.093$.

We had that $R \equiv \dfrac{r_{exp}}{r_{theory}}$

so, $R_{\text{Avignone spectrum}} = \dfrac{0.167 \pm .093}{0.44} = 0.38 \pm 0.21$

and, $R_{\text{Davis spectrum}} = \dfrac{0.167 \pm .093}{0.42} = 0.40 \pm 0.22$.

These represent a 3.0 to 2.7 standard deviation departure from unity, assuming that the σ_r calculated above is representative of a normal distribution.

In the same way as for the ccp experiments the allowed values of Δ and $\sin^2 2\theta$ are plotted in Fig. 12 for $R = .38 \pm .21$.

Consistency Checks

1. We have mentioned previously another experiment at the 11.2 meter position which measured the ccp process. This positron

LOW ENERGY NEUTRINO INTERACTIONS

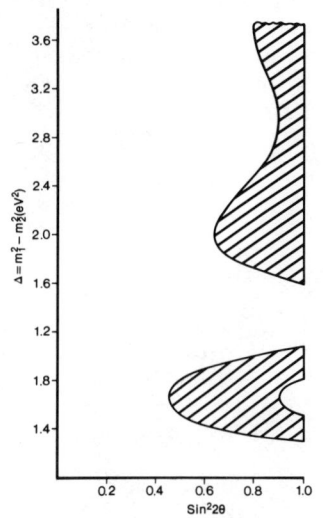

Fig. 12. $\Delta(\text{eV}^2)$ vs. $\sin^2 2\theta$ for $R = 0.38 \pm 0.21$.

spectrum has been used to obtain a $\bar{\nu}_e$ spectrum for $E_{\bar{\nu}_e} > 4$ MeV (Fig. 13). The value for R deduced using this spectrum, extrapolated below 4 MeV, is $0.47 \pm .24$, a 2.2 standard deviation effect. If neutrino oscillations occur, with the parameters implied by the deuteron experiment, then the extrapolation of the neutrino spectrum to lower energies would be in error and the value of R reduced.

2. In Table IV we list the individual ccd and ncd rates, and other previously mentioned reactor results compared to the predicted rates using the Avignone spectrum, the Davis spectrum and the measured $\bar{\nu}_e$ spectrum at 11.2 meters.

Unlike the insensitivity of R to the reactor neutrino spectrum, all other ratios of experimentally determined rates to predicted rates are markedly dependent on the spectrum and normalizations. For this reason, we consider the precise values of these other ratios listed in Table IV to be of less significance. They can however be used to test consistency with R.

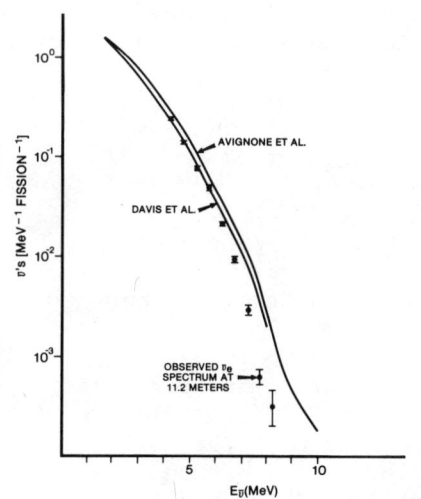

Fig. 13. The observed $\bar{\nu}_e$ spectrum at 11.2 meters compared to Avignone and Davis predictions.

a. We note that since our measurement of the neutrino spectrum is only sensitive to $\bar{\nu}_e$ it should enable us to correctly predict the ratio for the charged current branch. Table IV indicates that the preliminary prediction for this ratio using the measured spectrum is 1.3 standard deviations from the expected value of unity. If the difference can be attributed to a normalization error between the two experiments it would have no effect on the ratio R. If, however, the difference is due to a statistical fluctuation and we therefore choose for the charged current the most likely value consistent with the two experiments, then the ratio R would become 0.62 ± 0.16. We note in this case that whereas R has increased, its error has diminished reflecting the greater precision of the prediction based on the measured $\bar{\nu}_e$ spectrum.

Table IV. Summary of Results for the Ratio $\dfrac{\bar{\sigma}_{expt.}}{\bar{\sigma}_{th.}}$

Distance from Core Center (Meters)	Reaction	Neutrino Detection Threshold (MeV)	Ratio		
			Avignone Spectrum	Davis Spectrum	Measured $\bar{\nu}_e$ Spectrum (preliminary)
11.2	ncd	2.2	.83 ± .13	1.10 ± .16	1.3 ± .22
11.2	ccd	4.0	.32 ± .14	.44 ± .19	.61 ± .29
11.2	ccp	4.0	.68 ± .12	.88 ± .15	≡ 1.0
11.2	ccp	6.0	.42 ± .09	.58 ± .12	≡ 1.0
6	ccp	1.8	.65 ± .09	.84 ± .12	-
6	ccp	4.0	.81 ± .11	1.02 ± .15	1.19 ± .27
8.7	ccp	3.0	.68 ± .15	.87 ± .14	-

b. Allowed regions Δ vs. $\sin^2 2\theta$ can be drawn for each of the ratios listed in Table IV. For the Avignone spectrum there is an overlapping region consistent with all the experiments at 11.2 meters but not with the > 4 MeV data at 6 meters. We note that small changes in the normalization of the 6 meter data could give agreement. This yields

$$0.5 \leq \sin^2 2\theta \leq 0.8 \quad (32° > \theta > 22°)$$

and $\quad 0.7 \leq \Delta(eV^2) \leq 1.0$

We find that the Davis spectrum yields no overlapping region at the level of one standard deviation. On the other hand it appears to predict more precisely the observed neutral current branch of the deuteron experiment.

c. If oscillations occur with these approximate parameters, then the observed spectrum at 11.2 meters should show evidence of spectral changes. In Fig. 14 we plot the ratio of the observed 11.2 meter data to the Avignone prediction as a function of neutrino energy. If oscillations do not exist and further if the Avignone spectrum is the correct one, then the ratio should be 1.0 independent of energy. For comparison, the predicted ratio for $\Delta = 1$ eV^2 and $\sin^2 2\theta = 1$ is plotted. The same comparison is made, this time to the Davis spectrum in Fig. 15.

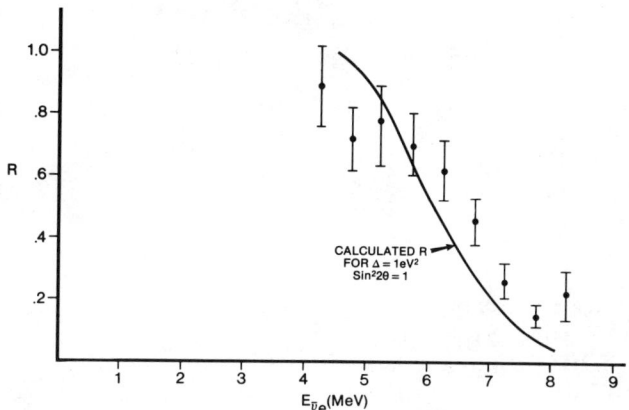

Fig. 14. The ratio of 11.2 meter ccp measured neutrino spectrum to that predicted by Avignone.

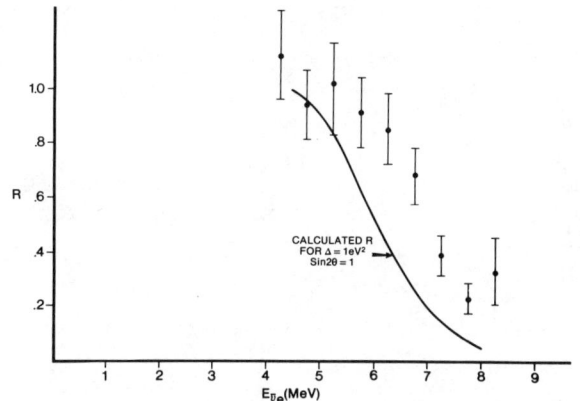

Fig. 15. The ratio of 11.2 meter ccp measured neutrino spectrum to that predicted by Davis.

Conclusions

The results of a reactor experiment comparing the observed rates of the charged current and neutral current interactions of reactor neutrinos with deuterons gives an indication of neutrino instability at the 2 to 3 standard deviation level.

References

1. S. Willis et al., Phys. Rev. Lett. 44, 522 (1980).
2. H.H. Chen, F. Reines et al., "A Study of Neutrino Electron

Elastic Scattering at LAMPF", LAMPF Experiment No. 225.
3. S. Blankenship - UCI internal report, UCI-10P19-102 (1976).
4. S. Blankenship - UCI internal report, UCI-10P19-104 (1976).
5. R. Davis, Jr., and D.S. Harmer, Bull. Amer. Phys. Soc. (2) $\underline{4}$, 219 (1959).
6. A. Soni and D. Silverman, private communication.
7. F. Nezrick and F. Reines, Phys. Rev. $\underline{142}$, 852 (1966).
8. F. Reines, H.S. Gurr, and H.W. Sobel, Phys. REv. Lett. $\underline{37}$, 315 (1976).
9. F. Boehm, private communication.
10. UCI internal reports - UCI-10P19-141 (1979), UCI-10P19-132 (1978), UCI-10P19-126 (1977).
11. S. Blankenship, Georgia Institute of Technology, private communication.
12. E. Pasierb, H.S. Gurr, J. Lathrop, F. Reines and H.W. Sobel, Phys. Rev. Lett. $\underline{43}$, 96 (1979).
13. R.M. Ahrens and L. Gallaher, Phys. Rev. $\underline{D20}$, 2714 (1979) and private communication (1980). S.L. Glashow, J. Iliopoulos and L. Maiani, Phys. Rev. $\underline{D2}$, 1285 (1970).
14. J.H. Munsee and F. Reines, Phys. Rev. $\underline{177}$, 2002 (1969).

DISCUSSION (Chairman: L. Wolfenstein)

J. LO SECCO, Cambridge: What is the effect of neutrino interactions on the non deuteron material inside the anticoincidence shield, including effects that may produce a pulse in the He_3 counter, but perhaps make no neutrons?

H. SOBEL, Irvine: Background from neutrino interactions on all other materials is less than one event per day.

R.G.H. ROBERTSON, Argonne: What is the sensitivity of your ^3He detectors to direct detection of antineutrinos?

H. SOBEL, Irvine: Small.

A.K. MANN, Philadelphia: Why not use deuterated scintillator?

F. REINES, Irvine: We looked at $\nu_e + d \rightarrow n + n + e^+$ in $\sim$ 1967 (Jenkins et al.). Now as then it is difficult primarily because it requires pure (i.e. no proton) deuterated scintillators.

M. CHEN, Cambridge: I have a trivial statistical question concerning how to combine your data set # 1 and # 2. I would use the % err. square as the statistical weight of each data point:

$$\text{mean} = \frac{\Sigma x_i/\sigma_i^2}{\Sigma \frac{1}{\sigma_i^2}} \quad \text{where } \sigma_i = \frac{\text{error}_i}{x_i}$$

and I got

$$<R_{exp}> = \frac{\frac{.206}{(\frac{.12}{.206})^2} + \frac{.114}{(\frac{.144}{.114})^2}}{(\frac{.206}{.12})^2 + (\frac{.114}{.144})^2} = .19 \pm .1$$

and $R = \frac{<R_{exp}>}{<R_{th}>} = 0.43 \pm .23$ instead of your number $R = .38 \pm .21$

If you do not use (percentage error)^2 as weight, then you should put the err. on the expected value i.e. .44 with the same percentage error and you will get the same result as I got above. The deviation is less than 2.5 σ from being 1.

T. BOWLES, Los Alamos: In computing the two-dimensional region of A versus θ is the size of the reactor core taken into account and how large is the effect?

H. SOBEL, Irvine: When we integrate over the spectrum with a threshold of 4 MeV for the reaction, the effect is small. In looking at oscillations derived from the spectra of $\bar{\nu}_e + p \rightarrow n + e^+$, the effect is about 15% and is taken into account.

M. ROOS, Helsinki: The errors you quoted on R were symmetric. However, since negative values of R are unphysical, I would expect rather asymmetric errors. Thus the number of standard deviations that you are away from unity may still be quite different in your final analysis.
In replay to a question about what the values are for the Irvine $\bar{\nu}_e e^- \rightarrow \bar{\nu}_e e^-$ cross-section using the Davis-Vogel spectrum, I can answer it. Davis and Vogel have communicated the parameters of the spectrum to me so that Liede and I (Nucl. Phys. B 1980) were able to use them in our global fit of $\sin^2\theta_W$. It turns out that the Irvine low energy cross-section is 1σ higher than the best fit value and the high energy cross-section is 2σ higher. The best fit including the Irvine data and the Davis-Vogel spectrum leads to $\sin^2\theta_W = 0.238$ whereas the fit without these data yields $\sin^2\theta_W = 0.233$. Since the Irvine cross-sections do not include the uncertainty in the spectrum I would rather recommend to use 0.233.

S. FREEDMAN, Stanford: What are the theoretical estimate to the errors in the ν-spectra? (Will be answered by Boehm).

LEPTON CONSERVATION

Felix Boehm

California Institute of Technology
Pasadena, Calif. 91125, USA

INTRODUCTION

A discussion of lepton conservation is conveniently based on the definition of the lepton numbers, N_e, N_μ and N_τ given in the following table

Leptons	e^-, ν_e	$e^+, \bar{\nu}_e$	μ^-, ν_μ	$\mu^+, \bar{\nu}_\mu$	τ^-, ν_τ	$\tau^+, \bar{\nu}_\tau$
N_e	+1	-1				
N_μ			+1	-1		
N_τ					+1	-1

In studying a weak interacting process we seek answers to some of the following questions:

1) How well are the lepton numbers N_e, N_μ, and N_τ individually conserved?

2) How well is the sum $N_e+N_\mu+N_\tau$ conserved?

3) Is there evidence for non-conservation of N_e, or N_μ, or N_τ while conserving the sum, $N_e+N_\mu+N_\tau$ as well as the product $N_e \cdot N_\mu \cdot N_\tau$ (multiplicative law)?

We wish to review here recent experimental progress relevant for these questions. The discussion will be divided in the following manner: a) results from (semileptonic) nuclear beta decay, where the leptonic current is presumably purely electronic, b) results from (purely leptonic) muon decay, c) results from neutrino induced reactions and their significance for possible neutrino oscillations.

BETA DECAY

Lepton non-conservation in the electron-leptonic current can be introduced with the help of three mixing parameters, η, ξ, δ, accounting for lepton number and helicity non-conservation. In the usual notation[1] the lepton current is given by

$$\ell_i^* \sim \bar{e}0_i \{[(1+\gamma_5)+\eta(1-\gamma_5)]\nu_e + [\xi(1+\gamma_5)+\delta(1-\gamma_5)]\bar{\nu}_e\},$$

where e and ν_e are the electron and neutrino wave functions, 0_i are the Dirac matrices, and $i=V,S,A,T,P$ stand for vector, etc. interactions. The first bracket corresponds to $N_e=+1$, and the second to $N_e=-1$. The role of the parameters η, ξ, and δ becomes more evident if the current is rewritten with helicity indices, L and R:

$$\ell_i^* \sim \bar{e}0_i [\nu_{eL} + \eta\nu_{eR} + \xi\bar{\nu}_{eL} + \delta\bar{\nu}_{eR}],$$

where $\nu_{eL} = (1+\gamma_5)\nu_e$, etc. Introducing the beta-decay coupling-constants C_i the first two terms can also be written as

$$\nu_{eL} + \eta\nu_{eR} \sim C_i + C_i'\gamma_5, \text{ with } C_i = 1+\eta \text{ and } C_i' = 1-\eta.$$

From a study of asymmetry parameters in parity experiments in beta-decay[1] the limits on η and δ are given by

$$|\eta| \leq 0.1, \quad |\delta| \leq 0.1.$$

Inverse beta decay, using wrong handed neutrinos ($\bar{\nu}_{eR}$ instead of ν_{eL}) in the reaction $\bar{\nu}_{eR} Z \to e^-(Z+1)$ furnishes[1] $\delta \leq 0.2$.

Double beta decay of the neutrino-less type (Fig.1a) requires violation of both, lepton number and handedness, and thus samples the parameter δ. The two-neutrino beta-decay (which is lepton number allowed) (Fig.1b) has been observed based on indirect (geochemical) evidence in the reactions $^{82}Se \to ^{82}Kr$, $^{128}Te \to ^{128}Xe$, and $^{130}Te \to ^{130}Xe$, as reviewed in ref.2.

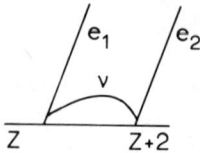

Fig.1a: Neutrinoless (o-ν) double beta decay

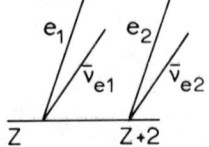

Fig.1b: Two neutrino (2-ν) double beta decay

LEPTON CONSERVATION

Up to 1979 the best limit of δ, $\delta < 3 \times 10^{-4}$, came from Cleveland et al[3] who combine their observed half-life limit for ^{82}Se, $T_{1/2}(0-\nu) \geq 3.1 \times 10^{21}$y, with the geochemical value[2] of $T_{1/2}(2-\nu) = 2.8 \pm 0.9 \times 10^{20}$y, taking into account the relative two-fermion and four-fermion phase space. In a recent report, Moe and Lowenthal[4], using a cloud chamber, present direct evidence for 2-ν double beta-decay in ^{82}Se. Based on the observation of 20 e$^-$e$^-$-pairs they find a shorter half-life of $T_{1/2} = (1.0 \pm 0.4) \times 10^{19}$y. This new half-life combined with the limit of $T_{1/2}(0-\nu)$ of ref.3 leads to a branching ratio $(0-\nu)/(2-\nu) < 0.003$, and thus furnishes a limit for the mixing parameter δ

$$\delta < 6 \times 10^{-5} \quad (68\% \text{ confidence level}).$$

Inasmuch as this limit depends strongly on the 2-ν decay rate, calculations of the 2-ν matrix-elements, as those by Haxton and Stephenson[5], are of great importance.

To conclude this chapter we wish to point out that available data from nuclear beta-decay experiments leave sizable margins for possible right-helicity couplings. These margins are particularly large for Fermi couplings (coupling constants C_V, C_V', C_S, C_S'), for which the free fits of ref.6 furnish

$$C_V'/C_V = 0.8^{+0.4}_{-0.1}, \quad \text{or} \quad |\eta_V| < 0.3,$$

$$C_S/C_V < 1.2 \quad , \quad C_S'/C_V < 1.0.$$

There are few restrictions on ImC_S and ImC_S' from published experiments. A fresh look into this neglected field should be valuable.

MUON DECAY

Stringent limits for the violation of the lepton numbers N_e and N_μ were set by two important experiments reported in Neutrino-79. The Los Alamos experiment[7] gives

$$(\mu^+ \to e^+ \gamma)/(\mu^+ \to e^+ \nu \nu) < 1.9 \times 10^{-10} \quad (90\%).$$

A recently proposed experiment[8] may improve the sensitivity of this test by two orders of magnitude. The SIN experiment[9] gives a limit for the μ-e conversion

$$(\mu^- Z \to e^- Z)/(\mu^- \text{capt.}) < 7 \times 10^{-11} \quad (90\%),$$

and

$$(\mu^-, Z, N \to e^+, Z-2, N+2)/(\mu^- \text{capt.}) < 9 \times 10^{-10} \quad (90\%).$$

The nucleus Z was ^{32}S.

Recently, Abela et al.[10] have studied the above reaction on ^{127}I. By searching for radioactive decay of ^{127}Sb this experiment yields the following limit,

$$(\mu^-, {}^{127}I \to e^+, {}^{127}Sb)/(\mu^- capt.) < 3 \times 10^{-10} \quad (90\%).$$

Improvements of the sensitivity of this experiment by two orders of magnitude appear possible[11].

Following the preliminary results reported in Neutrino-79, the Los Alamos muon-number conservation experiment by Willis et al.[12] has obtained a final result for the limit of the multiplicative conservation law. In order for the reaction

$$\mu^+ \to e^+ \bar{\nu}_e \nu_\mu$$

to occur, both N_e and N_μ must be violated, while the product, $N_e N_\mu$ remains conserved. The detector reaction for $\bar{\nu}_e$ is $\bar{\nu}_e p \to n e^+$. The lepton-number allowed reaction

$$\mu^+ \to e^+ \nu_e \bar{\nu}_\mu$$

was observed by means of the detector reaction $\nu_e d \to p p e^-$, and the following limit was obtained,

$$(\mu^+ \to e^+ \bar{\nu}_e \nu_\mu)/(\mu^+ \to e^+ \nu_e \bar{\nu}_\mu) < 0.065 \quad (90\%).$$

A similar test of N_e and N_μ in a reaction where $N_e \cdot N_\mu$ is conserved was submitted to this Conference by a Amsterdam-CERN-Hamburg-Moscow-Rome collaboration[13]. Using the CERN SPS the inverse muon decay

$$\nu_\mu e^- \to \mu^- \nu_e = 0.98 \pm 0.18 \text{ times V-A theory}$$

was observed to occur with V-A strength, while the forbidden reaction

$$\bar{\nu}_\mu e^- \to \mu^- \bar{\nu}_e$$

could not be seen. A limit of

$$(\bar{\nu}_\mu e^- \to \mu^- \bar{\nu}_e)/(\nu_\mu e^- \to \mu^- \nu_e) < 0.09 \quad (90\%)$$

is presented in this contribution.

NEUTRINO INDUCED REACTIONS AND NEUTRINO OSCILLATIONS

The oscillations[14] of weak interaction eigenstates, ν_e, ν_μ, ν_τ ("physical neutrinos") into each other or into some other states clearly constitute a violation of lepton number conserva-

LEPTON CONSERVATION

tion. For these oscillations to occur it is neccessary to assume that the neutrinos ν_e, ν_μ, ν_τ are superpositions of pure states ν_1, ν_2, ν_3, (eigenvalue of mass matrix) and that ν_1, ν_2, ν_3 have finite rest masses. This superposition can be written as

$$\begin{pmatrix} \nu_e \\ \nu_\mu \\ \nu_\tau \end{pmatrix} = \begin{pmatrix} U \end{pmatrix} \begin{pmatrix} \nu_1 \\ \nu_2 \\ \nu_3 \end{pmatrix}, \qquad (1)$$

where U is a unitary matrix characterized by four free parameters (three Euler angles, one phase). As for any mixed quantum mechanical system it follows that ν_e, ν_μ, ν_τ are not stationary states, thus have no definite mass.

In gauge model descriptions[15] the pure states ν_1, ν_2, ν_3 are given by either Dirac (four component) neutrinos or by Majorana (two component) neutrinos. Either description leads to

Flavor Oscillations: $\nu_e \leftrightarrow \nu_\mu$, $\nu_\mu \leftrightarrow \nu_\tau$, $\nu_\tau \leftrightarrow \nu_e$.

The Majorana description also leads to

Particle-Antiparticle Oscillations: $\nu_e \leftrightarrow \bar{\nu}_{eL}$, etc.

where $\bar{\nu}_{eL}$ is a wrong handed neutrino, presumably unable to undergo weak interaction.

It is convenient to discuss the neutrino oscillations in terms of two neutrino states. If only ν_e and ν_μ are invoked eq.(1) becomes

$$\nu_e = \nu_1 \cos\theta + \nu_2 \sin\theta$$
$$\nu_\mu = -\nu_1 \sin\theta + \nu_2 \cos\theta \qquad \text{mixing parameter } \theta. \quad (2)$$

In the analysis of oscillation experiments this two-neutrino description may prevail also for the general case of three neutrinos, as it is unlikely that the two pairs of mass differences (see below) or the mixing angles are equal or nearly equal.

The dynamics of a two state system is easily obtained since pure states $|\nu_\ell\rangle$ ($\ell=1,2$) evolve in time as $|\nu_\ell(t)\rangle \sim \exp(-iE_\ell t)$, with $E_\ell = (p^2 + m_\ell^2)^{1/2}$, it follows from eq.(2) that the composition of a neutrino beam is given by

$$|\nu_e(t)|^2 = 1 - \frac{\sin^2\theta}{2} \left[1 - \cos(E_2 - E_1)t\right].$$

With $E_2 - E_1 \simeq (m_1^2 - m_2^2)/2p$, the oscillation length Λ (in meters) is related to the neutrino kinetic energy E_ν (in MeV) and to the mass-square difference $|m_1^2 - m_2^2| \equiv \Delta^2$ (in (eV)^2) by

$$\Lambda(m) = 2.5 E_\nu (\text{MeV})/\Delta^2 (\text{eV})^2. \tag{3}$$

The parameters describing the oscillations thus are Δ^2 (∼frequency) and $\sin^2 2\theta$ (∼amplitude).

At a given time t, the counting rate in a detector at a distance d from the source where neutrinos are created is given (in the same units) by

$$Y(E_\nu, \Delta^2 d) = N(E_\nu)\sigma(E_\nu)\left[1 - \frac{\sin^2 2\theta}{2}\left(1 - \cos\frac{2.5\Delta^2 d}{E_\nu}\right)\right], \tag{4}$$

where $N(E_\nu)$ is the neutrino spectrum from the source and $\sigma(E_\nu)$ is the cross section of the detector reaction.

Evidence for oscillations can be obtained by searching for the "new" state assuming that the process proceeds in a specific channel, or by studying the disappearance of the original state. In the latter experiment all channels contribute.

Below we present the results from several recent experiments. We begin with the following specific channel:

$\overset{(-)}{\nu}_\mu \to \overset{(-)}{\nu}_e$ oscillations

Evidence has been sought for the occurence of $\overset{(-)}{\nu}_e$ induced reactions in an originally pure $\overset{(-)}{\nu}_\mu$ beam. The quoted values of Δ^2 are for full mixing and 90% confidence level.

Gargamelle-CERN (78)[16], ν_e and $\bar\nu_e$, $\Delta^2 < 1 (\text{eV})^2$
Los Alamos (80)[17], ν_e, $\Delta^2 < 0.7 (\text{eV})^2$
Aachen-Padova-CERN (80)[18], ν_e, $\Delta^2 < 0.9 (\text{eV})^2$
Serpukhov (80)[19], ν_e, $\Delta^2 = 1.4 \pm 0.4 (\text{eV})^2$ (68%)
Brookhaven (80)[20], ν_e, expected sensitivity: $\Delta^2 < 0.3 (\text{eV})^2$

The results of refs. 16 and 17 are depicted in Fig.2, showing log Δ^2 versus $\sin^2 2\theta$. The Serpukhov experiment[19] gives a non-zero value for Δ^2. If their data is interpreted with a smaller mixing angle, the resulting Δ^2 is correspondingly larger. (For example, the pair $\sin^2 2\theta = 0.2$ and $\Delta^2 = 3.5 \pm 1.0 (\text{eV})^2$ represents a solution). An excess by two standard deviations of the number of electron-neutrinos was observed when compared with the number calculated using absolute cross-sections for 2-5 GeV neutrinos. Clearly, the oscillation parameters implied by this experiment are in conflict with those of refs. 16,17,18 (see Fig.2).

$\nu_e \to$ anything

Several experiments performed at the CERN beam dump have focused attention to the question: is there a reduction of the ν_e/ν_μ ratio? These results are discussed in full in the lecture by Dydak at this conference. Below we briefly summarize the implications for possible oscillations.

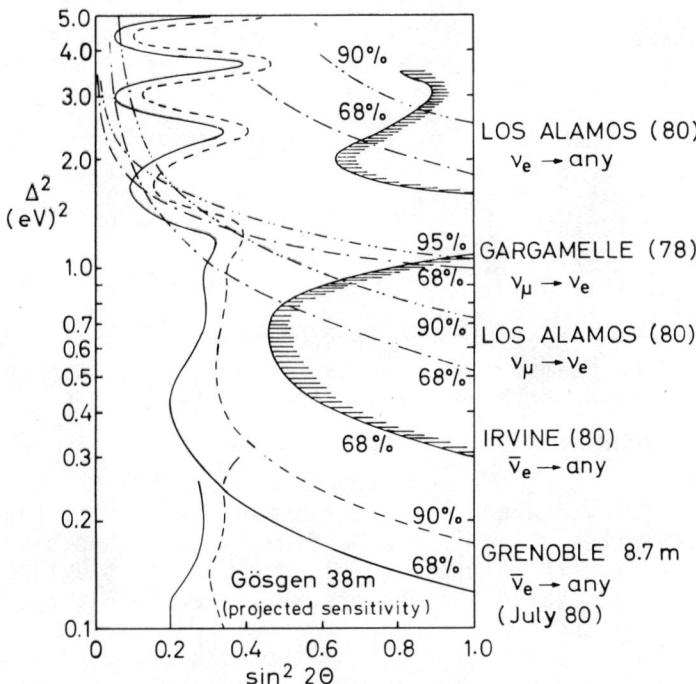

Fig.2: The values of the oscillation parameters $\Delta^2 = m_1^2 - m_2^2$ vs. $\sin^2 2\theta$ are plotted for several experiments. (For references see text.) In each case the regions to the right of the curve can be excluded, the curve representing the confidence limit indicated, except for the curve marked Irvine (80), where the region to the right of the curve is allowed and that to the left is excluded.

CDHS-CERN[21]. The ν_e/ν_μ ratio is depleted by a factor of 1/2, independent of energy, implying $\Delta^2 \sim 100(eV)^2$ and large mixing.

CHARM-CERN[22]. The ν_e/ν_μ ratio is lower than expected and the data is not inconsistent with oscillations with large mixing angle.

BEBC-CERN[23]. Using a narrow band 200 GeV neutrino beam no depletion is found, consistent with no oscillations.

Brookhaven Proposal[24]. A proposed experiment aims at studying attenuation of the number of ν_e's. The sensitivity is $\Delta^2=1-400(eV)^2$ and $\theta=15-45^\circ$.

At present the CERN beam dump results appear to be somewhat in a state of flux. In a recent review, DeRujula[15] summarizes the results as follows: "The indications are that the relevant mixing angles are comfortably large, $\sim 20^\circ$, and that the relevant mass differences Δ^2 may be in the range of hundreds of $(eV)^2$".

$\bar{\nu}_e \to$ anything

Reactor neutrinos have been successfully engaged to search for a depletion of $\bar{\nu}_e$ beams. Advantage is taken of the distance and energy dependence of the depletion (see eq.(4)).

The Irvine experiment

As described in the Conference report by H. Sobel the experiments[25] investigating the charged current reaction $\bar{\nu}_e d \to e^+ nn$ give evidence for an attenuation of the $\bar{\nu}_e$ flux by more than a factor of two. The results are interpreted as evidence for neutrino oscillations with oscillation parameters indicated in Fig.2 (The region to the right of the curve marked "Irvine"(80)" is the allowed region.) Further, a reanalysis[26] of the data from the reaction $\nu_e p \to e^+ n$ at 11.2 m from the reactor core shows a reduction of the neutrino rate at the upper region of the positron spectra and thus reenforces the findings of the deuteron experiment. The oscillation parameters consistent with both experiments are

$$\text{Irvine, } \sin^2 2\theta \sim 0.6; \ \Delta^2 \sim 1(eV)^2 .$$

The Grenoble Experiment

The results reported[27] to this Conference by the Caltech-Grenoble-Munich collaboration do not support the Irvine findings. Below we shall review these experiments. A detailed account will appear in publication[28].

The Grenoble experiment makes use of the research reactor of the Institut Laue Langevin (ILL). The reactor's core consists of a single fuel element, highly enriched in ^{235}U. The reactor power is 57 MW. With regard to studying oscillations its chief advantage is the small ("pointlike") core dimension and the good definition of the fission product distribution (^{235}U fission only).

The neutrino detector is set up at 8.7 m from the core. The detector reaction is $\bar{\nu}_e p \to e^+ n$. The detector, desribed in Neutrino-79[29] consists of 30 proton rich liquid scintillator target cells (total volume 375 liters), serving also as neutron moderators, interlaced with four ^{3}He wire chambers serving as neutron detectors. The system is surrounded with an active liquid scintillator veto. A valid event is a positron signal followed by a neutron signal in the ^{3}He detector. Emphasis was put on efficient light collection (good positron energy resolution) and low background in the ^{3}He counters. Proton recoil pulses, originating from secondary fast neutrons in the liquid scintillator are suppressed by pulse shape discrimination. With a neutrino source strength of 10^{19}/s the $\bar{\nu}_e$ flux in the detector is 1.2×10^{12}/cm^{2}s. The detector efficiency is measured to be $(21.5 \pm 0.5)\%$ and the neutrino event rate is about 1.5/h, with a signal-to-noise of better than 1:1. Reactor associated background (change in the singles rate in both, the target and the ^{3}He counters) is absent.

Fig.3 shows the observed correlated e^+ spectrum for reactor-on and reactor-off, as available at the time of the Conference. The positron spectrum ($E_{e^+} = E_\nu - 1.8$ MeV), corrected for correlated and accidental background is shown in Fig.4. This spectrum must now be compared to the calculated spectrum from ^{235}U fission.

Calculations of neutrino spectra from fission products are due to Avignone[30] and to Davis and Vogel et al.[31]. For ^{235}U fission the Avignone spectrum is about 30% higher than the Davis-Vogel spectrum (Fig.4), the difference being principally attributed to the different nuclear model assumptions used to calculate the unknown short lived beta decays. Two recent developments tend to favour Davis-Vogel: (1) a recent on-line electron spectrum measured[32] at the ILL reactor, and (2) a similar measurement at the Oak Ridge reactor[33]. Both experiments agree with the Davis-Vogel spectrum to within 10%. For the comparison of the Grenoble data it will therefore be assumed that the Davis-Vogel spectrum is valid. The experiment gives an integrated yield

$$\int Y(exp) / \int Y(D.-V.) = 0.89 \begin{array}{l} \pm\ 0.04\ \text{(statistical error)} \\ \pm\ 0.14\ \text{(systematic and theoretical errors).} \end{array}$$

Fig.5 shows the experimental spectrum divided by the Davis-Vogel spectrum (for which an uncertainty of 10% has been taken). Since the agreement is good we conclude that there is no evidence for oscillations. A chi square test with the Davis-Vogel spectrum using a wide range of oscillation parameters yields the curves contained in Fig.2. (To the left of the curves for 68% and 90% confidence limits lies the allowed region.) If $\sin^2 2\theta$ is sufficiently small, various solutions are possible. For example, the oscil-

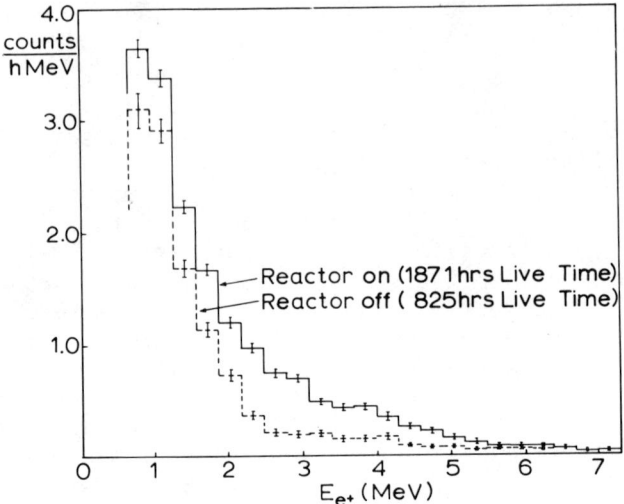

Fig.3: Results of the Caltech-Grenoble-Munich experiment. The positron spectra ($E_{e^+}=E_\nu-1.8$ MeV) are shown for "reactor on" and "reactor off". The energy bins are 0.302 MeV.

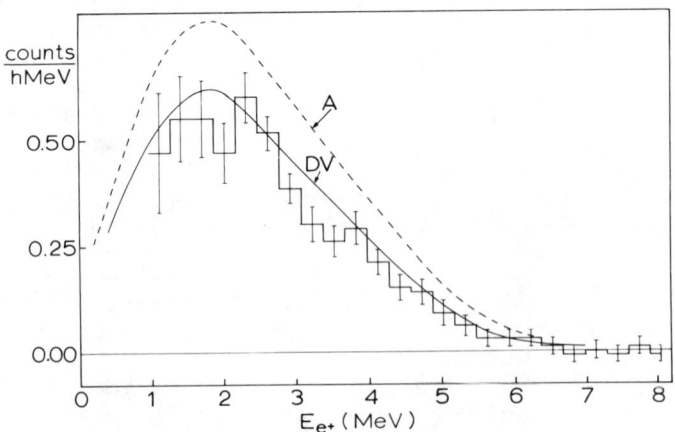

Fig.4: Positron spectrum (difference of the spectra in Fig.3) associated with reactor neutrinos. The error bars are statistical errors. The calculated spectra A (ref.30) and DV (ref.31) corrected for detector efficiency and resolution are also shown.

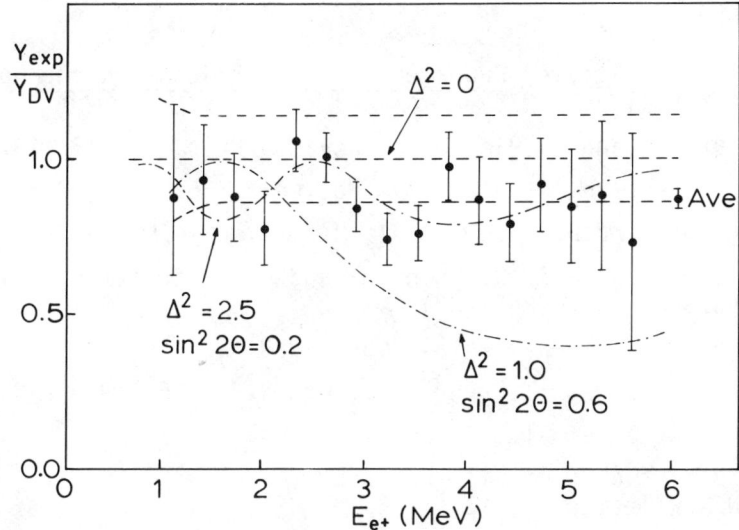

Fig.5: The experimental positron spectrum divided by the calculated spectrum for no oscillations is shown. The horizontal band represents the uncertainties of the calculations as well as those attributed to systematic errors in the experimental data. The curved lines give the calculated spectra for two sets of oscillation parameters. The set $\Delta^2=1(eV)^2$, $\sin^2 2\theta=0.6$ corresponds to the Irvine findings.

latory curve in Fig.5 shows the case of $\sin^2 2\theta=0.2$, $\Delta^2=2.5(eV)^2$ which obviously is an allowed solution. The Irvine solution ($\sin^2 2\theta=0.6$, $\Delta^2=1.0$) is also shown for comparison.

Possible uncertainties in the spectral distribution of the reactor neutrinos are best resolved by measuring the spectrum at several positions. Measurements at 38 m and 65 m are now planned by a Caltech-Munich-SIN group[34] at the Gösgen power reactor, allowing a sensitivity of $<0.03(eV)^2$ for full mixing.

In conclusion, there exist disagreements between the results of the Irvine and the Grenoble reactor experiments and also between CERN beam dump experiments and Serpukhov. Further work is needed to settle the issue[35]. Finally, it should be emphasized[14,36] that oscillation studies are the most sensitive tool for lepton mixing. If it is confirmed that $\Delta^2 < 1(eV)^2$, then one can estimate that the $\mu \to e\gamma$ process will have a rate of
$(\mu \to e\gamma)/(\mu \to e\nu\nu) \sim (3\alpha/32\pi)(\Delta^2/M_W^2) < 10^{-47}$ and the double beta decay rate is $(o-\nu)/all < 10^{-32}$.

Stimulating discussions with my colleagues of the Caltech-Grenoble-Munich team, and with R.P. Feynman, H. Fritzsch, P. Ramond and P. Vogel are gratefully acknowledged.

REFERENCES

1. R.J. Blin-Stoyle, *Fundamental Interactions and the Nucleus*, North Holland (1973)
2. D. Bryman and C. Picciotto, Rev.mod.Phys. $\underline{50}$, 15 (1978)
3. B.T. Cleveland et al., Phys.Rev.Lett. $\underline{35}$, 757 (1975)
4. M.K. Moe and D.D. Lowenthal, Report, UCI IOP19-143 (1979)
5. W.C. Haxton, G.J. Stephenson, and D.R. Strottman, Bull.Am. Phys.Soc. $\underline{25}$, 502 (1980)
6. H. Paul, Nucl.Phys. $\underline{A154}$, 160 (1970)
7. J.D. Bowman et al., Phys.Rev.Lett. $\underline{42}$, 556 (1979)
8. J.D. Bowman et al., Proposal LAMPF 444 (1978)
9. A. Badertscher et al., Phys.Lett. $\underline{79B}$, 371 (1978); SIN Newsletter $\underline{12}$, 13 (1979)
10. R. Abela et al., Phys.Lett.B, to be published
11. L. Simons, private communication, SIN (1980)
12. S.E. Willis et al., Phys.Rev.Lett. $\underline{44}$, 522 (1980)
13. M. Jonker et al., Preprint CERN-EP/80-30
14. S.M. Bilenky and B. Pontecorvo, Phys.Reports $\underline{41}$, 226 (1978), and earlier references quoted therein
15. For review and recent references see A. DeRujula et al. Report, CERN TH2788 (1980), A. Zee, Phys.Lett $\underline{93B}$, 389 (1980); H. Fritzsch, in *Fundamental Physics with Neutrons and Neutrinos*, p.117, Ed. T. von Egidy, Inst. Physics, Bristol and London (1978)
16. J. Blietschau et al., Nucl.Phys. $\underline{B133}$, 205 (1978)
17. P. Nemethy et al., Contribution to this Conference, see also ref. 12
18. H. Faissner et al., Contribution to this Conference, J.K. Bienlein et al., Phys.Lett. $\underline{13}$, 80 (1964)
19. A. Zaitzev, S. Zeldovich et al., ITEP-IHEP, Contribution to this Conference
20. Harvard-Michigan Collaboration, H. Sulak, private communication
21. CERN-Dortmund-Heidelberg-Saclay collaboration quoted in ref.23, see also Lecture by F. Dydak at this Conference
22. M. Jonker et al., Contribution to this Conference
23. Aachen-Bonn-CERN-Demokritos-London-Oxford-Saclay Collaboration, CERN/EP/NPC-N 80 Draft

24. B. Cortez et al., Contribution to this Conference
25. F. Reines, H.W. Sobel, and E. Pasierb, Report UCI-IOP19-144 (revised June 80), Phys.Rev.Lett., to be published
26. H.W. Sobel et al., Contribution to this Conference
27. F. Boehm et al., Contribution to this Conference
28. Caltech-ILL, ISN-Grenoble-TU Munich Collaboration, F. Boehm, J.F. Cavaignac, F. v. Feilitzsch, A.A. Hahn, H.E. Henrikson, D.H. Koang, H. Kwon, R.L. Mößbauer, B. Vignon, and J.L. Vuilleumier, to be published
29. F. Boehm et al., Neutrino-79, University of Bergen, Vol. II, p. 185 (1980)
30. F.T. Avignone and Z.D. Greenwood, Univ. Sout Carolina, Preprint (1980)
31. B.R. Davis et al., Phys.Rev. $C19$, 2259 (1979)
32. K. Schreckenbach et al., Phys.Lett., to be published
33. J.K. Dickens, private communication, Oak Ridge Natl. Laboratory (1980)
34. F. Boehm et al., Letter-of-Intent to SIN (1980)
35. An analysis based on earlier data has been submitted to this conference by V. Barger et al., COO-88-135 (1980)
36. H. Primakoff, Neutrino-79, University of Bergen, Vol. II, p. 1002 (1980)

DISCUSSION (Chairman: M. Baldo Ceolin)

J. LO SECCO, Michigan: Tell us more about the Serpukhov result.

A. ROZANOV, Moscow: The result of the Serpukhov experiment is not statistically significant to establish the existence of neutrino $\nu_\mu \to \nu_e$ oscillations, because the effect is 51 ± 24 events, i.e. only 2 standard deviations from zero. This means that there is no real contradiction with the Gargamelle and the Los Alamos experiments. In the Serpukhov paper it was clearly pointed, that the authors do not insist on $\nu_\mu \to \nu_e$ oscillation interpretation of their experiment.

K. WINTER, CERN: You conclude that there is disagreement between the CERN Beam Dump result and the Narrow Band Beam. As these are completely different sources of ν_e and ν_μ I do not see the contradiction. Could you please comment on that?

W. VENUS, Didcot: The BEBC-TST collaboration also has a new preliminary result for the ratio of observed to expected ν_e events in a Wide Band Beam. The ratio (observed events)/(expected events), assuming no oscillations etc. is 1.20 ± 0.24. The error is due mainly to the fact that the Wide Band Beam is, of course, less well understood than the Narrow Band Beam. The result is from a reanalysis of old data (1977 WB test run).

M. LUSIGNOLI, Roma: I would like to point out that comparing limits on $\nu_\mu \to \nu_e$ and $\nu_e \to$ any ν oscillations may be misleading in a three neutrino world. In fact, there are papers in the literature stressing the possibility that one may have large $\nu_e \leftrightarrow \nu_\tau$ oscillations with almost no $\nu_\mu \leftrightarrow \nu_e$ oscillations.

F. BOEHM, Pasadena: It is true, but in a situation in which you have no effect it is better to use the simplest scheme of analysis.

A.K. MANN, Philadelphia: Please comment on the direct measurement in the reactor of the e^- spectrum from the decay product of fission products. Will this give a very good measurement of the neutrino spectrum?

M. SHAPIRO, Washington: If I understood correctly, you said the Irvine report of this morning was based upon reanalysis of older data. I thought the new result was based in large part on new data (+ improved statistics).

F. DYDAK, Heidelberg: You quoted a paper from the CERN-Dortmund--Heidelberg-Saclay collaboration submitted to this Conference, claiming the observation of neutrino oscillations with large mixing and $\Delta^2 \sim 100$ eV^2. This is plane wrong. There exists neither a claim for such an observation nor has a paper been submitted to this Conference.

LEPTON CONSERVATION

D. PERKINS, Oxford: The CERN NB beam result (BEBC Ne) refers to the observed expected number of ν_e events in a very well-understood beam: 70 e^- events expected, 72 observed: no evidence for oscillation. The ratio of e/μ events in the CERN Beam Dump experiment is about 0.5; but one cannot conclude anything about oscillations from this. The source of prompt neutrinos is unknown; they come out of a beam dump, and if prompt ν_e and ν_μ are made in equal numbers, one might infer a discrepancy (this would be the case if they were from charmed hadron decay).

GRAND UNIFICATION OF QUARKS AND LEPTONS FRON ONE OF PREONS

Jogesh C. Pati

International Centre for Theoretical Physics, Trieste, Italy and
Department of Physics, University of Maryland
College Park, Maryland 20732, USA

ABSTRACT

Violation of baryon, lepton and, in general, fermion number is central to the hypothesis of quark lepton unification in a gauge context. Three of its characteristic signatures are proton decay, n-n̄ oscillation and neutrinoless double β decay. In 1974 and 1975 it was shown that within maximal gauging the proton may decay via four alternative modes (i.e. proton → one or three leptons or antileptons) satisfying $\Delta F = -2, 0, -4$ and -6 some of which may coexist; the deuteron may decay into pions and neutrinoless double β decay may occur in the context of spontaneous gauge symmetry breaking. It is now observed that n-n̄ oscillations (which are related to deuteron decays into pions) can coexist with proton decay of especially $\Delta F = -4$ variety ($p \to e^+\pi^0$) and both these processes may possess measurable strength so as to be amenable to forthcoming searches.

INTRODUCTION

The hypothesis of grand unification[1-3] stands at present primarily on its aesthetic merits. It gives the flavour of synthesis in that it provides a rationale for the existence of quarks *and* leptons by assigning the two sets of particles to one multiplet of a gauge symmetry G. It derives their forces - weak, electromagnetic as well as strong - through one principle - gauge unification. It provides a reason for the quantization of electric charges.

With quarks and leptons in one multiplet F of a local spontaneously broken gauge symmetry G, baryon and lepton number conservations cannot be absolute. To see this simply, take for illustration a multiplet F of three quarks of red, yellow and blue colours plus one lepton (i.e. $F = \{q_r, q_y, q_b, \ell\}$ and gauge the *maximal* symmetry

$SU(4)_{colour}$ in this space[2]. The gauge particle V_{15} associated with the 15th generator of this $SU(4)$ symmetry couples to the current $[\Sigma_{r,y,b} \bar{q}_i \gamma_\mu q_i - 3\bar{\ell}\gamma_\mu \ell] (2\sqrt{6})^{-1}$

The corresponding charge is proportional to $(B_q - 3L) = 3(B-L)$; here B_q denotes quark number, which is three times the baryon number B. This example serves to demonstrate the general result that once quarks and leptons are put in one multiplet and their maximal symmetry is locally gauged, some linear combination of baryon and lepton numbers (which in this example is B-L) must be among the generators of the local symmetry; it is thus conserved in the basic Lagrangian. Given that no massless vector particle can remain coupled to such a generator due to limits from the Eötvös-type experiments[4], however, the associated gauge particle (in this example V_{15}) must acquire a mass through spontaneous breakdown of the local symmetry. *Thereby the associated charge (in this example B-L) must be violated spontaneously.*[2]

Instead of gauging the maximal symmetry $SU(4)$, one might have chosen to gauge a subgroup of the maximal symmetry $SU(4)$. In this illustrative example the subgroup might have been $SU(2)$, which treats (q_r, q_y) and (q_b, ℓ) as doublets. In this case the gauge interactions of the basic Lagrangian would break B and L *explicitly* even prior to spontaneous symmetry breaking because one and the same gauge particle would couple for example to $\bar{q}_r \gamma_\mu q_y$ as well as to $\bar{q}_b \gamma_\mu \ell$ currents. An analogous situation is in fact what happens in some realistic models of grand unification, (e.g. $SU(5)$, see remarks later). The point of the remark made above, however, is that even if baryon and lepton numbers are conserved in the basic Lagrangian (and this holds automatically if we gauge the maximal symmetry of the quark-lepton-multiplet) either B or L or both must be violated spontaneously as the gauge particles of the quark-lepton symmetry acquire masses; this is in order that the theory may be compatible with the empirical fact based on Eötvös-type experiments. This line of reasoning had led Salam and myself to suggest in 1973 that the lighest baryon - the proton - must ultimately decay into leptons[2]. Theoretical considerations in the context of a number of models suggests a lifetime for the proton in the range of $10^{28} - 10^{33}$ years.[2,3,5,6]

Experiments are now underway to test proton stability to an accuracy one thousand times higher than before. In view of this, I shall concentrate primarily on the question of expected proton decay modes within the general hypothesis of quark-lepton unification and on the question of *intermediate mass scales* filling the grand plateau between 10^2 and 10^{15} GeV, which influence proton decay. The main point of my talk would be to stress that "maximal" symmetries[7] such as $SU(16)$ or its family extensions permit as a rule the existence of several intermediate mass scales lying within the grand plateau; *consequently they allow four major decay modes for the proton, some of which can even coexist* (this feature was first noted in Ref. 7). These are:*

$$\begin{aligned}
&\text{(i)} \quad p \to 3 \text{ leptons} + \text{(mesons)} \quad (\Delta F = 0; \; \Delta(B-L) = -4) \\
&\text{(ii)} \quad p \to \text{lepton} + \text{mesons)} \quad (\Delta F = -2; \; \Delta(B-L) = -2) \\
&\text{(iii)} \quad p \to \text{antilepton} + \text{(mesons)} \quad (\Delta F = -4; \; \Delta(B-L) = 0) \\
&\text{(iv)} \quad p \to 3 \text{ antileptons} + \text{(mesons)} \quad (\Delta F = -6; \; \Delta(B-L) = +2).
\end{aligned} \quad (1)$$

It is furthermore noted[8] that within such symmetries not only proton decay but also $\Delta B = 2$ n -- $\bar{n}$ oscillation[9] and $\Delta L = 2$ neutrinoless β decay can in general *coexist* with measurable strengths. In the last part of my talk I discuss the possibility that quarks, leptons *and* the associated gauge *and* Higgs particles are composites of more elementary objects – preons.

SPONTANEOUS VIOLATIONS OF B, L AND F; MAXIMAL SYMMETRIES

Much of what I have to say arises in the context of those unifying symmetries for which violations of baryon and lepton numbers and in general also of fermion number $F \equiv B_q + L = 3B + L$ arises only spontaneously rather than explicitly in the manner discussed in the introduction. Examples of this kind of symmetry are (i) the left-right symmetric subunification symmetry[2] $SU(2)_L \times SU(2)_R \times SU(4)_{L+R}$ and its unifying extensions[2], (ii) $[SU(4)]^4$, which operates on four flavours and six colours and (iii) $[SU(6)]^4$, which operates on six colours, etc. For these symmetries (B-L) is locally gauged, but fermion number F is still at best a global symmetry. Further extensions which put n left-handed fermions F_L and their n antiparticles F_L^c in one multiplet and gauge the *"maximal"* symmetry** $SU(2n)$, gauge not only (B-L) but also fermion number F as local symmetries. As an example, for a single family of two flavours and four colours including leptonic colour, n = 8, and thus the "maximal" symmetry G is $SU(16)$, which treats the 16 left-handed fields

---

*Here F stands for fermion number, which is +1 for a quark or a lepton; leptons comprise $(e^-, \nu_e, \mu^-, \nu_\mu)$ and mesons stand for π's and K's, etc.

**Such symmetries generate triangle anomalies, which are avoided by postulating that there exist a conjugate *mirror set of fermions* $F_{L,R}^m$ supplementing the basic fermions $F_{L,R}$ with the helicity flip coupling represented by the discrete symmetry $F_{L,R} \leftrightarrow F_{R,L}^m$. Thus by "maximal" symmetry we mean a symmetry which is maximal upto discrete symmetries such as the mirror symmetry.

$\{u_{r,y,b}, \nu_e; d_{r,y,b}, e^- | d^c_{y,y,b}, e^+; u^c_{r,y,b}, \nu^c_e\}_L$

as members of a single *16*-plet. Here c denotes charge conjugation.

To include the e, μ and τ families, whole attributing the distinction, which each family deserves in the sense of maximal gauging, one would need to gauge $[SU(16)]^3$ or the still extended symmetry * SU(48). Spontaneous symmetry breaking can permit the descent of these gigantic symmetries SU(48) or $[SU(16)]^3$ to the familiar low energy symmetry $SU(2)_L \times U(1) \times SU(4)^C$ via for example the diagonal symmetry ** $SU(16)_{e+\mu+\tau}$ so that interfamily universailty e -- μ -- τ appears only below some energy scale M_f; M_f need not be any higher than about 10^5 GeV. It turns out that these extended maximal symmetries such as $[SU(16)]^3$ or SU48) permit[10] signals for grand unification at low and intermediate mass scales ($\sim 10^4$-10^5 GeV and 10^8-10^{10} GeV) and thereby richer experimental possibilities[8] than for example SU(5) or[11] SO(10). In the discussions to follow, SU(16) will be used for simplicity *as a language* for maximal symmetry, though it may be viewed ultimately as part of the extended maximal symmetries such as $[SU(16)]^3$ or SU(48).

The gauge particles of SU(16) are symbolically listed in Fig. 1. Note that each of the gauge particles, prior to spontaneous symmetry breaking, carry *definite* baryon, lepton and fermion numbers. In fact there are two gauge particles V_{B-L} and V_F, which are coupled to B-L and F, respectively. As a consequence, these quantum numbers are conserved in the basic Lagrangian (as remarked before). They are violated in two distinct ways: (a) through spontaneously induced *mixings* of gauge particles carrying different B, L and F and (b) through spontaneously induced Yukawa transitions of the type $q \to \ell + \phi$, $q \to \bar{\ell} + \phi'$, or even $q \to \bar{q} + \phi''$, where ϕ's denote appropriate Higgs fields. In either case the violation is only spontaneou Two illustrative examples for case (a) and two for case (b) are shown in Figs. 2 and 3, respectively.

---

*These symmetries are no doubt gigantic, but if quarks and leptons are proliferated why not the associated gauge particles? As expresse elsewhere, the answer to proliferation must come by viewing quarks, leptons *and* the associated gauge particles as composites of more elementary objects - preons. From this point of view $[SU(16)]^3$ and $[SU(48)]$ are only effective gauge symmetries generated from a more economical basis of preons. See remarks at end.

**Note that for this diagonal symmetry $B = B_e + B_\mu + B_\tau$ and likewise for L and F.

GRAND UNIFICATION OF QUARKS AND LEPTONS

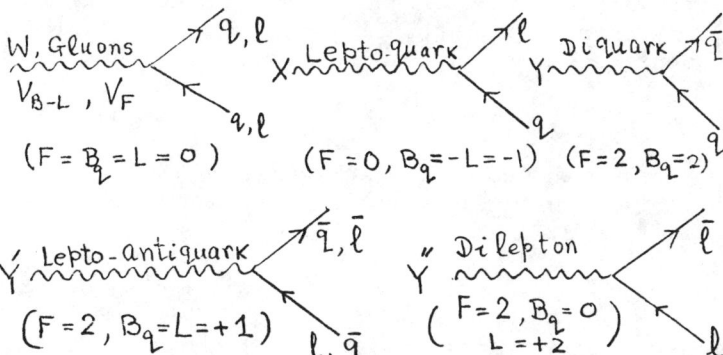

Fig 1. Gauge particles of SU(16); V_{B-L} and V_F couple to B-L and F.

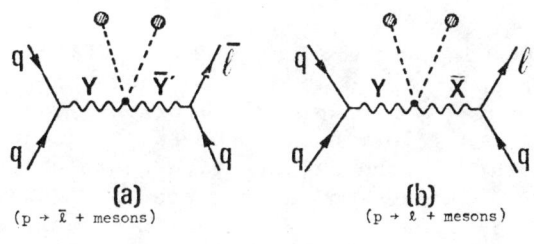

Fig. 2

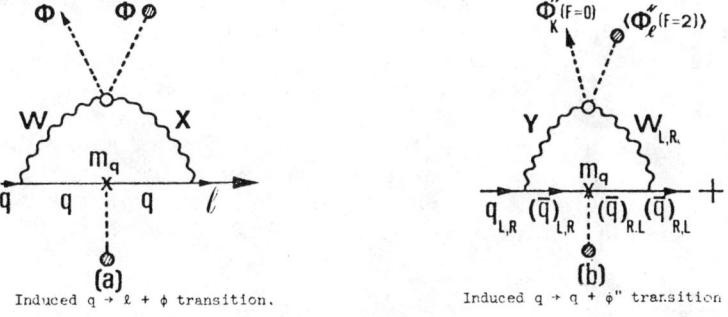

Fig. 3

Fig. 2(a) shows the mixing of the diquark gauge particle Y coupled to the $\bar{q}_L^c \gamma^\mu q_L$ current with the $\bar{Y}'$ gauge particle coupled to the $\ell_L^c \gamma_\mu q_L$ current; such a mixing induces the $\Delta F = -4$ (i.e. $\Delta(B-L) = 0$) transitions:

$$\text{proton} \to (e^+ \text{ or } \bar{\nu}_e) + \text{mesons etc.} \qquad (2)$$

The mixing of Y with the leptoquark gauge particle $\bar{X}$ coupled to the $\bar{\ell}_L \gamma_\mu q_L$ current shown in Fig. 2(b) induces the $\Delta F = -2$ (i.e. $\Delta(B-L) = -2$) transitions:

$$\text{proton} \to (e^- \text{ or } \nu_e) + \text{mesons etc.} \qquad (3)$$

The effective Yukawa transition $q \to \ell + \phi$ shown in Fig. 3(a) taken in third order[12,13] induces the $\Delta F = 0$ (i.e. $\Delta(B-L) = -4$) transitions

$$\text{proton} \to 3 \text{ leptons} + \text{mesons} . \qquad (4)$$

A similar mechanism involving $q \to \bar{\ell} + \phi'$ taken in third order would induce $\Delta F = -6$ decays (i.e. proton $\to 3\ell + $ mesons), while $\bar{q} \to q + \phi''$ (shown in Fig. 3(b)) taken in third order (see remarks later) induces $\Delta B = 2$ n -- $\bar{n}$ transitions. *It may be noted incidentally, that none of these mechanisms inducing the various proton decay modes and the n -- $\bar{n}$ transition depend in any way upon the nature of the quark charges.*

Instead of gauging the maximal symmetry SU(16), for which the violations of B, L and F are only spontaneous, one might have gauged one of its two special subgroups SO(10) or SU(5), which preserve the spirit of grand unification. These two subgroups violate B, L and F explicitly just as the illustrative example - the SU(2) subgroup does in the introduction and for the same reason. For SU(5) neither B-L nor F is locally gauged and B,L as well as F are violated explicitly*. For SO(10), B-L is locally gauged, but not F and here too B,L and F are violated explicitly. The explicit violations for SU(5) or SO(10) come about as follows. While Y's and Y''s coupled to $\bar{q}^c \gamma_\mu q$ and $\bar{\ell}^c \gamma_\mu \ell$ currents, respectively, are *distinct* gauge particles for the maximal symmetry SU(16), the subgroups SO(10) and SU(5) contain only the linear combinations $Y_s \equiv (Y + \bar{Y}')/\sqrt{2}$ but not the combinations $Y_a \equiv (Y - \bar{Y}')/\sqrt{2}$. Quite clearly exchanges of Y's would violate B,L and F explicitly in the second order of the gauge interactions.

The two cases of spontaneous versus explicit violations of B, L and F differ from each other conceptually as well as physically. Spontaneous violation would vanish at high temperatures exceeding the masses of the relevant gauge particles, where both Y and Y' or

---

*For the *minimal* SU(5) model, B-L turns out to be a global symmetry even though B,L and F are violated explicitly.

equivalently Y_s and Y_a would exist as degenerate massless particles. Explicit violation, on the other hand, would acquire its maximum strength at such high temperatures, since in this case only the Y_s gauge particles would exist as massless particles with the Y_a's being still absent. This distinction would play its most obvious role in the early stage of the Universe (see remarks later).

It is of interest to see the alternative routes for the spontaneous descent of SU(16) down to the low energy symmetry $SU(2)_L \times U(1) \times SU(3)_C$. The Higgs structure permitting the descents via such alternatives routes is given in detail in a recent paper by Salam, Strathdee and myself[8]. The three most obvious descents go via 1) SO(10) with respect to which the 16-plet remains irreducible, 2) the maximal chiral symmetry $SU(8)_I \times SU(8)_{II} \times U(1)_F$ where the two SU(8)'s operate in the spaces of eight fermions (F_L) and the eight antifermions (F_L^c), respectively, and $U(1)_F$ represents fermion number and 3) $SU(12)_q \times SU(4)_\ell \times U(1)_{|B|-|L|}$, where $SU(12)_q$ operates on (6 quarks + 6 antiquarks), $SU(4)_\ell$ on (2 leptons + 2 antileptons) and U(1) denotes $|B_q| - 3|L| = 3(|B| - |L|)$. These three alternative routes are shown in Fig. 4.

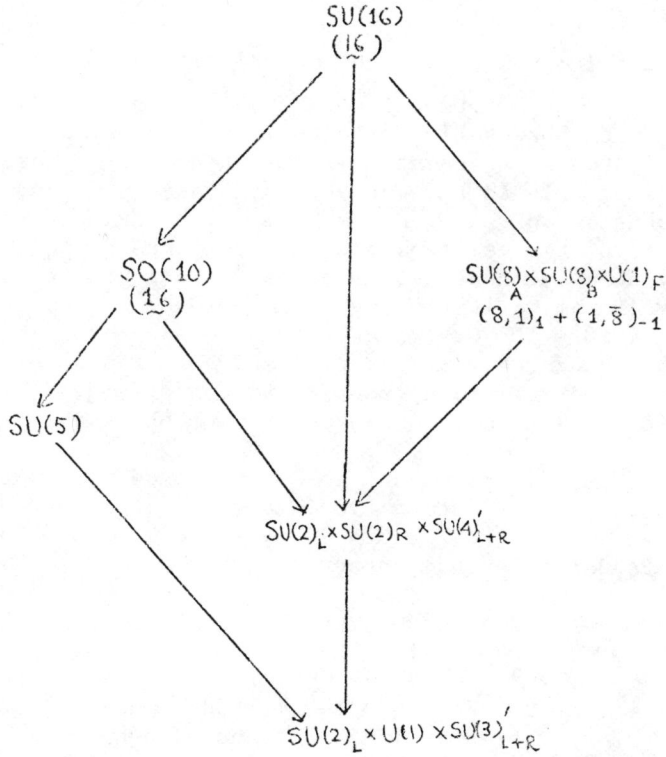

Fig.4: Three alternative routes for spontaneous descent of SU(16).

Note that in contrast to the first two routes, the third route separates quarks from Leptons at the very first stage of symmetry breaking. Complexions for proton decay as well as the strength of $n \text{---} \bar{n}$ oscillation depend crucially upon the route for spontaneous descent, which in turn depends upon the pattern of VEV of the Higgs multiplets. For instance, an adjoint 255 of Higgs by itself can take SI '6) to $SU(8)_I \times SU(8)_{II} \times U(1)_F$, while a four index multiplet $\phi^{\{AB\}}_{\{CD\}} \sim \{16 \times 16\}_{symm} \times \{16^* \times 16^*\}_{symm}$ by itself can break $SU(16)$ in general to $SU(2)_L \times SU(2)_R \times SU(4)^c_{L+R}$ or for a special case[8] into $SO(10)$. If 255 far exceeds $\phi^{\{AB\}}_{\{CD\}}$, the descent via $SU(8) \times SU(8) \times U(1)_F$ would be prominent; in this case the proton can decay via a variety of channels satisfying $\Delta F = -4$, $\Delta F = -2$ as well as $\Delta F = 0$; here $\Delta F = -2$ (i.e. $p \to e^- +$ mesons) and/or $\Delta F = 0$ (i.e. $p \to 3\ell +$ mesons) can even far supercede $\Delta F = -4$ mode (i.e. $p \to e^+ +$ mesons). If on the other hand $\phi^{\{AB\}}_{\{CD\}}$ far exceeds 255 and in particular the special $SO(10)$ chain materializes, then only the $\Delta F = -4$ mode ($p \to e^+ +$ mesons etc.) would be dominant. Thus predictions of $SO(10)$ (or $SU(5)$) can emerge as special cases within those of the maximal symmetry $SU(16)$. Finally, the third route with the intermediate $SU(12)_q \times SU(4)_\ell \times U(1)$ symmetry can result from an alternative pattern of VEV of the adjoint 255 plus $\phi^{\{AB\}}_{\{CD\}}$. The third route has the potentiality of yielding $\Delta F = -4$ proton decay ($p \to e^+ +$ mesons) *coexisting*[8] with $\Delta B = 2$ $n \text{---} \bar{n}$ oscillation with observable strength. This is because, for this route, the leptoquark X and the lepto-antiquark Y' gauge particles coupled respectively to the currents $\bar{q}\gamma_\mu \ell$ and $\bar{q}^c \gamma_\mu \ell$ acquire masses at the first stage of spontaneous symmetry breaking, while the di-quark Y gauge particles coupled to the currents $\bar{q}^c \gamma_\mu q$ acquire their masses only at a secondary stage of SSB. This generates the possibility (for $m_Y \sim 10^4$–10^5 GeV, $m_Y' \sim 10^{14}$ GeV and Y-$\bar{Y}$' mixing masses $\approx m_Y \sim 10^4$–10^5 GeV) that *both* $\Delta F = -4$ proton decay (i.e. $p \to e^+ \pi^0$ etc.) induced via Fig. 2(a) and $\Delta B = 2$ $n \text{---} \bar{n}$ oscillation induced via effective third order iteration of Fig. 3(b) followed by a quartic $\lambda\phi^4$ interaction can occur with observable strength.[8]

It would be only premature to speculate on which of these routes if any is preferred by Nature. One must wait for the forthcoming experiments searching for proton decay as well as $n \text{---} \bar{n}$ oscillation to provide guidance in this matter.

CHARACTERISTIC MASS SCALES FOR THE RELEVANCE OF ALTERNATIVE PROTON DECAY MODES AND n-$\bar{n}$ OSCILLATION

The $\Delta F = 0$ proton decay ($p \to 3$ leptons + mesons) arises in the third order of the effective Yukawa transition $q_i \to \ell_i + \phi$ (Fig.3(a)) followed by the quartic interaction $\lambda_{ijk\ell}\phi_i\phi_j\phi_k\phi_\ell$ of the Higgs fields subject to one component ϕ_ℓ having a non-zero VEV ($<\phi_\ell> \sim m_{W_L,R}/g$). The $\Delta F = -2$ and the $\Delta F = -4$ decays are induced through gauge mixings as shown in Figs. 2(a) and 2(b). The corresponding amplitudes are symbolically given by:*

$$\text{Amp}(\text{"udd"} \ \ell_1 \ell_2 \ell_3)_{F=0} \quad \frac{g^4 \, m_q (m_{W_L}/g)^3}{m_X^2} \quad \frac{\lambda \ <\phi_\ell>}{m_{\phi_1}^2 \, m_{\phi_2}^2 \, m_{\phi_3}^2} \quad (5a)$$

$$\text{Amp}(\text{"ddd"} \to e^-)_{\Delta F = -2} \sim g^2 \, (\Delta_{XY}/(m_X^2 m_Y^2)) \, m_{W_L}^2 \quad (\lesssim \Delta_{XY} \lesssim_{W_L} M_Y), \quad (5b)$$

$$\text{Amp}(\text{"uud"} \to e^+)_{\Delta F = -4} \sim g^2 (\Delta_{YY'}/(m_y^2 m_y^2)) \, (\Delta_{YY'} \lesssim m_y^2 \text{ or } m_{y'}^2, \text{ whichever is smaller}). \quad (5c)$$

The $\Delta F = -6$ decay ($p \to 3\bar{\ell}$ + mesons) will have an amplitude similar to that of the $\Delta F = 0$ decay except that m_X will be replaced by $m_{Y'}$. Taking effective masses of the Higgs particles** $\phi_{1,2,3}$ to be ~ 3 to 10 GeV and λ to be of order , the $\Delta F = 0$ amplitude would lead to proton decay with a partial width of order $(10^{28} - 10^{33}$ years$)^{-1}$ if $m_X \sim (3 \times 10^4 - 10^5)$ GeV.

The $\Delta F = -2$ (as also $\Delta F = 0$) proton decay violates[14] $SU(2)_L \times U(1)$. Thus the X-Y mixing (mass)^2 Δ_{XY} must be proportional to a VEV $\lesssim m_{W_L}$. This is why Δ_{XY} is chosen to lie between $m_{W_L}^2$ and $m_{W_L} m_Y$. Taking $\Delta_{XY} = m_{W_L}^2$ (or $m_{W_L} m_X$) and $m_X \sim 10^5$ GeV (as for $\Delta F = 0$), we see that $\Delta F = -2$ amplitude would have the "canonical" value $\approx 10^{-29}$ GeV^{-2} corresponding to a proton decay partial width $\sim (10^{30}$ years$)^{-1}$ for $m_Y \approx 3 \times 10^{11}$ GeV (or $\approx 10^{13}$ GeV). If Δ_{XY} has its maximal value $\sim m_{W_L} m_Y$, the *characteristic mass scale for this case would be given by* $M \equiv (m_X^2 m_Y)^{1/3}$, *which must be* $\approx 10^{10.3}$ *GeV for the maplitude to be of order* 10^{-29} GeV^{-2} (see (5c).

---

* A more careful representation of the $\Delta F = 0$ amplitude which includes the fact that two quarks must transform into leptons + Higgs fields via loop diagram (Fig. 2(a)) and the third via a tree diagram is to be found in Ref. 12. Here we are interested in the order of magnitude argument.

**These Higgs particles $\phi_{1,2,3}$ have $SU(2) \times SU(3)^{col}$ quantum numbers identical to those of the quarks; the component ϕ_ℓ with non-vanishing VEV being distinct from them (see Ref. 12 for details).

The $F = -4$ proton decay amplitude would have the "canonical" strength $\sim 10^{-29}$ GeV^{-2} if for example $m_Y \approx m_{Y'} \approx \Delta_{YY'}^{1/2} \sim 10^{14.5}$ GeV, or alternatively, if $m_Y \approx m_{Y'} \sim 10^{12}$ and $\Delta_{YY'} \approx 10^{9.5}$ GeV, etc.

In summary a gauge mass pattern

$$m_X \approx 10^4 - 10^5 \text{ GeV}, \quad m_Y \approx m_{Y'} \approx 10^{12}$$

$$\Delta_{YY'}^{1/2} \approx 10^{9.5} \text{ GeV and } \Delta_{XY}^{1/2} \sim m_{W_L} m_X \tag{6}$$

would permit the possibility that $\Delta F = 0, -2$ *and* -4 proton decay modes coexist and be relevant to forthcoming proton decay searchers. It is worth noting that none of the mechanisms for proton decay outlined above depends in any way on the nature of quark charges.

Finally the $\Delta B = 2$ n-$\bar{n}$ oscillations can occur through third order of effective transitions of the form $q_i \to \bar{q}_i + \phi_i''$ (Fig. 3(b)) followed by quartic scalar interaction $\lambda'_{ijk\ell} \phi_i'' \phi_j'' \phi_k'' \phi_\ell$ together with VEV for one component * ϕ_ℓ'' being non-zero (see Fig. 5).

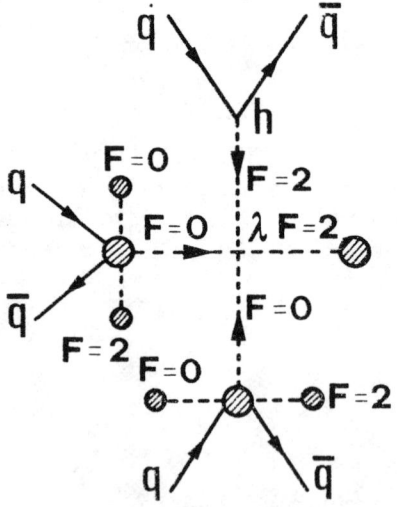

Fig. 5: $3q \to 3\bar{q}$ transition leading to n-$\bar{n}$ oscillation.

---

* The $\{\phi''\}$ multiplet transforms as *136* under SU(16); the ϕ_ℓ'' component transforms as $(1,3,10) \subset 136$ under the $SU(2)_L \times SU(2) \times SU(4)^C$ subgroup.

GRAND UNIFICATION OF QUARKS AND LEPTONS

This is analogous to the mechanism for $\Delta F = 0$ proton decay except that the mediating gauge particle is the diquark Y (see Fig. 3(b)) rather than the leptoquark X (see Fig. 3(a)). For a reasonable choice of the strength of quartic coupling constants ($\sim \alpha$), and Higgs masses (~ 100 GeV), the Y mass needs to be 10^4 GeV for the n -- $\bar{n}$ oscillation to have measurable strength; this corresponds to an oscillation period for free neutrons $\tau_{n\bar{n}} \approx 10^5 - 10^6$ sec. (or the partial width for say deuteron $\to$ pions $(10^{30}$ years$)^{-1})^{18}$. Such a light Y can in general arise, as mentioned before, through the third route of spontaneous descent (i.e. $SU(16) \to SU(12)_q \times SU(4)_\ell \times U(1)$, see Fig. 4) which makes Y' and X superheavy ($\sim 10^{14}$ GeV), but Y much lighter. Even with such a light Y, $\Delta F = -4$ decays ($p \to e^+_\pi \pi^0$ etc.) involving Y-$\bar{Y}$' mixing would not become too fast because for this case $\Delta_{YY'} \approx m_Y^2$, so that Amp $(uud \to e^+) \approx g^2/m_{Y'}^2$, this would have the canonical strength for $m_{Y'} \sim 10^{14}$ GeV. In other words, for the spontaneous descents as in the third route (i.e. $SU(16) \to SU(12)_q \times SU(4)_\ell \times U(1)$), *$\Delta F = -4$ proton decay as well as $\Delta B = 2$ n-$\bar{n}$ oscillation can coexist* * *with measurable strengths.* (Note however that for this case $\Delta F = 0$ proton decay (i.e. $p \to 3$ leptons + mesons) would be strongly suppressed, because the leptoquark gauge particle X is now superheavy. The $\Delta F = -2$ proton decay ($p \to e^-$ + mesons etc.) may or may not be suppressed depending upon whether $\Delta_{XY} \lesssim m_{W_L} m_Y$ or $> m_{W_L} m_Y$.

In summary, the different complexions for proton decay and $\Delta B = 2$ n-$\bar{n}$ oscillation if found to occur with measurable strengths for the present level of experimentation, would symbolize different characteristic mass scales (M_c) for the hierarchy of grand unification. These characteristic mass scales ** are listed below:

$\Delta F = -4$ ($p \to e^+_\pi \pi^0$ etc.) .... $\to M_c \sim 10^{14}$ GeV (if $m_Y = m_{Y'}$
$= \Delta_{YY'}$)
$\Delta F = -2$ ($p \to e^- \pi^+ \pi^+$ etc.) .... $\to M_c \sim 10^8 - 10^{12}$ GeV
(see discussions
$\Delta F = 0$ ($p \to 3$ leptons + pions) .... $\to M_c \sim 10^4 - 10^5$ GeV
$\Delta F = -6$ ($p \to 3$ antileptons + mesons) .. $\to M_c \sim 10^4 - 10^5$ GeV
$\Delta B = 2$ (n -- $\bar{n}$ oscillation) .... $M_c \sim 10^4 - 10^5$ GeV .

(7)

---

* Such a coexistence was not permissible within the mechanism considered in Ref. 15, which led to $\Delta B = 2$ n-$\bar{n}$ oscillation, but no proton decay. See Ref. 8 for a comparative discussion on this point.

**Note that the characteristic mass scales as obtained above with explicit mechanisms are not too different from those obtained in Ref. 14 from simplifying operator dimensional analysis.

Question 1: Can the vastly different mass scales exhibited in (6) or (7) coexist in accordance with the constraints of renormalization group equations for the running coupling constants and the observed values of $\sin^2\theta_W$ and α_S so as to permit the coexistence of alternative decay modes of proton with or without n -- $\bar{n}$ oscillation?

It has been discussed elsewhere[10,13,8] that the coexistence of such mass scales - in particular the mass pattern exhibited in Eq. (6) and *therefore the coexistence of $\Delta F = 0$, -2 and -4 proton decay modes - is indeed permissible* if we allow basic family distinctions[16] for quarks and leptons and gauge their maximal symmetries such as $[SU(16)]^3$ or $SU(48)$ as mentioned in the introduction. Some of the intriguing experimental consequences of gauging maximal symmetries, which allow family distinctions at the basic level, are that (a) they obviate the necessity for a long extending "desert" or grand plateau in physics and (b) they permit family non-universality in the form of couplings of gauge particles to *differences* of currents of the electronic and the muonic families rather than to their sums to appear at modest energy scales; these scales need be no higher than about 10^4-10^5 GeV. *In other words, such symmetries would make high energy machines of the future fertile - rich with the discoveries of new physics- rather than be sterile.*

Finally, the realization of n-$\bar{n}$ oscillation together with $\Delta F = -4$ proton decay ($p \to e^+\pi^0$ etc.) at an observable level, as discussed above, requires a mass pattern; $m_Y \sim 10^4$-10^5 GeV, $\Delta_{YY'} \sim m_Y$, $m_{Y'} \sim m_X \sim 10^{14}$ GeV. Such a mass pattern is permissible within the constraints of renormalization group equations (RGE) for a four-family symmetry of the type $[SU(16)]^4$ descending in a manner analogous to the third route (i.e. $[SU(16)]^4 \to SU(12)_q \times [SU(4)_\ell] \times U(1) \to [SU(12)_q \times SU(4)_\ell]_{e+\mu+\tau+\tau'} \times U(1))$. For a one or two family symmetry however, such a mass pattern is not allowed, while for a three family symmetry $[SU(16)]^3$, the perturbative solutions of RGE turn out to be independent of M_X and $M_{Y'}$. Thus within this picture, a coexistence of $\Delta F = -4$ proton decay with n-$\bar{n}$ oscillation is allowed but amusingly enough such a coexistence appears to need the existence of at least three but possibly four families.

Question 2: Does the coexistence of the alternative proton decay modes ($\Delta F = -4, -2, 0$ and 6) with or without n -- $\bar{n}$ oscillation, symbolizing different characteristic mass scales, conflict with the cosmological generation of baryon excess for a Universe, which is assumed to be initially matter-antimatter symmetric?

Weinberg[16] noted that the answer to this question is in the affirmative, if the violations of B, L and F are explicit. The argument is this: since the $\Delta F = 0, -2$ and -6 proton decays are mediated by intermediate mass scales $M_I(\sim 10^4$ or 10^{12} GeV$)<<M\sim 10^{14}$ GeV,

such processes would have relatively fast rates $\sim \alpha T$ or $\alpha^2 T$ in the early Universe at temperatures in the range $M_I \ll T \ll M$. Thus these processes with rates exceeding the rate of expansion of the Universe would be in thermal equilibrium and would wipe out any baryon excess generated in earlier epochs, due to the $\Delta F = -4$ processes, unless a specific linear combination $B + aL$ is conserved ($a \neq 0$). This wo31d imply that only one of the three modes $F = 0$, -2, or -6 can coexist with the $\Delta F = -4$ mode, but not all, and furthermore there should be *no* $\Delta B = -2$ n-n̄ oscillation.

These arguments apply, however, only if the violations of B,L and F are explicit rather than spontaneous. For the latter case the violations disappear for temperatures $T > M_I$. These violations do appear at a later epoch (i.e. at temperatures $T \vee M_I$); however, the associated gauge particles acquire their masses simultaneously and the rates of the B,L, F violating processes ($\Delta F = 0$, -2, -6, etc.), damped by gauge masses, are lower at such epochs than the expansion rate of the Universe. This keeps *all* of these processes out of thermal equilibrium, as and when they are operative. Consequently, the baryon excess generated in an earlier epoch $T > M_I$ by the $\Delta F = -4$ process, if any, is not wiped out by any or all of the other processes occurring at later times. *We conclude that there is no conflict between the coexistence of $\Delta F = 0$, -2, -4 and -6 proton decay modes and even $B = 2$ n-n̄ oscillation, on the one hand and the generation of baryon excess on the other, if the violations of B,L, F are spontaneous.* This is one of the crucial differences between explicit versus spontaneous violations of B,L, F as indicated before.

A SUMMARY

(1) Violation of baryon, lepton and in general fermion number is central to the hypothesis of quark lepton unification. Three of its characteristic signatures are proton decay, n-n̄ oscillation and neutrinoless double decay. Within maximal symmetries permitting intermediate mass scales (i) proton may decay via alternative modes satisfying $\Delta F = 0$ ($p \to 3\ell$ + mesons), $\Delta = -2$ ($p \to \ell$ + mesons), $\Delta F = -4$ ($p \to \ell$ + mesons) and $\Delta F = -6$ ($p \to 3\ell$ + mesons), some of which can coexist; (ii) n-n̄ oscillation 9 (related to deuteron decay into pions[7]) can *coexist*[8] with proton decay especially of the $\Delta F = -4$ variety (i.e. $p \to e^+ \pi^0$ etc.) and both of these processes may possess measurable strength so as to be amenable to forthcoming searches[17,18] and (iii) coexistence of any two or even all of the proton decay modes with or without n-n̄ oscillation poses no conflict with the cosmological generation of baryon excess. Maximal gauging with spontaneous rather than explicit violations of B,L and F plays an essential role in the realization of these features.

(2) Non-maximal symmetries like $SO(10)$, $SU(8) \times SU(8) \times U(1)_F$ and $SU(5)$ are contained within $SU(16)$ and arise effectively as special cases through spontaneous descent of the maximal symmetry $SU(16)$. Their predictions on proton decay are therefore naturally contained within the predictions of $SU(16)$. Unlike the case of $SU(16)$ none of these subsymmetries ($SO(10)$ or $SU(5)$ etc.*) however easily permit of the coexistence of proton decay and $n-\bar{n}$ oscillation.

(3) It is a reasonable expectation within most models that the lifetime of the proton will lie within the range of 10^{28}–10^{33} years making forthcoming searches fully sensitive to all these models.

(4) Neither proton decay into the modes enumerated in this paper nor $n-\bar{n}$ oscillation are tied in any way to the nature of quark charges.

(5) Quark-lepton unification-symmetry gauged in its maximal form, permits in general several intermediate mass scales (~ 10 to 10^{12} GeV) filling the grand plateau between 10^2 and 10^{15} GeV. Proton decay and $n-\bar{n}$ oscillation can provide a window to these intermediate scales. In particular, the observation of the $\Delta F = 0$, or -6 mode (i.e. $p \to 3\ell$ or $3\ell +$ mesons) and/or $n-\bar{n}$ oscillation at any level within the conceivable future will strongly suggest the existence of new physics in the 10 to 100 TeV region and thereby motivate building high energy machines in this range. For this reason, second and third generation experiments for proton decay and $n-\bar{n}$ oscillation must be planned to look for all modes listed above as possible rare processes, in case they are not found in the first generation experiments.

(6) Coexistence of any two of the proton decay modes $\Delta F = 0$, -2 or -6 or the mere existence of $\Delta B = 2$ $n-\bar{n}$ oscillation with proton decay of the $\Delta F = -4$ variety would strongly suggest that the associated violations of B,L, F are spontaneous rather than explicit. This is in order that cosmological generation of baryon excess may survive.

(7) Observation of proton decay will strongly support the idea that quark matter and leptonic matter are similar[2] in their composition, though this has no bearing on the question of whether quarks and leptons represent the ultimate constituents of matter.

---

* L.N. Chang and N.P. Chang, Phys. Lett. *92B*, 103 (1980) have considered such a coexistence within $SU(5)$; their model however would appear to conflict with the generation of baryon excess, as B, L and F are violated explicitly in this model (see text).

GRAND UNIFICATION OF QUARKS AND LEPTONS

Why preons?

The grand unification hypothesis attempts to remove a certain degree of arbitrariness from particle physics by linking the basic particles - quark and leptons - and their three forces - weak, electromagnetic and strong. The arbitrariness persists however if we view quarks, leptons and the associated gauge and spon-0 Higgs particles as elementary in that these are vastly proliferated. The degrees of freedom needed for some alternative grand unification symmetries are listed below:

	SU(5)	SO(10)	SU(16)
Fermions	$15_e + 15_\mu + 15_\tau$	$16_e + 16_\mu + 16_\tau$	$16_e + 16_\mu + 16_\tau$
Gauge bosons	24	45	255
Higgs	$24 + 5 + ...$	$45 + 126 + ...$	$255 + 136 + ...$

In addition to these degrees of freedom the scalar masses and quartic and Yukawa couplings are new parameters, unrelated in principle to the gauge coupling constant. If the attractive technicolour ideas are invoked, elementary Higgs fields may be dispensed with, but a new proliferation of fermions and gauge bosons would be needed. This kind of proliferation and arbitrariness runs counter to one's intuitive notion of elementarity and suggests that quarks, leptons as well as technifermions and the associated gauge and Higgs particles must define only a stage in one's quest for elementarity. They must be composities[19,20] of a more elementary set of objects, which collectively we name as *"preons"*. Quarks and leptons appear pointlike at present centre-of-mass energies ($\lesssim$ 100 GeV), but at approximately higher energies depending upon their sizes they must reveal a composite nature. From (g-2) and other experiments their size r_0 (at least for leptons*) is known to be less than 10^{-16} cm.

In general one may conceive of several layers of increasing elementarity** - preons, pre-preons and pre-pre-preons - within quarks and leptons with sizes decreasing progressively from say 10^{-17} or 10^{-18} cm. down to perhaps Planck size (M_{Planck}^{-1}) $\sim 10^{-33}$ cm. We shall by convention call the objects, which belong to the layer just preceding quarks and leptons, preons.

---

* For quarks, following deep inelastic scattering experiments with momentum transfers $|Q| \lesssim 10$ GeV, one may infer that quark sizes are less than 10^{-15} cm.

** This would correspond to having no long extending "desert" or grand plateau in physics.

With composite quarks and leptons, it appears natural to assume in the interest of economy that the associated proliferated gauge and Higgs particles are composites as well and that the quark-lepton gauge symmetry generating the electroweak strong forces is only an effective low energy symmetry valid at energies below the inverse size of the composites. A compelling motivation for this point of view is given later.

Two simple alternative pictures for preons, which we wish to consider, arise depending upon the level at which the flavour and colour attributes have a well defined meaning.

(I) *Flavour and colour attributes arise at the preon level:* Here preons already carry flavour and colour quantum numbers even though pre-preons or pre-pre-preons in the sense defined above may not; the preons are thus *directly* responsible for the manifestation of these quantum numbers at the quark lepton level. Quarks and leptons in this case may be viewed as composites for example of three* sets of preons: the flavons $f_i = (u,d)$, the chromons** $C_\alpha = r,y,b,\ell)$ and a third set of entities S_k, which are provisionally called*** "somons". Each set is characterized by specific values of the binding charges (i.e. transformation properties with respect to the symmetry of the "primordial" preon binding force, see later). The third set of S_k may contain as few as a single member. Quarks and leptons are composites of the form $(f_i C_\alpha S_k)$ and are neutral with respect to the binding charges, whatever the nature of these charges may be. With two flavons, four chromons and a single S, the composite $f_i C_\alpha S$ can represent the six quarks + 2 leptons in one family. Additional families may arise dynamically without involving new ingredients (see later), or alternatively, the set $\{S_k\}$ having more than one member may provide the family quantum number, or else the set of flavons may consists of more than two members.

---

* Minimally two may have sufficed if we take spin 1/2 for one set (e.g. the flavons) and spon 0 for the other (the chromons) as in Ref. 2. The appearance of a third set appears to be desirable in realistic models.

** The fourth chromon ℓ goes with the idea of lepton number being the fourth colour[2]. In general, for a preonic model, leptons may differ from quarks by more than one attribute.

*** The origin being the Sanskrit word "soma-ras", meaning the nectar in God's drink. If the set of S particles having say three members S_ℓ, S_μ and S_τ turn out to provide the family quantum numbers, they may be more appropriately named as "familons".

(II) *Flavour and colour attributes arise at the quark lepton level:* Here preons consists of a few sets of objects each set possessing well defined binding charges, but they do not possess well defined flavour and colour attributes. These quantum numbers arise dynamically only at the composite level of quarks and leptons*.

For any of these preonic models, some of the important questions which one faces are these:

(1) What is the nature of the "priomordial" preon binding force, which binds preons to make quarks and leptons? Can such a force be generated retaining the spirit of grand unification (i.e. non-proliferation of fundamental coupling constants) and also non-proliferation of fundamental building blocks - in this case preons? A related question: Do the electroweak strong gauge forces coexist with the preon-binding force at the basic preonic level or are they generated effectively only at the composite quark-lepton level?

(2) Why do we see for example only the fCS composites for models of class I (and some analogous composites for models of class II)? What about other three-body composites such as fCS, fCS, ffS, etc? Do these form if so with what masses and what sizes?

(3) Why do the three-body composites, identified with observed quarks and leptons, carry only spin 1/2. What excludes the formation of analogous composites with relatively low mass but higher spins?

(4) With spin-0 Higgs particles viewed more attractively as preonic composites, can have some understanding of the desired hierarchical pattern for the spontaneous breakdown of quark-lepton gauge symmetries?

(5) Does the observed family structure for quarks and leptons arise dynamically, or does it arise due to the existence of new basic ingredients (i.e. additional members within the set of S particles or the set of flavons)?

The single most important question among these is the first one. To obtain an insight into the nature of the preon binding force, we shall insist on two points:[21] (a) non-proliferation of fundamental coupling constants (i.e. grand unification of the basic forces) and (b) non-proliferation of preons (if the number of preons needed is comparable to or exceed the number of quarks and leptons, that would go against one of our primary motivations for preons - economy).

---

* In general if photon is a preonic composite, even electric charge need not have a well defined meaning at the preonic level.

It has been observed[21] that ordinary "electric" type forces*
(abelian or non-abelian) arising within a non-abelian simple or semi-
simple grand unifying symmetry are inadequate to bind preons to make
quarks and leptons of small size r_0 ($<10^{-16}$ cm) without proliferating
preons unduly.

To arrive at this conclusion, we shall follow the conventional
perturbative renormalization group approach for the evolution of all
effective gauge coupling constants g_i down to such momenta where
they are small (i.e. $g_i^2/4\pi < 0.3$ say). The argument briefly is
this: For quarks and leptons to have small size, the preon binding
force F_b must be strong or superstrong ($g_{eff}^2/4\pi \gtrsim 1$) at some
appropriately short distance $\lesssim 10^{-17}$–10^{-18} cm corresponding to
running momentum $Q \gtrsim 1$ to 10 TeV. This says that the symmetry G_b
generating the preon binding force must lie outside of the familiar
quark-lepton symmetry $G_0 = SU(2) \times U(1) \times SU(3)^c$, since the forces
associated with the latter symmetry are weak ($g_i^2/4\pi << 1$) at such
short distances.

Now there are two possibilities:

(i) First assume that the preon binding forces F_b *as well as*
the known quark lepton force F_k arise from a grand unifying symmetry
G_p operative at the preon level. In this case, either the symmetry
G_p is of the form $G_k \times G_b$ with G_k generating the known forces F_k
and G_b generating the preon binding forces F_b (G_k and G_b are related
to each other by discrete symmetry so as to permit a single gauge
coupling constant), or the unifying symmetry G_p breaks spontaneously
as follows:

$$G_p \xrightarrow{SSB} G_k \times G_b \times [\text{possible } U(1) \text{ factors}].$$

In the second case G_k need not be related to G_b by discrete symmetry.
In either case G_k contains the familiar $SU(2) \times U(1) \times SU(3)^c$
symmetry and therefore the number of attributes N_k on which G_k
operates is *at least* 5. This corresponds to having two flavons (u,d)
plus three chromons (r,y,b). To accommodate the leptonic chromon ℓ
and possibly also the somon s, N_k may need to be at least 7, but con-
servatively we shall take $N_k \geq 5$.

Now consider the size of G_b. If the effective coupling constant
$\bar{g}_b$ of G_b and that of $SU(3)^c$ are to be equal at the grand unification
mass M, and simultaneously if $\bar{g}_b^2/4\pi$ ($Q \sim 1$ to 10 TeV) $>> \alpha$ ($Q \sim 1$ to 10
TeV), it follows that the size of G_b must be larger than $SU(3)^c$.

---

* By "electric" type forces, we mean forces whose effective coupling
strength ($\bar{g}^2/4\pi$) is of order $\alpha \simeq 1/137$ at the grand unification
mass scale.

Thus minimally G_b is SU(4). To make singlets of G_b, quarks and leptons would have to be four preon composites if G_b is SU(4); this is however not acceptable if all preons carry spin 1/2. Thus minimally G_b is SU(5).

For the case under consideration, where both the binding symmetry G_b and the known symmetry G_k are equally basic, preons must in the first place carry well defined transformation properties with respect to both these symmetries and in the second place they must be non-trivial with regard to both these symmetries*. Since each of G_b and G_k requires for their operations a space, which is minimally five dimensional, it follows that the number of preons N_p needed (under the hypothesis alluded to above) is minimally

$$N_k \times N_b \geqslant 5 \times 5 = 25$$

$$N_p \geqslant N_k \times N_b \geqslant 25.$$

This number already exceeds or is close to the number of quarks and leptons which we need at present, which** is 24. Such a proliferation of preons defeats from the start the very purpose of economy for which they were introduced. In other words, the alternative of a non-abelian grand unifying symmetry G_p generating simultaneously the preon binding as well as the known electroweak forces do meet our twin requirements of proliferation of fundamental coupling constants and non-proliferation of preons.

There is a second alternative. Assume that only the binding symmetry G_b is basic at the preonic level; the symmetry G_k of known forces derives its origin as an effective symmetry at the composite quark-lepton level. G_b may in this case be either non-abelian or abelian. The case of G_b being non-abelian does not however appear to lead to any simple or satisfactory solution.***

---

* They have to be non-trivial with regard to G_b in order that they may experience the binding force; they must also be non-trivial with regard to G_k in order that quarks and leptons be non-trivial with regard to G_k (this is assuming that the preons do carry well-defined transformation property with regard to G_k in the first place).

** To be precise we presently need 21 four component plus 3 two component quark-lepton entities, assuming that the top flavour will be discovered.

*** To see this, assume that there are N preons. One possibility is that $G_p = G_b = $ SU(N). In this case quarks and leptons would be singlets of SU(N), if they are N preon composites. If $N \geqslant 3$, we would need to supplement SU(3) binding with additional global and local symmetries to make many singlets, thereby proliferating preons.

We are thus led to suggest that the primordial preon binding force is essentially abelian in character adn that the known electroweak strong forces are derived through the formation of composite spin 1 W's, Z's and gluons.

Within the class of abelian forces, it appears especially appealing to postulate[21] that *the primordial force is just electromagnetism generated by dual electric and magnetic charges*[22] of an $U(1) \times U(1)$ symmetry[23] generating a single photon. The preons possess not only electric but also magnetic charges; their binding force is magnetic in nature. The two types of charges are related to each other by the familiar Dirac-like quantization conditions for charge monopole or dyon systems (eg/4π = 1/2 N, or $(e_i g_j - e_j g_i)/4\pi$ = 1/2 N_{ij}, N = 1,2,3,...), which imply that the magnetic coupling strength is superstrong. Note that such a binding force, unlike the case of non-abelian G_h deviates necessity of proliferation of preons. Quarks and leptons do not exhibit even a trace of this superstrong force at present energies because they are extraordinarily small* in size ($r_0 \lesssim 10^{-17}$ to 10^{-18} cm) and they are *neutral* with respect to the magnetic charge.

There is a variant of this idea. Preon binding force is based on dual abelian charges as above; they are coupled however to a new (perhaps massless) gauge particle A' distinct from the ordinary photon**. In this case one may regard both dual charges (g and h) to be manifest only at the preon (and pre-preon) levels but hidden at the quark-lepton level. A model of this latter kind was recently presented in Ref. 8, in which the flavons are assigned charges (g,0), chromons (0,h) and somons (-g,-h). (This charge pattern corresponding to an ordinary U(1) U(1) symmetry was suggested in 1975 (Ref. 19).)

To illustrate these idea two specific models of preons are worth considering, each of which is based on the idea of abelian dual binding charges. They differ from each other in that the flavour colour attributes are defined for one and not for the other at the preon level.

(*Model I*): *A model of preons with dual charges with preons possessing flavour and colour attributes*

$$f_i^a = (u_{1/2}, d_{-1/2})^a, \quad c_\alpha^b = (r_{1/6}, y_{1/6}, b_{1/6}, \ell_{-1/2})^b$$

$$S = \{s_0^k\}^c$$

---

* See further discussions regarding constraints from (g-2) and other experiments in for example S.J. Brodsky, S.D. Drell, SLAC-PUB 2534(198

** In this case, photon must be a composite like W, Z and gluons.

Here the superscripts a,b,c denote the magnetic binding charges and the subscripts the electric charges of the preons. A minimal assignment $(a,b,c) = g_0(+1, +3, -4)$, where $g_0^2/4\pi \sim O(137)$, guarantees that quarks and leptons, interpreted as $fiC_\alpha S$ composites are neutral with respect to magnetic charge *and* that these are the only magnetically neutral three-body composites (apart from $\overline{fCS}$). This provides a reason why fCS, fCS, ffS, etc. which are magnetically charged, are not among the low-lying spectrum.

For the variant, where the primordial spon-1 gauge particle is A', distinct from photon, and *both* charges g and h are hidden at the quark-lepton level a,b,c would specify both g and h. One special assignment is this: $a = (g,0)$, $b = 0,h)$ and $c = (-g.-h)$ with $gh/4 = 1/2$ (integer).

(Model II): A model of preons, with dual charges, without flavour and colour attributes

Models of this class have been proposed by I.G. Koh, A.M. Rodriguez-Vargas and myself.[24] Assume that there exist three sets of preons, as before, but each set possessing only two members; the two members differ from each other in respect of their electric charges, but possess the same binding charge:

$$\{preons\} = \begin{matrix} P_{1/3}^a & P_{1/3}^b & P_{1/3}^c \\ R_0^a & R_0^b & R_0^c \end{matrix}$$

As before, the superscript denotes the binding charges and the subscript the electric charge. Assign minimally $(a,b,c) = g^0 (+1, +3, -4)$, or alternatively for the variant $a = (g,0)$, $b = (o,h)$ and $c = (-g,-h)$. Among three index combinations, only (abc) or $(\overline{abc})$ are neutral with respect to binding charges. Quarks and leptons may now be identified with the following unique and *distinct* three preonic composites which are magnetically neutral:

$$P^a P^b P^c = e^+$$

$$R^a R^b R^c = \nu$$

$$R^a P^b P^c, R^b P^c P^a, R^c P^a P^b = u_{red,yellow,blue}$$

$$P^a R^b R^c, P^b R^c R^a, P^c R^a R^b = \bar{d}_{red,yellow,blue}$$

The antiparticles may be constructed similarly. Both flavour and colour have emerged dynamically at the composite level. Note that the preons $P^{a,b,c}$ and $R^{a,b,c}$ do not possess well defined flavour

and colour transformation properties. This model resembles one
feature of the Harari-Shupe model (based on just two rishons),
in that the specific correlation between electric charges of the
preons and the composite is the same in our case as in the Harari-
Shupe model. The difference lies in that the model presented here
permits quarks and leptons to be *neutral* with respect to the binding
charges; the HS model does not (regardless of whether the binding
force is abelian or non-abelian), which would appear to be a
difficulty of the HS model. Furthermore, there is in our case a
reason for the generation of a three valued colour degree of freedom
at the composite level; this is due to the fact that there are
three sets of preons with *distinct* binding charges (a,b,c) but the
same electric charge. A corresponding reason is missing in the
HS model.

The generation of effective gauge symmetries like [SU(16) or
$SU(2)_L \times SU(2)_R \times SU(4)^c] \times$ (possible U(1) factors), where the
U(1)'s denote symmetries generated by traces of $\bar{f}f$, $\bar{C}C$ and $\bar{S}S$
currents has been discussed in Refs. 8 and 25.

We now observe one interesting possibility for the construction
of three families + three mirror families for spin 1/2 preons. Take
$f_{L,R}$ $C_{L,R}$ and $S_{L,R}$ to have magnetic charges proportional to +1,
+3 and -4 (for example). Naively threr are two attractive pairs
(fS) and (CS). Assume that (fCS) composites, either attractive pair
(fS) or (CS) moves in the direction of motion of the (massless)
composite with the third preon C or f moving in the opposite
direction and that *the repelling members f and C do not move together*. This has the amusing consequence that there are just three
distinct configurations for generating spin 1/2 (f_L $C_{L,R}$ $S_{L,R}$)
composites which couple to composite W_L with left helicity and three
additional configurations which couple with right helicity. Similar
remarks apply to couplings of W_R, etc. This may provide a reason
for the existence of three families plus three mirror families (the
latter are needed for cancellation of anomalies within maximal
symmetries like (SU(16).

Apart from magnetically neutral composites representing
standard quarks, leptons and their gauge particles, we expect
magnetically charged relatively heavy fermions, gauge and Higgs
bosons should also form as preonic composites. These will be
subject to the strong magnetic force as well as to the effective
electroweak strong forces. These may thus play the role* of
technifermions, technibosons and extended technigauge bosons. Note
this extra dimension in composite spectrum can arise without the
introduction of any new attribute at the preonic level.

---

* It is amusing to note that $(\alpha_m/\alpha_c) \sim (137)/(0.2) \sim 10^3$, which
corresponds to the ratio of $(\Lambda_T \Lambda_C) \sim (300 \text{ GeV})/(0.5 \text{ GeV})$.

Thus ideas of grand unification including those of technicolour may find their support through a simple economical picture of preons or pre-preons governed by a single primordial force. Even with a composite picture for quarks, leptons and gauge bosons, proton must decay, though the characteristic mass scales M can differ drastically (in that it may be much lower) compared to the case of elementary quarks, leptons and gauge bosons.

REFERENCES

1. J.C. Pati and Abdus Salam, "Lepton hadron unification" (unpublished) reported by J.D. Bjorken in the proceedings of the 15th High Energy Physics Conference held at Batavia, Vol. 2, p. 304, September (1972); J.C. Pati and Abdus Salam, Phys. Rev. D8, 1240 (1973).
2. J.C. Pati and Abdus Salam, Phys. Rev. Letters 31, 661 (1973); Phys. Rev. D10, 275 (1974); Phys. Letters 58B, 333 (1975); J.C. Pati, Proceedings of the Seoul Symposium (1978).
3. H. Georgi and S.L. Glashow, Phys. Rev. Letters 32, 438 (1974).
4. T.D. Lee and C.N. Yang, Phys. Rev. 98, 1501 (1955).
5. H. Georgi, H. Quinn and S. Weinberg, Phys. Rev. Letters 33, 451 (1974).
6. T. Goldman and D. Ross, Cal. Tech. preprint (1979).
7. J.C. Pati, Abdus Salam and J. Strathdee, Il Nuovo Cimento 26A, 77 (1975); J.C. Pati, Proceedings of the Second Orbis Scientiae, Coral Gables, Florida, p. 253, January (1975); J.C. Pati, S. Sakakibara and Abdus Salam, ICTP, Trieste, preprint IC/75/93, unpublished.
8. J.C. Pati, Abdus Salam and J. Strathdee, ICTP, Trieste, preprint IC/80/180.
9. V.A. Kuzmin, Pisma Zh. E Ksp. Teor. Fiz. 13, 335 (1970); S.L. Glashow, Cargese Lectures (1979); R.N. Mohapatra and R.S. Marshak, VPI-HEP-90/1 and 80/2; L.N. Chang and N.P. Chang, CCNY-HEP-80/5 (1980); J.C. Pati, Abdus Salam and J. Strathdee, work reported at 20th Int. Conf. on H.E. Physics, Madison, Wisconsin (July 1980) and preprint to appear.
10. B. Deo, J.C. Pati, S. Raypoot and Abdus Salam, preprint to appear. See Ref. 15 for partial discussions on $\Delta F = 0$-decay.
11. H. Fritzsch and P. Minkowski, Ann. Phys. (N.Y.) 93, 193 (1975); H. Georgi, Proceedings of the Williamsburg Conference (1974).

12. B. Deo, J.C. Pati, S. Rajpoot and Abdus Salam, preprint to appear. See Ref. 15 for partial discussions on hierarchy.
13. J.C. Pati and Abdus Salam, "Quark-lepton unification and proton decay", ICTP, Trieste, preprint IC/80/72, to appear as invited talks by J.C. Pati in the Proceedings of the Grand Unification workshops held at Erice (March 1980) and Durham, New Hampshire (April 1980).
14. S. Weinberg, Phys. Rev. Letters 43, 1566 (1979); F. Wilczek and A. Zee, Phys. Rev. Letters 43, 1571 (1979); See F. Wilczek and A. Zee, preprint UPR-0135 T; R.N. Mohapatra and R. Marshak, preprints VPI-HEP-80/1,2; S. Glashow (unpublished).
15. R.N. Mohapatra and R. Marshak, Ref. 9.
16. S. Weinberg, preprint HUTP-80/A023; H.A. Weldon and A. Zee, preprint (1980); V. Elias and S. Rajpoot, ICTP, Trieste, preprint IC/78/159.
17. See L. Sulak, Proceedings of the Erice Workshop (March 1980) and M. Goldhaber, Proceedings of the New Hampshire Workshop (April 1980) and 1980 ν Conference (Erice), June (1980) for a description of the Irvine-Brookhaven-Michigan set up; The Harvard-Purdue-Wisconsin experiment is the other parallel experiment of comparable magnitude in progress (private communications, D. Cline and C. Rubbia). The Tata Institute - Osaka collaboration working at Kolar Goldmines, India, has recently reported two promising candidates for proton decay (S. Miyakee, Proc. of Neutrino Conf. held at Erice (June 1980). The present best limit on proton lifetime is given by F. Reines and M.F. Crouch, Phys. Rev. Letters 32, 493 (1974).
18. See for example R. Wilson, Proc. of New Hampshire Workshop (April 1980) and M. Baldo-Ceoline, Proc. of 1980 ν Conference (Erice, 1980).
19. J.C. Pati and Abdus Salam, Phys. Rev. $D10$, 275 (1974); Proceedings of the EPS International Conference on High Energy Physics, Palermo, June (1975); p. 171 (Ed. A. Zichichi) J.C. Pati, Abdus Salam and J. Strathdee, Phys. Letters $59B$, 265 (1975).
20. Several other authors have also worked on composite models of quarks and leptons with an emphasis on classification rather than gauge unification of forces. K. Matumoto, Progr. Theor. Phys. 52, 1973 (1974); O.W. Greenberg, Phys. Rev. Letters 35, 1120 (1975); H.J. Lipkin, Proc. EPS Conf. Palermo, June (1975), p. 609 (Ed. A. Zichichi); J.D. Bjorken, C.H. Woo and W. Krolikowski (all unpublished); E. Nowak, J. Sucher and C.H. Woo, Phys. Rev. $D16$, 2874 (1977); H. Terezawa, preprint, Univ. of Tokyo, September (1979); INS-Rep-351 (a list of other references may be found here). H. Harari, Phys. Letters $86B$, 83 (1979) and M.A. Shupe, Phys. Letters $86B$, 78 (1979) have recently proposed the most economical model of all, but with a number of dynamical assumptions, whose bases are not clear.

21. J.C. Pati, University of Maryland preprint TR-80-095 (April 1980). "Magnetism as the origin of preon binding", to appear in Phys. Letters.
22. P.A.M. Dirac, Proc. Roy. Soc. (London) A*133*, 60 (1931); Phys. Rev. *74*, 817 (1948); J. Schwinger, Phys. Rev. *144*, 1087 (1966); D*12*, 3105 (1975); Science *165*, 757 (1969); D. Zwanziger, Phys. Rev. *176*, 1489 (1968); R.A. Brandt, F. Neri and D. Zwanziger, Phys. Rev. Letters *40*, 147 (1978); G. 't Hooft, Nucl. Phys. B*79*, 276 (1974); A.M. Polyakov, JETP Letters *20*, 194 (1974); C. Montonen and D. Olive, Phys. Letters B*72*, 117 (1977); P. Goddard, J. Nuyts and D. Olive, Nucl. Phys. B*125*, 1 (1977).
23. The formalism may follow that of D. Zwanziger, Phys. Rev. D*3*, 880 (1971).
24. I.G. Koh, J.C. Pati and A.M. Rodriguez-Vargas (to appear).
25. J.C. Pati, Abdus Salam and J. Strathdee, ICTP, Trieste, preprint IC/80/183.

BARYON CONSERVATION (Experiments)

Maurice Goldhaber

Brookhaven National Laboratory
Upton, New York 11973

The organizers of this conference were perhaps subconsciously aware that the discussion of proton stability started at about the same time as the concept of the neutrino, 50 years ago. During most of the half century since Weyl, Stückelberg and Wigner first postulated proton stability, nearly everybody believed in it as an absolute truth.[1]

About a quarter of a century after Weyl we realized that proton stability is an empirical question and asked: "Is the proton really stable?" Of course, experimentally we cannot prove absolute stability. But we can at least give quantitative limits of stability or find the actual decay of the proton. This was our attitude. By now everybody has gotten used to the idea that so-called "elementary particles" can decay, but I still remember the shock I felt when we found at the Cavendish in 1934 that the free neutron weighed definitely more than the hydrogen atom and that it therefore should decay by β emission with a half life which we then could estimate roughly from β-decay systematics to be about half an hour. In order to avoid confusion with the β-decay of the neutron when we talk of nucleon decay I shall usually use proton decay as a generic term by which I often mean to include the decay of neutrons which have been stabilized against β-decay by nuclear binding.

The first results on limits for proton stability were obtained in 1954:[2] two methods were discussed, a nuclear method and a counting method. The nuclear method permits us to obtain a lifetime limit for a nucleon because of the fact that nuclear shells exist. Only rarely, if one of the loosely bound nucleons would decay, would the nucleus be left unexcited. Usually some excitation energy is left in the nucleus which leads to a de-excitation which can be detected in various ways. The mode of de-excitation which we considered at first was induced fission which would look like spontaneous fission. Since spontaneous fission had been looked for in thorium but not found, we could give a limit for nucleon decay of $\sim 10^{20}$ y. The best limit for spontaneous fission of ^{232}Th was obtained later by Flerov. If we multiply his limit by ~ 200, the approximate number of particles in Th, we get an approximate limit for nucleon decay of $\sim 2 \times 10^{22}$ y. To a very high approximation it would not matter how the nucleon decays. For instance, if a neutron would decay into three neutrinos which could be described as "disappeared without a trace", it would still leave a hole behind which is equivalent to a high average nuclear excitation and this would lead to fission.

The nuclear method was extended by Peter Rosen[3] who proposed the use of more versatile radiochemical methods. Essentially one uses the fact that an excited nucleus resulting from proton decay can emit 0, 1, 2, ... nucleons and one can calculate the relative probabilities roughly from the shell model. For instance, in ^{130}Te the decay of a neutron, if it does not leave the nucleus too highly excited, will lead to ^{129}Te which decays into ^{129}I and ultimately into ^{129}Xe. Typically the early experimental results were all parasitic, i.e., the experiments were originally done for different reasons. In this case people were interested in double-β- decay and looked for ^{132}Xe from the isotope ^{132}Te. They also got limits on ^{129}Xe and from this Evans and Steinberg[4] deduced a limit of 1.6×10^{25} y, again independent of the decay mode. And then Fireman, et al,[5] and Steinberg et al[5] in 1977 did a deliberate or "dedicated" experiment. They had a large amount of potassium where the ^{39}K can, after the decay of one nucleon and the loss of another one, ultimately end up as ^{37}Ar, for which there are very sensitive methods of detection developed by Ray Davis. They obtained a limit for nucleon decay $>2.2 \times 10^{26}$ y. Thus nuclear methods have gone from $\sim 10^{20}$ to $\sim 10^{26}$y. The counting methods in which certain charged particles of some minimum energy are detected were largely pursued by Reines and his collaborators and have reached $\sim 10^{30}$ y as a limit for nucleon decay.[6]

There is a proposal for a nuclear method which might go well beyond 10^{26} y, but it is a difficult one. It is due to Charles Bennett of Princeton University who takes samples of mica from a deep mine so that they have not been exposed to too many cosmic rays. If a nucleon were to decay, π mesons would be emitted sometime and reabsorbed in the nucleus leading to spallation. The spallation gives heavier tracks which etching brings out and these etched tracks can then be counted. It is very laborious; he has now samples from fairly deep down and he thinks in principle if he gets very old samples he could reach lifetimes close to those obtained by the present counting methods. This is probably somewhat optimistic but it is interesting that methods which are independent of decay modes may go pretty far.

If Weyl had asked himself how can a proton decay, he would not have been able to write down a single equation conserving charge, energy, spin and momentum because he speculated on proton conservation before the positron was discovered. But look at the embarrassment of riches which we have now. Fig. 1 shows all the known particles whose masses are below that of a nucleon. (I hope you will forgive me for not putting the neutrino masses at their "correct" place.) Therefore in principle the nucleon could decay in many ways, involving at least one fermion to conserve spin, and one or more other particles. Of course proton decay could only happen if the charges of, say, the proton and the π^+ were exactly equal because otherwise charge conservation alone would stop it.[7] On the other hand, if we ever find nucleon decay then we could conclude that the charges are equal. The present limits for the differences between π and nucleon charges, or positron and nucleon charges, etc., are of the order of 10^{-19} of an electron charge.

While it is against my taste to talk of experiments in the future tense, it is unavoidable in this field. I shall therefore now talk about the experiments which are either in progress or planned. It is a very laborious task because proposals, some tentative, have been reaching me in a haphazard manner, mostly in the last few days, some only this afternoon, and so I hope you forgive me if it is not perfectly organized. Until very recently most experiments were parasitic to neutrino experiments for the very simple reason that to detect neutrinos with their small cross sections people built very massive counters and such counters can also be useful for the very long lifetime looked for in proton decay. Just as an illustration, one ton of matter, say, has roughly 10^{30} nucleons, so if the lifetime of the proton were 10^{30} years only one would decay per year in a ton of matter and therefore to push this limit much further one needs many tons of matter. In the United States there are four different attempts to measure the proton lifetime in progress. I shall talk first about these and then tell you of plans outside the United States. Dr. Miyake will talk later in detail about an Indian-Japanese collaborative effort which is in progress. Now why do we all make this effort at this time? While the older generation of theoretical physicists felt it in their bones that the proton is stable, the younger generation has a visceral feeling that it decays, as we have just heard from Professor Pati.

The lifetimes which are predicted by the Grand Unified Theories (GUTS) are very seductively close to the old experimental limit and it looks as if just a little extra effort should bring us to the promised land. So, very many groups are now in the

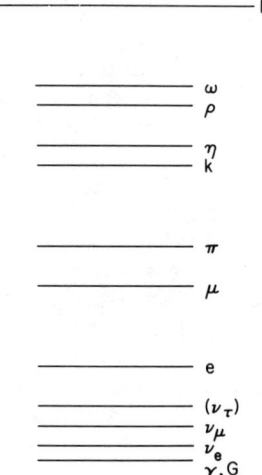

Fig. 1. Schematic presentation of particles of mass lower than that of a nucleon.

process of preparing or proposing experiments, and for the first time instead of being parasitic to neutrino experiments these proposed experiments are dedicated to proton decay studies, and by being dedicated they emphasize sometimes a different technique than one most useful in neutrino research. But of course the situation is immediately turned around: as soon as you build these very large counters there are many new neutrino ideas which have to be tested and the neutrino becomes parasitic to the proton, so let us perhaps drop this nasty word and say from now on that these experiments will be often symbiotic.

I hope you will forgive me if I start out with the experiment I am best acquainted with, the one with which I am connected, the Irvine-Michigan-Brookhaven search for proton decay. We tried to find a place where it would not be too expensive to dig a very large hole because we wanted to do an experiment with a sufficient amount of material that we could test some of the predictions for the proton lifetime, e.g., those based on SU(5) and related higher groups which predict lifetimes well below 10^{33} y. We hope to do an experiment which can reach, or very closely reach, this limit. The place we chose was one known to Reines and collaborators for a long time. It is a salt mine in Fairport Harbor, Ohio, not far from Cleveland, and it is a place where Mary K. Gaillard comes from, a good omen!

If one wants to use a very large amount of material at a reasonable cost, water seems a good choice. We intend to use Čerenkov radiation from water. If a proton decays, the energetic particles would produce Čerenkov radiation and if we can detect the radiation efficiently, and if we can reconstruct the events, then we should be able, by using several thousand tons of water, to reach the present theoretical limits.

The Čerenkov spectrum in water decreases steeply as the wavelength increases, but the ultraviolet part gets very rapidly absorbed in water, leaving only the visible part of the spectrum and therefore some people are using a different method, a wave shifter, which shifts the ultraviolet part of the spectrum to the visible where the water does not absorb appreciably, and where the phototubes can detect it easily. We chose instead to stay with the Čerenkov radiation alone because we believe we can in this way define the positions better.

After 10 meters of water the Čerenkov light left is essentially in the visible region where water is very transparent. The mean free path at the maximum transmission ($\sim$300 nanometers) is over 40 meters in pure water. Our phototubes have a wavelength sensitivity which matches the surviving photon spectrum closely.

To test the ideas, a water tank was built which has a 10-meter baseline. It was built at Michigan University and tested by some members of our group there. The test is made by allowing a coincidence between a cosmic ray μ meson traversing the water (defined by a coincidence between two counters) and phototubes recording Čerenkov radiation from the μ meson. We could predict 2.5 photoelectrons at 35 feet. Actually three photoelectrons were seen. The water was carefully purified and the amount of water circulating per unit time was scaled so that we should be able to achieve the same kind of purification for the large volume of water to be used in the final arrangement. The big hole which had to be dug at the mine is now essentially finished.

Our plans were discussed in detail at the Bergen ν '79 meeting.[8]

We believe we can take care of backgrounds by using just the inner part of the water cube as the fiducial volume, with the outer part as a sort of anticoincidence by software, except that one cannot stop neutrinos from initiating, inside the fiducial volume, events which might sometime mimic proton decay. It was therefore very important that the results of a Gargamelle neutrino experiment, where the PS neutrinos had roughly the same spectrum as the atmospheric ones, were made available to us through the kindness of G. Martin and M. Pohl. We therefore expect in two years of running only about one background count due to the neutrinos simulating a proton decay. That would correspond to a limit close to 10^{33} y.

The digging in the salt mine was done by a continuous digger which has been developed by an English company called DOSCO. At present the hole is being prepared for the final putting down of the floor and then we will install a liner to hold the water. We hope that by late spring of next year, barring the usual holdups or lack of money, etc., we can start taking data.

One experimental effort by a University of Pennsylvania group[9] is getting data right now; this was due to the happy circumstance that they built some time ago a neutrino detector in the Homestake Mine where Ray Davis' chlorine detector is located. In fact their detector surrounds his solar neutrino detector. The Pennsylvania neutrino detector has symbiotically developed into a proton decay detector. They use wavelength shifters and look for μ-e decay, where a μ^+ could either be created directly from proton decay or indirectly as a decay product of the presumably more copious π^+'s. To reduce the data they use some theoretical branching ratio predictions.

They believe from a Monte Carlo study of the energy distribution of events expected from decays which lead to μ-e decay that none of their observed candidates are due to proton decay and therefore they can give a limit of $\gtrsim 2 \times 10^{30}$ y for the proton lifetime. They will now go from about 150 tons to 500 tons of water and hope to push the limit to $\sim 10^{31}$ y.

The Harvard-Purdue-Wisconsin Group[10] is building a one kiloton water detector in an old silver mine in Park City, Utah, which is the place where Keuffel and his group used to do their cosmic ray research. It has a depth of 1800 meter water equivalent; they also use a water Čerenkov counter, but with a wavelength shifter and they distribute their phototubes throughout the counter volume. In this way they will obtain good energy resolution.

A University of Minnesota group uses an old iron mine at Soudan, Minnesota.[11] They use very cheap counters which can be built in small units and as more money becomes available they can build more and more units. They make gas proportional counters from simple steel tubes embedded in ferro-concrete which they get cheaply and the first modules are in the mine. They want at first to build a 45-ton detector to just check the ideas. They expect to be able to recognize a number of different decay modes.

Tables 1 - 7 on planned proton lifetime experiments are based on questionnaires filled out for this conference by a number of research groups planning proton lifetime experiments outside the United States.

Recently there has been a renewed interest in n-n̄ oscillations.[12] An experiment will soon start at the Laue-Langevin Center at Grenoble[13] and another is being proposed for Oak Ridge.[14]

Table 1. P Lifetime Experiments

Collaborative Institutions	Frascati-Milano-Torino plus support from CERN (1st Generation Experiment)
Location	Mont Blanc Tunnel (Garage 17)
Depth	> 5000 metres of w.e.
Weight of Detector	150 tons
Method of Detection	1 cm iron plates interleaved with limited streamer tubes
(Partial) p Lifetime Limits Obtainable	$10^{31} - 10^{32}$ y depending on length of the run
Present Status and/or Time when (Preliminary) Results are Expected	Experiment in preparation. First data end of 1981. Exposure to neutrinos from unfocused 10 GeV protons at CERN.

Table 2. P Lifetime Experiments

Collaborative Institutions	Frascati-Milano-Rome-Torino (2nd Generation Experiment)
Location	In Fréjus or other European Tunnels
Depth	∼ 4000 m(w.e.)
Weight of Detector	A few kilotons
Method of Detection	Fine grain calorimeter
(Partial) p Lifetime Limits Obtainable	10^{32} - 10^{33} y depending on length of the run
Present Status and/or Time when (Preliminary) Results are Expected	Letter of Intent to the Italian authorities.

Table 3. P Lifetime Experiments

Collaborative Institutions	Orsay-Ecole Polytechnique-Saclay, plus Italian Groups
Location	Fréjus Tunnel
Depth	~ 4000 m(w.e.)
Weight of Detector	2 kilotons at start
Method of Detection	Calorimetry with 4mm section flash tubes plus Fe (3mm); Possibly C in H_2O
(Partial) p Lifetime Limits Obtainable	$\sim 10^{32}$ y
Present Status and/or Time when (Preliminary) Results are Expected	Negotiation on excavation on possible Franco-Italian laboratory in the Fréjus Tunnel, and tests on 5m long, 4 mm cross section plastic flash chambers.

Table 4.   P Lifetime Experiments

Collaborative Institutions	Istituto di Cosmo-Geofisica del CNR - Torino Institute for Nuclear Research-Moscow Laboratori Nazional Frascati - INFN - Frascati
Location	Laboratory of the Mt. Blanc Tunnel
Depth	$\sim$ 4,270 m(w.e.)
Weight of Detector	$\sim$ 60 tons of Liquid Scint. $\sim$ 40 tons of Iron
Method of Detection	Liquid scintillator detection and anticoincidence system } 100 Tons
(Partial) p Lifetime Limits Obtainable	5×10^{31} y
Present Status and/or Time when (Preliminary) Results are Expected	End of 1981 (20 tons are running at present).

Table 5. P Lifetime Experiments

Collaborative Institutions	Institute for Nuclear Research of the Academy of Sciences of the USSR, Moscow
Location	Salt mine, Artyemorsk, Ukraine
Depth	~ 600 m(w.e.)
Weight of Detector	100 ton liquid scintillator, surrounded by ~ 200 ton liquid scintillator anticoincidence.
Method of Detection	Liquid scintillator 128 PM-tubes
(Partial) p Lifetime Limits Obtainable	10^{31} years?
Present Status and/or Time when (Preliminary) Results are Expected	Only projected so far.

Table 6. P Lifetime Experiments

Collaborative Institutions	Institute for Nuclear Research of Academy of Sciences of the USSR, Moscow
Location	Baksan Valley, North Caucasus
Depth	~ 850 m(w.e.)
Weight of Detector	80 tons of liquid scintillator 220 tons of l.s. anticoincidence shield
Method of Detection	1200 liq. sc. detectors (Internal part of Baksan scintillator telescope)
(Partial) p Lifetime Limits Obtainable	For leading to μ–e decay only $\sim 5 \times 10^{30}$ years
Present Status and/or Time when (Preliminary) Results are Expected	Starting from 1 July 1980. The beginning of 1981.

Table 7. P Lifetime Experiments

Collaborative Institutions	Tata Institute of Fundamental Research, Bombay, India Osaka City University, Osaka, Japan and ICR, University of Tokyo
Location	Kolar Gold Field, South India
Depth	$\sim$ 7600 m(w.e.)
Weight of Detector	150 tons. Fiducial volume $\sim$ 100 tons.
Method of Detection	PR counter array in crossed geometry
(Partial) p Lifetime Limits Obtainable	$\sim 10^{31}$ y
Present Status and/or Time when (Preliminary) Results are Expected	Within this year.

REFERENCES

1. For a review see, e.g., M. Goldhaber, P. Langacker, and R. Slansky, Science, in press.
2. F. Reines, C. L. Cowan, Jr., and M. Goldhaber, Phys. Rev. 96:1157 (1954).
3. S. P. Rosen, Phys. Rev. Lett., 34:774 (1975).
4. J. C. Evans, Jr., and R. I. Steinberg, Science, 197:989 (1977).
5. E. Fireman et al.; R. I. Steinberg et al., International Conference on Neutrino Physics, Elbus, USSR, June 1977.
6. F. Reines and M. F. Crouch, Phys. Rev. Lett. 32:493 (1974); J. Learned, F. Reines and A. Soni, ibid. 43:907 (1979).
7. G. Feinberg and M. Goldhaber, Proc. Nat. Acad. Sci., 45:1301 (1959).
8. ν '79 Conference, Bergen, Norway.
9. K. Lande and R. I. Steinberg, et al., private communication.
10. J. Blandino et al., Harvard-Purdue-Wisconsin Proposal.
11. R. Marshak et al., Minnesota Proposal.
12. See R. N. Mohapatra and R. E. Marshak, Phys. Lett., 94B:183 (1980) for earlier references.
13. M. Baldo-Ceolin, private communication.
14. R. Wilson, private communication.

COMMENTS ON "THE ELECTRON NEUTRINO MASS FROM ^{3}H-^{3}He β-DECAY [+]

14 eV $\leq m_{\bar{\nu}_e} \leq$ 46 eV (99 % C.L.)" (V.A. Lyubimov et al.)

K.E. BERGKVIST

Institute of Physics, University of Stockholm
Vanadisvägen 9
113 46 Stockholm, Sweden

GENERAL FEATURES OF RESULT 14 eV $\leq m_{\bar{\nu}_e} \leq$ 46 eV (99 % C.L.) (1)

1) Profond impact on current theory building – if assumed correct.

2) Systematic errors – perhaps even to the extent of making 14 eV → 0 <u>not obviously</u> completely excluded.

3) Almost fail-safe statement – at least for reasonable future. Proving that (1) is really wrong at the lower limit would imply setting new upper limit like, say

$$m_\nu \leq 10 \text{ eV} \qquad (2)$$

Such an accuracy does not seem achievable without some new break-through in basic procedure.

4) Confirmation of (1) – especially if m_ν is not too far from the upper limit – should not be too difficult, combining now available procedures.

[+] The year 1980 for the first time saw experimental claims for a finite neutrino mass. We reproduce here the overhead notes used by K.-E Bergkvist in a commentary talk following the presentation – by Dr. A. Rozanov – of the work of Lyubimov et al. at the Neutrino '80 Conference in Erice, Sicily, in June 1980. The talk by Dr. Bergkvist replaced an intended panel discussion which was cancelled in the absence at the Conference of any actual member of the Russian team behind the neutrino mass work. The instruction to Dr. Bergkvist was to take the role of the "Devil's advocate" in his comments on the Russian work. Ed.).

Presumably:

1) and 4) will strongly stimulate further efforts
2) will necessitate further and deeper analysis of details
3) will somewhat deter some people to enter, the interesting and realistically probed mass interval already used up in (1).

THE IMMEDIATE EXPERIMENTAL ACCURACY: HOW IT WAS IMPROVED

A. Big, iron-free multi-foci β-spectrometer with excellent properties especially with respect to keeping background from scattering in the walls of the tank down.

B. Predetermination of over-all spectrometer response function $R_{i,k}$ allowing both for backscattering in the source backing and for energy losses in the activity layer.

C. Inclusion of electrostatic method (Bergkvist 64, 72) allowing an increase in source area (i.e. in intensity) by appr. a factor of 10.

D. Detectors with background suppression

E. Endurance and dedication of team

Measures A. and B. the most specific for the work, but all points A. - E. probably essential.

CONCEIVABLE SYSTEMATIC ERRORS

A. The very parametrization of the measured β-spectrum:

$$N(p_i) = A \sum_k S(p_k, M_\nu, E_0) \left[1 + \alpha(p_{0k} - p_k)\right] R_{i,k} + \phi \quad (1)$$

is not readily fully understood.

First is it assumed that

$$p_{0k} = p_{0k'} = p_0 \ ?$$

Moreover, a parametrization capable of allowing for any amount of backscattering from the source would rather like

$$N(p_i) = A \left[1 + \alpha(p_0 - p_i)\right] \sum_k S(p_k, M_\nu, E_0) R_{i,k} + \phi \quad (1')$$

QUESTIONS

(i)   Are (1) and (1') both appropriate in actual application?
(ii) How much do the obtained χ^2 minima move in energy when all the conceivable experimental errors in $R_{i,k}$ are allowed for?
(iii) The determination of $R_{i,k}$ is such that charging-up of the

ELECTRON NEUTRINO MASS

Valine T layer will not be seen in the measured line profile. How to be sure the Valine layer does not acquire some potential of 5 - 10 volts with respect to the source backing?

(iv) Do the polarized ($v/c \simeq 0.3$) β-particles with certainty scatter in the source precisely as the test conversion electrons? (Client's inquiry)

B. SPECTRUM MODIFICATION DUE TO ATOMIC INTERPLAY IN THE β-PROCESS

β-spectrum assumed to be given by

$$S(p, M_\nu, E_0) = F(E)p^2 \left[W_1(E_0-E)\sqrt{(E_0-E)^2 - M_\nu^2} + W_2(E_0-E^*-E)\sqrt{(E_0-E^*-E)^2 - M_\nu^2} \right] \quad (1)$$

$W_1 = W_2 = 1$

QUESTIONS

(i) Suppose mean square energy spread in β decay energy in Valine layer maintains its T and T_2 values, i.e.

$$(\Delta E_{\beta+\nu})^2 = \langle\phi_i|(\Delta V)^2|\phi_i\rangle - (\langle\phi_i|\Delta V|\phi_i\rangle)^2 \simeq (27 \text{ eV})^2 \quad (2)$$

How does the position of the relevant χ^2 min. move when all distributions of $\Delta E_{\beta+\nu}$ compatible with (2) are allowed for?

(ii) Relaxing from just the $(27 \text{ eV})^2$, what variations in extracted ν_e mass can be obtained using proper insight in the atomic interplay?

C. OTHER CONCEIVABLE ORIGINS OF SPECTRUM DISTORTION

(i) Ghost intensity from activity migrated within the source arrangement.
(ii) Host of others, requiring detailed knowledge of every step in the experiment.

CONCLUDING REMARKS ABOUT RUSSIAN WORK

1. Clearly:
   Introduction of new methods ⟨ Importance ⟨ Change of paradigm
   of values in further efforts ⟨ of work ⟨ in ν physics
   on ν_e mass

2. No systematic effect which obviously is big enough to produce the same type of end-point modification. More detailed account of work <u>requested</u>!

3. In general, ν_e mass lower limits easier to claim than to prove really wrong — if so.

4. Next measures – in order of urgency
   a) Independent reproduction of experiment, with some differences to improve the intensity situation (hence reduce time required)
   b) Further exploration of atomic interplay (ask quantum chemists)
   c) Search for basically new methods of measuring m_{ν_e}
   d) Perhaps motivated to reconsider effects from nuclear structure (closelying states) and radiative corrections with $m \neq 0$.

ELECTRO-WEAK PHYSICS IN e^+e^- INTERACTIONS

Min Chen

Department of Physics and Laboratory of Nuclear Science
Massachusetts Institute of Technology
Cambridge, Massachusetts 02139 U.S.A.

1. INTRODUCTION

At the core of our present understanding of weak interactions is the mediation of the weak force between two point-like fermions via a weak intermediate vector boson, W^{+-} or Z^o with vector and axial vector couplings whose strength may depend on the type(s) of fermions. Electron-positron collisions provide a particularly clean way to study the nature of these fundamental constituents of matter and their interactions because of the point-like nature of leptons. Experiments here are free of the complications that arise in studies of hadron-hadron or lepton-hadron collisions, where the hadrons themselves have a complex internal structure which must be understood in detail before new information on the large q^2 interactions of the constituents can be extracted. Therefore e^+e^- collisions at high CM energies can be directly used to probe the very fundamental questions in weak interactions, e.g., how many quarks and leptons are there? Are there additional weak bosons with mass below that predicted by the standard models? How large are the weak interaction coupling constants?

2. The High Mass Lepton, Tau

Let me use the high mass lepton, tau as an example. For many years physists wondered why the sequence of charged leptons should stop at electrons and muons. Early searches for another sequential lepton decaying into e or muon plus neutrinos were carried out at Frascati[1] up to a cm energy of 3 GeV setting a lower mass limit of 1.15 GeV in 1975. Immediately following the discovery of the J particle[2] and the charmed particles,[3] an abnormal excess of events of the type $e^+ + e^- \to e^\pm + \mu^\mp$ and nothing else in the detector were seen at SPEAR,[4] providing the first evidence of a new weakly decaying particle with a mass of about 1.8 GeV, decaying into $\mu(e) + \nu + \bar{\nu}$. The mass was later determined from the threshold behavior of the production cross section (see Figure 1) to be 1.782 ± 0.002 GeV, i.e., slightly below the D meson mass of 1.868 GeV.

The spin of the tau is also determined from the threshold behavior of the production cross section. The cross sections $\sigma_{\tau\tau}$ in units of $\sigma_{\mu\mu}$ are calculated to be

$$= 1/4\, \beta^3 \qquad \text{If } s = 0$$
$$= \beta(3-\beta^2)/2 \qquad \text{If } s = 1/2$$
$$= \beta^2(\gamma^4 \pm 5\gamma^2 + 0.75) \qquad \text{If } s = 1$$

where $\beta = p_\tau/e_\tau$ and $\gamma = e_\tau/m_\tau$, and are shown in Figure 1 the energy dependence of the production cross section favors the hypothesis that the spin of the tau is 1/2. The weak interaction properties of the tau are revealed by studying its decay products. For example, from the electron momentum distribution[6] from tau $\to e+\nu + \bar{\nu}$ it was found that the V-A form is favored over the V+A form of the interaction, as in the case of muon decay. Various decay branching ratios have been measured and found to be in agreement with the expected properties of a new sequential lepton. The remaining questions are: "Is the tau, which is twice as heavy as the proton, also point-like?", and "What is the lifetime of the tau?" Both are studied at

ELECTRO-WEAK PHYSICS IN e^+e^- INTERACTIONS

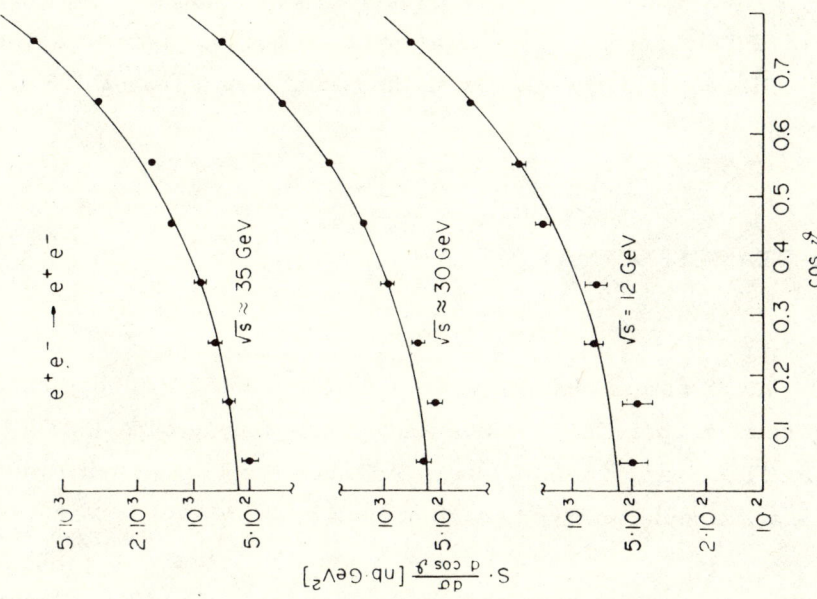

Fig. 2
Angular distribution of Bhabha scattering at PETRA energies measured by MARK J. The solid lines display the QED cross sections.

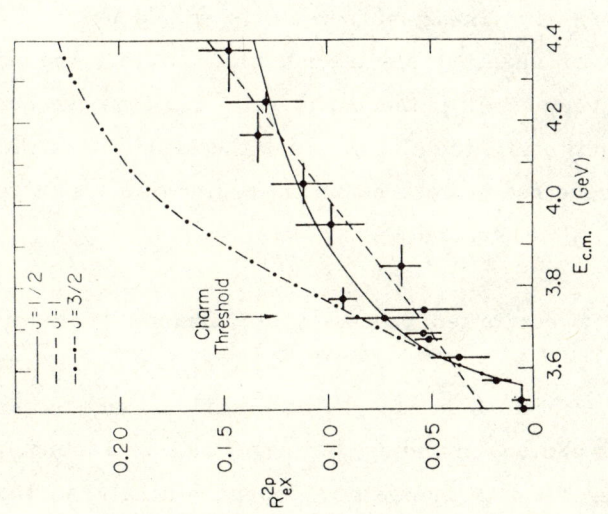

Fig. 1
Delco: Inclusive electron production in the two-prong events. The ratio of the electron to μ-pair production cross section is plotted versus CM energy, together with the prediction for spin 1/2, 1 and 3/2.

the newly completed e^+e^- storage ring at PETRA. Due to limited time today I will not try to cover the work on the weak decay of charmed particles performed[7] at SPEAR and DORIS, nor the work on bottom[8] carried out at CESAR and DORIS, but limit myself now to the recent results from PETRA.

3. ELECTRO-WEAK EFFECTS AT PETRA

PETRA[9] is the world's highest energy electron-positron colliding beam storage ring now in operation. There have been four groups (JADE,[10] MARK J,[11] PLUTO,[12] and TASSO[13]) using PETRA since 1979. Their results are supportive and complementary to each other.

At PETRA we study the reactions $e^+e^- \to \ell^+ + \ell^-$ for all the known charged leptons (ℓ = e, μ, τ) by measuring the dependence of the cross section on center of mass energy or scattering angle over a wide range of PETRA energies. These measurements enable us to compare the data with predictions of quantum electrodynamics, to test the universality of these leptons at very small distances, to set a limit on the charge radius of these particles and to test the electro-weak theories. Up to the present time these reactions have been measured at the center of mass energies $\sqrt{s}$ = 12, 13, 17, 22, 27.4, 30, 31.6, 35, and 36.6 GeV.

The lowest order processes for Bhabha scattering or μ (τ) pair production are of order α^2. The next higher order processes[14] consist of radiative corrections due to photon emissions, and the interference between the virtual photon (including $\ell^+\ell^-$ and hadron vacuum polarization) and the lower order terms. Since they depend strongly on the properties of the experimental set up (such as energy and angular resolution) the experimental results are presented here with the contributions from radiative corrections removed.

ELECTRO-WEAK PHYSICS IN e^+e^- INTERACTIONS

A measured cross section, say $\frac{d\sigma}{d\Omega}$, is corrected for radiative effects δ_r and hadronic vacuum polarization δ_h and then compared to the lowest order QED cross section:

$$\frac{d\sigma}{d\Omega}(1 - \delta_r - \delta_h) = \frac{d\sigma_{QED}}{d\Omega}(1 + \delta)$$

A deviation δ can have two sources, a violation of QED or weak interaction effects.

(a) $e^+e^- \to e^+e^-$ (Bhabha scattering).

To check the radiative corrections, the distribution of the acollinearity angle between the outgoing $e^+ + e^-$ has been measured out to $\sim 120°$ and the cross section varies over four orders of magnitude. The data agree very well with the QED prediction folded with the experimental resolution. Figure 2 shows the angular distribution of Bhabha scattering measured by MARK J at $\sqrt{s}$ = 12, 30 and 35 GeV.

To express the agreement between the data and QED analytically, we modified the QED formula with a form factor of the photon $F = 1 \mp q^2/(q^2 - \Lambda_\pm^2)$. The resultant limits on $\Lambda_\pm$ are listed in Table I.

(b) $e^+e^- \to \ell^+\ell^-$ with $\ell = \mu$ or τ.

The differential cross section reads

$$\frac{d\sigma}{d\Omega} = \frac{\alpha^2}{4s}(1 + \cos^2\theta).$$

Possible deviations from QED will depend on s and can only be detected by measuring the absolute magnitude of the cross section. This can be done e.g., by comparing μ-pair and τ pair production to small angle Bhabha scattering.

The resultant $e^+e^- \to \mu^+\mu^-$ and $\tau^+\tau^-$ cross sections, as a function of the cm evergy $\sqrt{s}$ and $\cos\theta$ are plotted in Figures 3 and 4 respectively together with the QED predictions. The deviation of the measured cross-section σ from QED prediction σ_{QED} due to a charge radius is described with a form factor F such that

$$\sigma = \sigma_{QED} F^2(s) \qquad \text{where } F(s) = 1 \mp s/(s - \Lambda^2 \pm)$$

The resulting cut-off parameters, Λ, at 95% confidence level, are listed in Table I.

Table I. Cut-off Paramters of ee, $\mu\mu$ and ee in GeV

	JADE	MARK J	PLUTO	TASSO	
e^+e^-:					
$\Lambda+$	132	97	80	150	GeV
$\Lambda-$	113	157	234	136	
$\mu^+\mu^-$:					
$\Lambda+$	137	192	116	80	
$\Lambda-$	96	139	101	118	
$\tau^+\tau^-$:					
$\Lambda+$	---	127	74	115	
$\Lambda-$	---	100	65	76	

The lifetime of the tau was recently measured by PLUTO and TASSO. The TASSO's limit is

$$\tau = 1.4 \times 10^{-12} \text{ sec (95% C.L.)}$$

to be compared with the predicted value of 2.8×10^{-13} sec., assuming $\mu-\tau$ universality. This limit can be converted into a lower limit for the strength of the weak coupling constant of the τ, g_τ in units of that of the electron or muon, g_e, i.e., $g_\tau > 0.46\ g_e$. This limit shows that the weak coupling constant of the τ, g_τ is greater than $0.46\ g_e$, where g_e is the weak coupling constant of the electron.

ELECTRO-WEAK PHYSICS IN e^+e^- INTERACTIONS

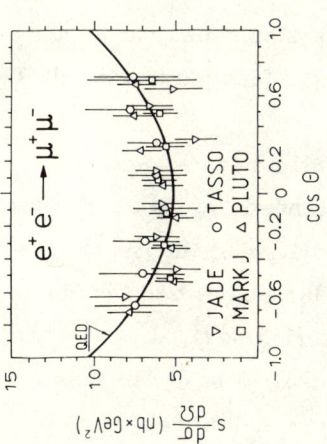

Fig. 4

Angular distribution of μ-pair production measured by PETRA experiments at energies between 27 and 35 GeV.

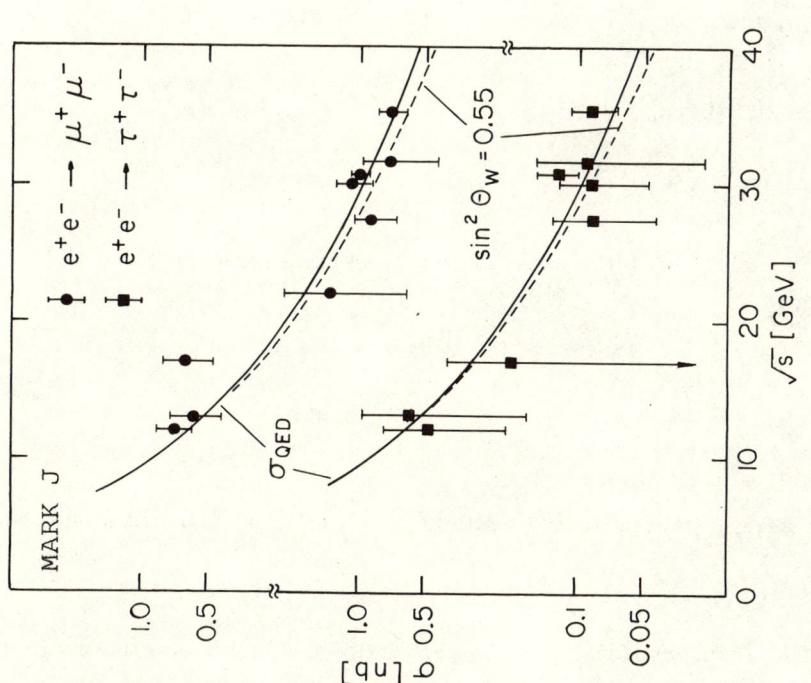

Fig. 3

τ and μ-pair cross section as a function of energy. Also indicated is the QED cross section (solid line) and the limits (dashed line) corresponding to $\sin^2\theta_W = 0.55$.

The electro-weak theory of Glashow, Weinberg, and Salam (GWS)[15] has become the "standard" model through its remarkable successes. The theory incorporates a single neutral gauge boson Z^o whose coupling is expressed in terms of a single free parameter $\sin^2\theta_w$.

There have been suggestions [16] that the standard model can be naturally extended to models based on an enlarged symmetry group SU(2) x U(1) x G with more than one Z^o, without changing any of the predictions which have been made for the low q^2 range covered by lepton-nucleon scattering data. In order to be consistent with low energy data, the lowest mass neutral boson in these models have a mass smaller than the Z^o mass in the standard model.

The deviation of the multiple Z^o models from the standard model can be approximately described by an additional term $C j_{em}^2$ in the effective Hamiltonian of the standard model. The effective Hamiltonian contains three parameters[17] h_{VV}, h_{VA}, and h_{AA}, where the parameter C only modifies[18] h_{VV} and $\sin\theta_w$.

$$h_{VV} = \frac{1}{4}(1-4\sin^2\theta_w)^2 + 4C, \qquad C > 0$$

$$h_{VA} = \frac{1}{4}(1-4\sin^2\theta_w)$$

while the terms containing the axial vector current remain the same as in the standard model,

$$h_{AA} = g_A^2 = \frac{1}{4}.$$

In the case where C vanishes, h_{VV} reduces to g_V^2 as expected in the standard model.

Since the term containing C is proportional to the electro-magnetic currents which couple to charge and is parity conserving, it contributes neither to the neutrino scattering processes, nor to polarized e-D scattering. On the other hand the reactions $e^+e^- \to \ell^+\ell^-$ ($\ell = e, \mu, \tau$) at present PETRA energies $\sqrt{s} = 35$ GeV begin to be sensitive to the effects of an additional Z with a mass below that of the standard model Z^0 and provide an opportunity to set stringent limits on C.

From the limit of C, limits on the masses of Z^0's in specific models can be obtained. As an illustration, we consider the case of two Z^0's with masses m_1 and m_2. In models based on $SU(2) \times U(1) \times U(1)$, with one doublet of charged and two neutral gauge bosons, we have the relation[18]

$$C = \cos^4\theta_W \left(\frac{m_Z^2}{m_1^2} - 1\right)\left(1 - \frac{m_Z^2}{m_2^2}\right),$$

while in models based on $SU(2) \times U(1) \times SU(2)$, with two doublets of charged and two neutral gauge bosons, we have[20]

$$C = \sin^4\theta_W \left(\frac{m_Z^2}{m_1^2} - 1\right)\left(1 - \frac{m_Z^2}{m_2^2}\right).$$

With $\sin^2\theta_W = 0.23$, for the same mass m_1 and m_2, C in the $SU(2) \times U(1) \times U(1)$ model is about eleven times greater than the C in $SU(2) \times U(1) \times SU(2)$ model. These two models represent approximately the extreme cases of the strength of coupling constants of these types of models.

The cross-sections for $e^+e^- \to \ell^+\ell^-$ ($\ell = e, \mu, \tau$) have been calculated in terms of $\sin^2\theta_W$ and the parameter C. Using the measured values of $\sin^2\theta_W = 0.23$, we make a one parameter fit to all of the ee, $\mu\mu$, $\tau\tau$, and the charge asymmetry of the $\mu\mu$ data which is shown in Table II while the normalization is allowed to vary within $\pm 3\%$ due to the uncertainty in the luminosity. The MARK J experiment found $-0.04 < C < 0.027$, at the 95% confidence level.

The upper limit on C can be converted in a limit in the $m_1 - m_2$ plane for the two models,[19,20] as shown in Figure 5. In all these models there is a constraint that $m_1 \lesssim m_Z \leq m_2$. Therefore, the two lines $m_1 = m_Z$ and

$m_2 = m_Z$ are the natural boundaries of all the allowed region. The data put severe limits on the SU(2) x U(1) x U(1) model, constraining the masses of the two Z's in a small region. The limits on the SU(2) x U(1) x SU(2) are less stringent but significant.

In the framework of the single Z models, $C = 0$, h_{VV} reduces to g_V^2, $h_{VA}^2 = g_V \cdot g_A$ and h_{AA} to g_A^2. With the assumptions that the known charged leptons act universally as point-like charges and the weak interactions depend only on vector and axial vector currents, constraints on g_V and g_A are obtained by fitting all the e^+e^-, $\mu^+\mu^-$, and $\tau^+\tau^-$ data along with the $\mu^+\mu^-$ asymmetry. The MARK J experiment found $g_V^2 = -0.07$ and $g_A^2 = 0.23$, with $\chi^2 = 34$ for 38 degrees of freedom. These results including correlations can be converted into an allowed region in the $g_V - g_A$ plane. Since the predicted effects do not depend on the sign of g_V and g_A, the allowed region is four fold symmetric, as shown in Figure 6 for the 95% confidence level contour (which corresponds to an increase in χ^2 of 6 from the minimum value). This contour corresponds to the choice $m_Z = \infty$, which of all possible m_Z values gives the weakest overall constraints in the $g_V - g_A$ plane. Also shown in Figure 6 are the standard deviation limits deduced from the purely leptonic processes[21]:

$$\nu_\mu e^- \to \nu_\mu e^-$$

$$\bar{\nu}_\mu e^- \to \bar{\nu}_\mu e^-$$

$$\bar{\nu}_e e^- \to \bar{\nu}_e e^-.$$

Neutrino electron scattering data alone limit the possible values of g_V and g_A to two regions in the $g_V - g_A$ plane. The first solution is around $g_V = 0$, $g_A = -1/2$ and the second solution is around $g_V = -1/2$, $g_A = 0$. Our data find $\chi^2/\text{D.F.} = 35/38$ for the first solution.

Thus, combining the νe data with the results on purely leptonic interactions obtained in the MARK J experiment rules out the

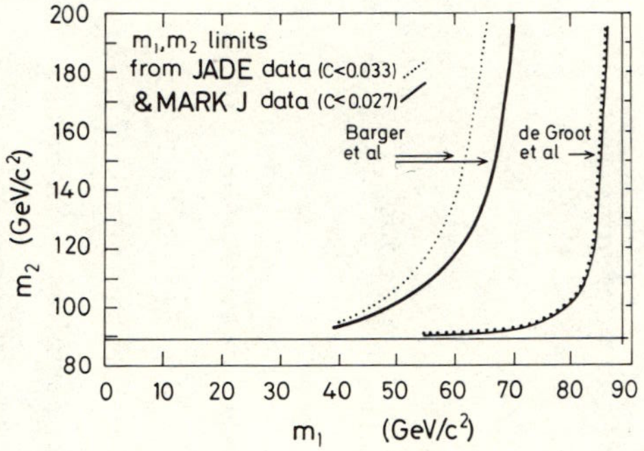

ELECTRO-WEAK PHYSICS IN e^+e^- INTERACTIONS

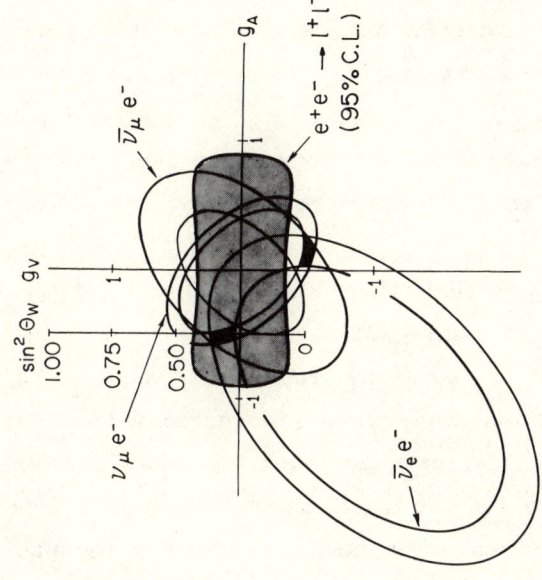

Fig. 6

Results obtained from neutrino experiments and the MARK J experiment expressed in terms of limits on g_V and g_A. The regions in between the concentric ellipses correspond to 1σ limits from the neutrino-electron scattering experiments. The two black areas indicate the two allowed regions for g_V and g_A from the combined neutrino data. The shaded area represents the 95% confidence limit contour from the MARK J experiment.

Fig. 5

Limits on m_1 and m_2 deducted from a limit on C for models with two Z^0 bosons.

second solution with more than 95% confidence. This confirms the conclusions drawn on the basis of deep inelastic neutrino nucleon scattering and polarized electron deuterium scattering data, without recourse of models of hadron production by the weak neutral current. We have also searched for new heavy leptons at PETRA, beyond the series e, μ, τ. Analagously to the τ-lepton we assume that a new heavy lepton HL couples universally to leptons and quarks according to the standard V-A weak interaction theory and has the following decay modes: (in %) $\tau^- \bar{\nu}_\tau \nu_{HL}$ (9.2), $\bar{\mu} \bar{\nu}_\mu \nu_{HL}$)10.6), $e \bar{\nu}_e \nu_{HL}$ (10.6) and hadrons + ν_{HL} (69.6). Owing to their large mass and low velocity, the decay products would be expected to have large angles with respect to the HL line of flight. This contrasts to the decay products of the τ at PETRA energies which are tightly collimated. Heavy lepton production is most easily recognized in most of the PETRA detectors for events in which one lepton decays into a muon and neutrinos and the other lepton decays into hadrons and neutrinos. No event was observed in the data. The limits on the mass of a new high mass lepton are listed in Table III.

SUPERSYMMETRY PARTICLES

In the framework of supersymmetric theories,[23] spin zero partners of the muon are expected to decay according to the reactions

$$\bar{s}_\mu \to \ell^- + \text{photino (goldstino)} \qquad \ell = \mu, e \qquad (a)$$

$$\bar{t}_\mu \to \ell^- + \text{antiphotino (antigoldstino)} \qquad (b)$$

where s_ℓ and t_ℓ are the spin zero partners of the lepton muon or electron associated with the left and right handed parts of the lepton field respectively, and the photino is the spin 1/2 partner of the photon and the goldstino is the goldstone fermion. Since s_ℓ and t_ℓ carry unit electric charge, once above threshold, they are expected to be produced in pairs in

e^+e^- annihilation according to the cross section:

$$\frac{d(e^+e^- \to s_\ell^- s_\ell^+ \text{ or } t_\ell^- t_\ell^+)}{d(\cos\theta)} = \frac{\pi \alpha^2 \beta^3 \sin^2\theta}{4s} \quad ; \quad \beta = ((1 - \frac{m}{E})^2)^{1/2}$$

Table II. $\mu^+\mu^-$ Charge Asymmetry

	MARK J	JADE	PLUTO		TASSO
$A_{\mu\mu}$ (%)	-5 ± 10	-3 ± 9	-8 ± 9	7 ± 10	-1 ± 12
Predicted by W-S model (%)	-5.5	-7.5	-6	-6	-6
E_{CM} (GeV)	30	35	30	30	30

Table III.   Search for Superheavy Sequential Lepton $e^+e^- \to$

Exp.	Method	M_L (GeV)
PLUTO	Single μ Recoiling Against Many Hadrons $E_{vis} > 3$ GeV; $\|P_{miss}\| > 2.5$ GeV	> 14
MARK J	Single μ ....., $N_c > 2$ $10\% < \frac{E_{vis}}{\sqrt{s}} < 50\%$; $\alpha_a > 30°$	> 15
TASSO	Single Charge Particle Recoiling Against Many Hadrons $E_{vis} \geq 8$ GeV; $N_c \geq 5$	> 13
JADE	$11 < E_{vis} < 32$ GeV ($\sqrt{s} = 35$ GeV) $\|\cos\theta_T\| < 0.75$, $\alpha_a \geq 45°$	> 17

Table IV.   Lower Limits on the Masses of the Supersymmetric Particles

	PLUTO	JADE	MARK J
Selection Criteria	2 electrons with $E_e < \sqrt{s}/10$ Acopl. $>15°$	2 electrons with $E_e < \sqrt{s}/6$ Acopl. $>10°$	2 muons with $E_\mu > \sqrt{s}/10$ Acopl. $>20°$
Limits on the Mass	$M(s_e, t_e)$ > 13 GeV	> 16 GeV or < 1 GeV	$M(s_\mu, t_\mu)$ > 15 GeV or < 5 GeV

which is characteristic of spin zero particle production. m is the mass of t_ℓ or s_ℓ, E is the beam energy, and θ is the scattering angle.

Because of the uniqueness of the decay reactions (a) and (b), the extremely short lifetime of s_ℓ and t_ℓ and the prediction that the interaction cross section of photino and goldstino are expected to be very small,[23] only muon or electron pairs are observed in the final state. Near threshold production of s_ℓ and t_ℓ the two residual leptons would be produced isotropically in space. Data from SPEAR place a lower limit of 3.5 GeV[24] on the mass of s_ℓ and t_ℓ. Thus, over the PETRA energy range of 12 to 36.7 GeV, an increase in the production of acoplanar muon or electron pairs should be observed if a new threshold is passed.

MARK J has looked for the muon partner while PLUTO and JADE searched for the electron partner. No events were observed. The resultant limits are shown in Table IV.

TOP SEARCH

At PETRA we have searched for top[25] by:
1. Looking for an increase of ~ 1 unit in R which is defined as the ratio of $\sigma_h / \sigma_{\mu\mu}$ (Figure 7), as the machine energy is raised.
2. Looking for a shoulder in the thrust distribution at thrust ≈ 0.7
3. Looking for excess of spherical events (Table V) over QCD background.

No signal above QCD is seen. We conclude the top threshold is above 36.6 GeV.

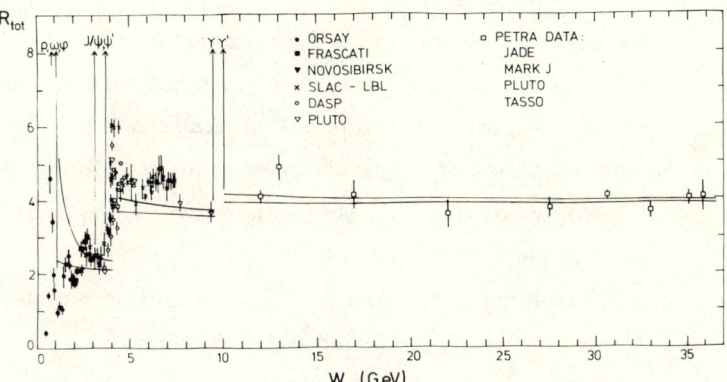

Fig. 7

Energy dependence of R, the ratio of hadronic to muon pair cross section as measured by various e^+e^- colliding beam machines. The curves are QCD predictions.

CONCLUSION

The test of electro-weak theories performed at PETRA leads to the following conclusions:

1. QED is valid up to $s = 1225$ GeV^2 and $-q^2 \sim 1000$ GeV.
2. Leptons, including the τ-lepton are point-like particles with a charge radius < than 2×10^{-16} cm.
3. The search for weak interaction effects not only leads to strong limits on neutral currents parameter, but also tests the theories at very large s and q^2.
4. The vector and axial-vector coupling can be uniquely determined for purely leptonic interactions.
5. For multi gauge boson models stringent restrictions are placed on coupling constants and neutral boson masses.
6. The existence of a new charged sequential heavy lepton is excluded with 95% confidence level for a mass lower than ~ 16 GeV.
7. The production of a scalar supersymmetric partner of the muon or electron is excluded with 95% confidence level for a mass lower than ~ 15 GeV.
8. The top threshold is above 36.6 GeV.

REFERENCES

1. M. Bernadini et al., Nuovo Cim. 17 (1973) 383.
   S. Orito et al., Phys. Lett. 48B (1974) 165.
2. J. J. Aubert et al., Phys. Rev. Lett. 33, (1974) 1404.
   J. E. Augustin et al., Phys. Rev. Lett. 33 (1974) 1406.
3. J. Siegrist et al., Phys. Rev. Lett. 36 (1976) 700.
4. M. L. Perl et al., Phys. Rev. Lett. 35 (1975) 1489;
   M. L. Perl et al., Phys. Lett. 63B (1976) 366.
5. W. Bacino et al., Phys. Rev. Lett. 41, 13 (1978).
6. W. Bacino et al., Phys. Rev. Lett. 42, 749 (1979).
7. D. L. Scharre, SLAC-PUB-2538.

8. J. Lee Franzini, SUNY Preprint
   H. Schröder, DESY 80/61.
9. PETRA Proposal, DESY, Hamburg (Feb., 1976).
10. JADE Collaboration, W. Bartel et al., Phys. Lett. 88B (1979) 171
11. MARK J Collaboration, Physics Reports 63, 337 (1980).
12. PLUTO Collaboration, DESY 80/01 and Phys. Letters 89B (1980) 120.
13. TASSO Collaboration, R. Brandelik et al., Particles and Fields (Z. Physik C) 4 (1980) 87.
14. F. A. Berends at al. DESY 80/66 and 80/73.
15. S. Weinberg, Phys. Rev. Lett. 19, 264 (1967);
    A. Salam, Elementray Particle Theory, ed. N. Svartholm (Almquist and Wiksell, Stockholm), p. 367;
    L. Glashow, Nucl. Phys. 22, 579 (1961).
16. H. Georgi and S. Weinberg, Phys. Rev. D17, 275 (1978).
17. P. Q. Hung and J. J. Sakurai, Phys. Lett. 69B, 323 (1977), and Phys. Lett. 88B 91 (1979).
18. E. H. de Groot and D. Schildknecht, Bielefeld preprint, BI-TP 80/08, May 1980.
    R. Bondy, Phys. Lett. 45B, 340 (1973); Phys. Lett. 55B, 227 (1975).
19. E. H. de Groot, G. J. Gounaris, D. Schildknecht, Phys. Lett., 90B 427 (1980);
    E. H. de Groot and D. Schildknecht, Bielefeld preprint BI-TP 80/22 (1980).
20. V. Barger, W. Y. Keung and E. Ma, Phys. Rev. Lett. 44, 1169 (1980).
    V. Barger, W. Y. Keung and E. Ma, Phys. Lett. 94B, 377 (1980).
21. R. H. Heisterberg et al., Phys. Rev. Lett. 44, 635 (1980);
    L. W. Mo, Contribution to Neutrino 80, Erice (1980), and private communication H. Faissner and H. Reithler.
22. Yu. A. Gol'fand and E. P. Likhtman, JETP Letters 13, 323 (1971).
    G. R. Farrar and P. Fayet, Phys. Lett. 89B 191 (1980).
23. P. Fayet, Phys. Lett. 86B, 272 (1979).
24. F. B. Heiel et al., Nucl. Phys. B138, 189 (1978).

25. M. Chen, Proceeding of Meson Conference, Brookhaven National Laboratory, May 1980.

H. Newman, contribution to International High Energy Physics, Madison, July 1980.

DISCUSSION (Chairman A.K. Mann)

H. FRITZSCH, Munich: Can you set a limit on the lifetime of the b-quark?

L. CHEN, Cambridge: About 10^{-10}s. It is more difficult than measuring the life time of the τ for two reasons: the B's are 3 times more massive therefore less relativistic; we cannot identify the B's as well as the τ events.

J. BERNABEU, Valencia: Forward-backward asymmetry can be generated by 2γ-exchange. How data are analyzed to extract weak interaction information?

M. CHEN, Cambridge: The asymmetry due to QED has been calculated by Bahrends et al. and Paschols et al. in details. All the data I showed have already been corrected before they were compared with weak interactions predictions.

A. DI GIACOMO, Pisa: Which is the value of α strong from your fits? Which is the value of R above 10 GeV?

M. CHEN, Cambridge: α_s from Markj is .23 ± .02(stat.) ± 0.04(sys.) and α_s = 0.17 ± .02(stat.) ± 0.03(sys.) from Tasso recently R = 4.0 ± .17 ± 10% systematic.

P. SCHMID, CERN: Is there evidence of a multicomponent effect in the overall multiplicity distribution due to 3-jet events?

M. CHEN, Cambridge: The overall multiplicity of 3-jet events is higher than what one expects. The multiplicity distribution of each jet of the 3-jet events is also measured.

G.A. SNOW, College Park: What are the future plans at Petra?

M. CHEN, Cambridge: Install low β insertion for all the intersections to increase liminosity by a factor of five. Double the R.F. and increase the energy eventually to about 44 GeV.

E. PASCHOS, Dortmund: Would you comment if the beams are polarized and comment on the plans to use the polarization in future experiments?

M. CHEN, Cambridge: You do see from time to time some indication of the polarization of the beams. That is, one sees an azymuthal dependence of the jet axis of the particles. We are prepared to use the polarization when it becomes available. However the polarization is a sensitive function of the operation of the beams and this should be understood first. A polarization monitor is being set up at Petra.

BEAM-DUMP EXPERIMENTS

F. Dydak

Institut für Hochenergiephysik der Universität
Heidelberg, Germany, and
CERN, Geneva, Switzerland

1. INTRODUCTION

In a beam-dump experiment, a beam of high-energy protons is dumped
into a dense block of heavy material, in order to absorb the hadronic
cascade as quickly as possible. The short decay path minimizes the
background of conventional leptons from the decay of known long-lived
particles (π, K, Λ, ...), thus facilitating the search for penetrating
stable particles, produced either directly in the proton-nucleus interaction, or originating from the decay of short-lived particles
(lifetime $\lesssim 10^{-11}$ s). Leptons produced in this way are referred to as
prompt.

Typical beam-dump experiments are designed to detect either prompt
muons or prompt neutrinos. Whereas electron- and muon-neutrinos are
separately observable, the measurement of charged leptons is restricted
to muons. In the case of neutrino detection, a decay tunnel and a long
passive muon shield was located in all past experiments between the
dumps and the detector. Hence, the acceptance for prompt neutrino detection was limited to a small cone in the very forward direction. In
the case of muon detection, a large portion of phase space can be explored with a well-designed experiment.

While the background from the decay of conventional long-lived
particles is small for electron-neutrinos, it constitutes the bulk of
the observed muon-neutrinos, making a precise determination of the
prompt muon-neutrino flux difficult. The measurement of prompt single
muons is even harder, since on top of the large conventional background
there is a further background from prompt muon pairs (mainly from vector
meson decays and electromagnetic pair creation) with one muon missing
the acceptance criteria.

The prompt electron-neutrino flux can best be determined by subtracting the calculated small background from conventional sources. The same method is also applicable to the case of muon-neutrinos, but the inherent systematic uncertainties in the calculation of the large conventional background seriously affect the prompt flux. Another method is the use of dumps with variable density, to permit separation of the prompt and conventional fluxes by extrapolation to infinite density. This method is in principle safe, but requires large statistics and a precise relative normalization of the runs at different dump densities.

First exploratory experiments have reported evidence for prompt electron-neutrinos[1-3], and for prompt single muons[4], originating from the interactions of 400 GeV/c protons with heavy nuclei. The observed fluxes of prompt single leptons have been attributed to the production and subsequent semileptonic decay of charmed D-mesons. The experimental results are summarized elsewhere[5].

The purpose of this talk is to review the results of a second series of proton beam-dump experiments, carried out in 1979 at BNL with 28 GeV/c protons, at FNAL with 350 GeV/c protons, and at CERN with 400 GeV/c protons. The aims of these experiments were: i) to check the equality of the prompt ν_e, $\bar{\nu}_e$, ν_μ, and $\bar{\nu}_\mu$ fluxes expected from the $D\bar{D}$ production model, and in particular to establish the prompt ν_μ and $\bar{\nu}_\mu$ fluxes by use of the extrapolation method, ii) to clear up some discrepancy in the prompt electron-neutrino flux[5] between the CERN-Dortmund-Heidelberg-Saclay (CDHS) counter experiment[2], and the BEBC[3] and Gargamelle[1] bubble-chamber experiments, iii) to explore the energy dependence and the differential cross-section for charmed-particle production, and, of course, iv) to look for new and unexpected phenomena beyond charm production. A summary of early results has been given by Wachsmuth at the 1979 Lepton-Photon Conference[6].

2. THE BNL EXPERIMENTS

At the Brookhaven AGS, three beam-dump experiments have been performed with 28 GeV/c protons. The protons were targeted on a solid copper block placed in front of the muon shield of the neutrino beam line. In order to reduce the measurement to one comparing relative rates, a comparable number of conventional neutrino events was taken in a bare target beam, produced by a thin target at a position normally occupied by the neutrino production target, 60 m upstream of the copper-dump position. The difference in the conventional rates is then expected to be dominated by the differences in the decay path and the geometrical acceptance rather than by the details of the hadron production spectra.

Three experiments took data in the same exposure: i) the 7 ft bubble chamber, filled with a heavy Ne-H$_2$ mixture, run by the Rutgers-Stevens-Columbia (RSC) Collaboration[7]; ii) a counter set-up consisting of thin-plate optical spark chambers interspersed with planes of plastic scintillator, run by the Columbia-Illinois-Brookhaven (CIB) Collaboration[8]; and iii) a totally active liquid scintillator neutrino detector with drift chambers, run by the Brookhaven-Harvard-Oak Ridge-Pennsylvania (BHOP) Collaboration[9]. The more important parameters of the experimental situation are listed in Table 1.

The RSC and CIB Collaborations do not find a significant difference in the number of observed and expected neutrino events. Their negative result is, however, within errors consistent with an indication of an excess of neutrino events, reported by the BHOP group. Out of a total of 104 observed neutrino events, they claim an excess of 48 ± 10 (stat.) ± 12 (syst.) events, compared to the expected number. This 3σ excess is the only hint for prompt neutrino production, since none of the experiments observes other differences in the events from the beam-dump and bare-target exposures that could point to non-conventional sources of neutrinos. Hence, the existence of prompt neutrino events produced in 28 GeV/c proton interactions still deserves an unambiguous confirmation by experiment.

Table 1

Parameters of the BNL beam-dump experiments

	RSC	CIB	BHOP
Type of experiment	7 ft bubble chamber	Counter set-up	
Effective No. of protons	1.3×10^{18}	4.7×10^{18}	4.9×10^{18}
Distance from dump	43 m	∼ 70 m	105 m
Fiducial mass	2.8 t	5 t	11 t
Sensitivity[a]	22	53	54
Proton momentum	28 GeV/c		

a) In units of (10^{18} protons on target) × (tons of fiducial mass at 105 m distance from the dump).

3. THE FNAL EXPERIMENT

The CalTech-FNAL-Rochester-Stanford (CFRS) Collaboration has recently presented new results on prompt muon production in the collisions of 350 GeV/c protons with iron nuclei[10]. One of the salient features of their apparatus is a variable-density target calorimeter, which serves as a proton dump. It consists of iron plates interspersed with scintillator sheets. In the compacted version, its high density (5.9 g/cm^3) minimizes the conventional π and K decay muon background. The separation of the prompt muon rate and the conventional muon background rate is done by extrapolation to infinite density. The second important component of their set-up is a muon spectrometer with very large acceptance, located downstream of the target calorimeter. The large muon acceptance is particularly relevant for the detection both of single prompt muons and muon pairs.

Their experiment was the first to establish a prompt 1μ^+ signal, obtained in the region $0.8 < p_T < 2.5$ GeV/c and $10 < p_\mu < 60$ GeV/c (small x_F region), induced by 400 GeV/c protons on iron nuclei[4]. At this time, the muon spectrometer was located on axis, and the hole in the centre of their magnet toroids excluded the measurement of muons with small p_T. The results from this first experiment have been reviewed on several occasions[11].

In a modified version of their apparatus, the magnet toroids were placed off axis, to avoid the hole and thereby have good acceptance also for the low p_T region. For muon momenta above 30 GeV/c, the muon acceptance is well understood, and not much different for μ^+ and μ^-. This enables a comparison of the prompt 1μ^+ and 1μ^- rates in the forward direction.

The rates of the 1μ^+ events, as well as the rates of dimuon events with a triggering positive muon (2μ^+ events), are shown in Fig. 1 as a function of the inverse target density. The intercept at zero inverse density is the apparent prompt rate. After corrections for π and K decays downstream of the calorimeter, and for dimuon events with an undetected second muon, the prompt 1μ rates for muon momenta > 30 GeV/c are obtained. These are $(4.7 \pm 1.3) \times 10^{-6}$ prompt 1μ^+ and $(6.1 \pm 1.0) \times 10^{-6}$ prompt 1μ^- per interacting proton. In a model of $D\bar{D}$ production with subsequent semileptonic decay one expects equality of prompt 1μ^+ and 1μ^-. This expectation is consistent with the measured ratio 1μ^-/1μ^+ = 1.3 ± 0.4. This result is interesting in view of an analogous result for prompt ν_μ and $\bar{\nu}_\mu$ production in the very forward direction (see Section 4), where the measured prompt $\bar{\nu}_\mu$ flux is found to be smaller than the prompt ν_μ flux. In the same kinematical domain, one expects the 1μ^-/1μ^+ ratio to be equal to the $\bar{\nu}_\mu/\nu_\mu$ ratio. While the muon experiment has good acceptance for $x_F \gtrsim 0.2$, the neutrino experiment is sensitive only for $x_F \gtrsim 0.8$. Also the acceptance in the p_T of

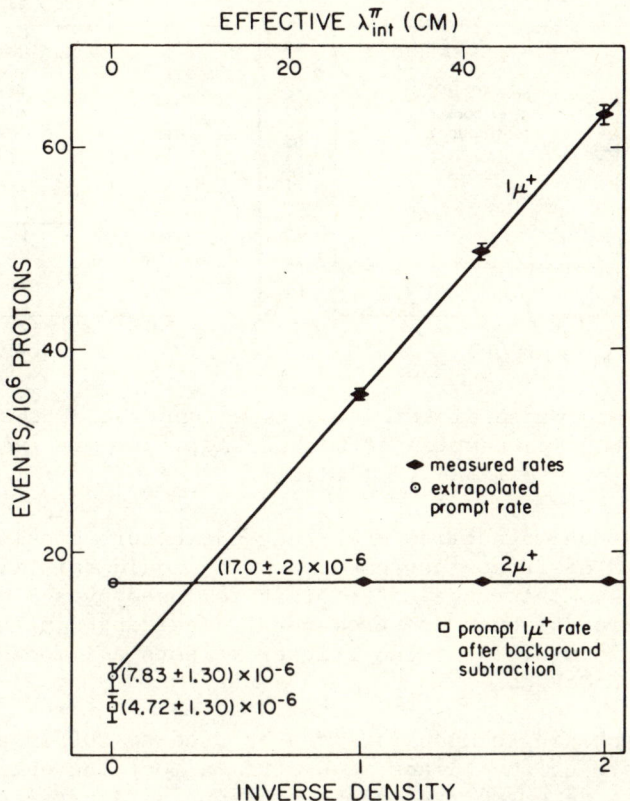

Fig. 1 The measured $1\mu^+$ and $2\mu^+$ rates from the off-axis configuration of the CFRS detector, as a function of the inverse target density. The lines are least squares fits to the measured rates. The intercept of the $1\mu^+$ line measures the apparent prompt $1\mu^+$ rate. The value of the $1\mu^+$ rate after background subtraction is also shown.

the lepton is different: the muon experiment accepts almost all p_T, but the neutrino experiment is limited to small p_T ($\lesssim 0.1$ GeV/c). The p_T distribution of prompt $1\mu^-$ and $1\mu^+$ shows no evidence for an asymmetry at low p_T (see Fig. 2). Since the p_T kick from charmed-particle decay (~ 0.5 GeV/c) would give an asymmetry also at larger p_T, the smearing of ~ 0.3 GeV/c in p_T due to multiple scattering in iron does not matter. To summarize, the difference in the experimental results is due either to the different domains in x_F, or to experimental errors (a statistical fluctuation being unlikely but not excluded).

Recently, the total prompt muon production rate for almost the entire forward hemisphere has been measured by the CFRS group[12] with a new version of its apparatus. The target calorimeter is now

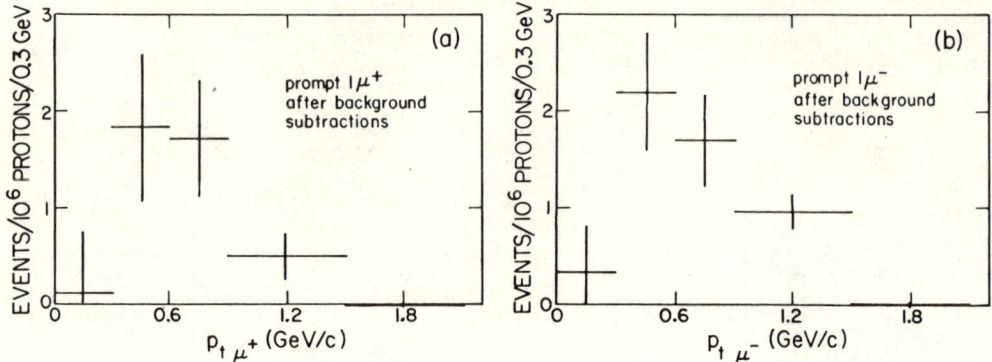

Fig. 2 The prompt 1μ distribution as a function of the transverse momentum, after subtraction of background, for positive (a) and negative (b) single muons (CFRS data).

followed by a non-magnetized muon-range detector, consisting of a total of 4.5 m of iron, instrumented with liquid scintillator and spark chambers. The range detector is followed by a 3.5 m magnetized iron toroidal muon spectrometer. Requiring a penetration depth of ≥ 5.75 m of iron, the trigger selects all muons with momentum above 8 GeV/c.

The prompt single muon rate can be used to obtain a charm production cross-section. Since muons from almost the entire forward hemisphere are detected, the cross-section is rather insensitive to model assumptions on the inclusive cross-section for charm production. Using the parametrization for uncorrelated $D\bar{D}$ production,

$$E \frac{d^3\sigma}{dp^3} \propto (1 - |x_F|)^n e^{-bp_T} ,$$

an average semileptonic branching ratio of 8%, and a linear A-dependence, the CFRS group obtains a charm production cross-section of $\sigma = 22 \pm 9$ μb/nucleon[12]).

4. THE CERN EXPERIMENTS

4.1 The Experimental Situation

As in the first beam-dump experiments carried out at CERN in 1977, 400 GeV/c protons were incident on a thick copper target. An improvement on the previous experiments is the use of dumps with different densities. Neutrino data were taken with dumps of density 1 and ⅓, with about 70% of all protons used for density 1.

While the density-1 target consisted of a solid copper block, the density-⅓ target was built of copper disks of 2 cm thickness,

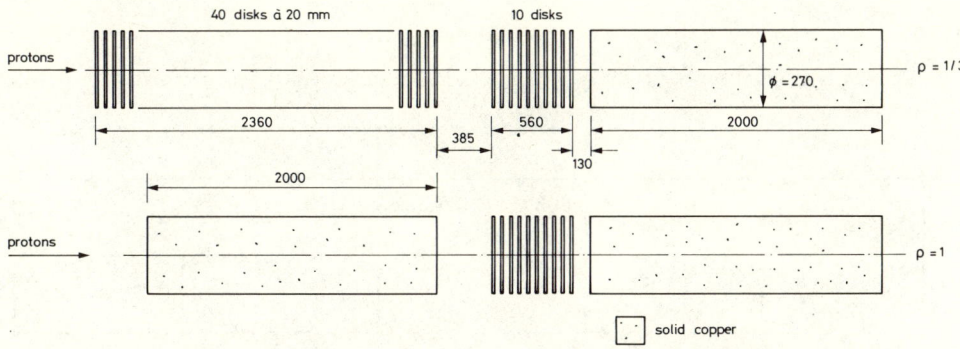

Fig. 3 Layout of the density-1 and density-$\frac{1}{3}$ dumps used in the 1979 CERN beam-dump experiments.

4 cm apart. The detailed layout of the dumps, which had to meet some constraints of the existing neutrino target area, is shown in Fig. 3. The effective average dump densities agree within 1% with those of infinitely large and homogeneous dumps.

The proton intensity per burst was measured with a beam current transformer. The effective number of protons after dead-time correction was used for the normalization of the runs.

The potentially most dangerous source of systematic uncertainties in the determination of the prompt muon flux is proton beam scraping upstream of the dump. Such proton interactions may increase the conventional muon-neutrino flux considerably, since the secondaries have a long decay path. In order to ensure that proton beam scraping had no significant effect, eight ionization chambers were installed along the ejected proton beam line. Their response was recorded continuously during data taking, and indicated that the losses were typically less than 1×10^{-4}. Although the monitors did not cover the full azimuthal angle around the beam line, it is believed that the effect of proton beam scraping was negligible. This is supported by the observation of both the CDHS and CERN-Hamburg-Amsterdam-Rome-Moscow (CHARM) groups that cutting out a portion of the data with the largest response of the scraping monitors had no significant influence on the results.

Owing to some material about 1 m upstream of the dumps (target monitor, vacuum flanges, etc.) the conventional background was increased by $(5.8 \pm 0.6)\%$ in the case of the density-1 exposure. All prompt and non-prompt rates given below are corrected for this effect.

The main characteristics of the detectors taking part in the 1979 CERN beam-dump exposure are: i) the excellent $e^{\pm}$ detection and measurement capability of BEBC, filled with a heavy Ne-H_2 mixture,

Table 2

Parameters of the 1979 CERN beam-dump experiments

	ABCDLOS	CDHS	CHARM
Type of experiment	Bubble chamber (BEBC)	Counter set-up	
Effective No. of protons in density-1 exposure	1.12×10^{18} a)	0.66×10^{18}	0.70×10^{18}
Effective No. of protons in density-$\frac{1}{3}$ exposure	0.31×10^{18}	0.25×10^{18}	0.26×10^{18}
Distance from dump	820 m	890 m	910 m
Fiducial mass	13.3 t b)	492(465) t c)	100 t
Sensitivity[d] in density-1 exposure	17.5	326(308)	67
Proton momentum	400 GeV/c		

a) Including the 1977 exposure.
b) Weighted average of the 1977 and 1979 fiducial masses.
c) 492 t for 1μ analysis, 465 t for 0μ analysis.
d) In units of $(10^{18}$ protons on target$) \times$ (tons of fiducial mass at 890 m distance from the dump).

operated by the Aachen-Bonn-CERN-Demokritos-London-Oxford-Saclay (ABCDLOS) Collaboration; the excellent $\mu^{\pm}$ detection and measurement capability and the large fiducial mass of the CDHS neutrino detector[13]; the good muon detection and the ability to recognize and measure electromagnetic energy within a hadronic shower in the fine-grain CHARM detector[14]. The more important parameters of the experimental situation are listed in Table 2, and a schematic layout is shown in Fig. 4. The angular acceptance of the detectors is limited to an angle of less than 1.8 mrad with respect to the forward direction.

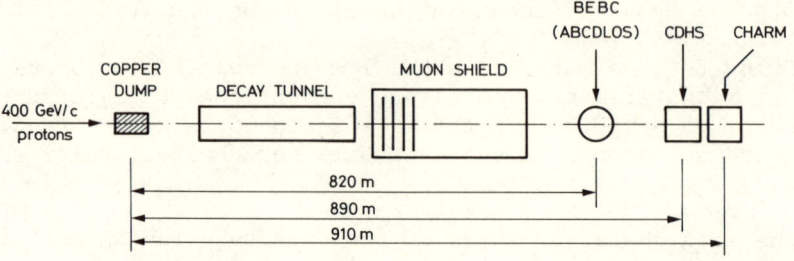

Fig. 4 Schematic layout of the 1979 CERN beam-dump experiments

In order to facilitate the comparison of experimental results obtained with different effective numbers of protons, fiducial masses, and distances from the dump, all event numbers given below refer to 10^{18} protons on target and 1 t of detector, at the position of the CDHS apparatus. The correction of the flux dilution as a function of distance is done proportionally to the distance squared.

4.2 The Prompt Muon-Neutrino Flux

Because of the large background from conventional π and K decays, the measurement of the prompt muon-neutrino flux needs large event statistics, and can therefore best be done by the CDHS experiment. Figures 5a and b show the $E_{tot} = E_H + p_\mu$ spectra of $1\mu^-$ and $1\mu^+$ events measured in the density-1 exposure[15]. The data are corrected for losses due to geometrical acceptance and the requirement of a minimum muon momentum of 5 GeV/c. The spectra are steeply falling with energy. The $1\mu^-$ spectrum obtained by the CHARM Collaboration[16] is consistent with the CDHS $1\mu^-$ spectrum (Fig. 5a).

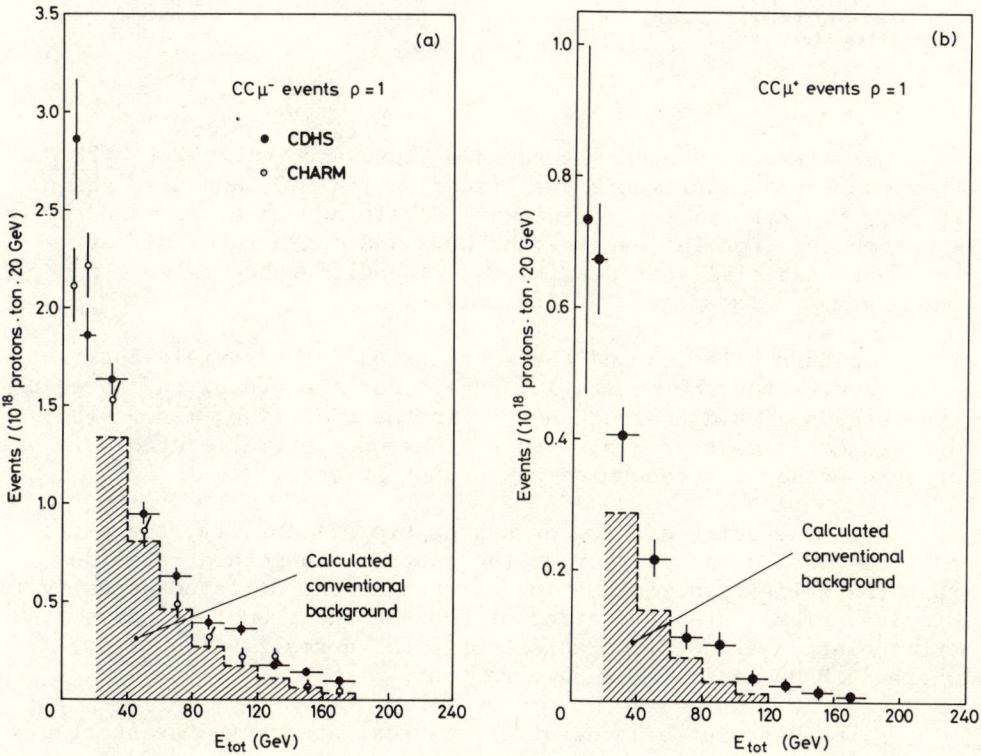

Fig. 5 Over-all rates of $1\mu^-$ (a) and $1\mu^+$ (b) events, normalized to 10^{18} protons and 1 t of detector at the CDHS position, as a function of E_{tot}.

Table 3

1μ event rates of the CERN beam-dump experiments for $E_{tot} > 20$ GeV, normalized to 10^{18} protons on target and 1 t of detector at the CDHS position[a]

	ABCDLOS	CDHS	CHARM
All 1μ⁻ (density 1)	4.90 ± 0.55	4.57 ± 0.13	4.15 ± 0.28
All 1μ⁻ (density ⅓)	10.2 ± 1.5	10.92 ± 0.33	9.87 ± 0.71
Prompt 1μ⁻ (extr.)	–	1.17 ± 0.26 ± 0.32	1.11 ± 0.56
Prompt 1μ⁻ (subtr.)	1.70 ± 0.61	1.33 ± 0.13 ± 0.49	1.12 ± 0.28 ± 0.34
All 1μ⁺ (density 1)	1.25 ± 0.27	0.91 ± 0.06	1.23 ± 0.15
All 1μ⁺ (density ⅓)	1.85 ± 0.62	2.51 ± 0.16	1.77 ± 0.35
Prompt 1μ⁺ (extr.)	–	0.05 ± 0.12 ± 0.07	0.95 ± 0.30
Prompt 1μ⁺ (subtr.)	0.65 ± 0.30	0.38 ± 0.06 ± 0.08	0.71 ± 0.16 ± 0.06
Prompt 1μ⁻ + 1μ⁺ (extr.)	–	1.23 ± 0.29 ± 0.39	2.06 ± 0.64
Prompt 1μ⁻ + 1μ⁺ (subtr.)	2.52 ± 0.72	1.70 ± 0.14 ± 0.56	1.83 ± 0.32

a) If two errors are quoted, the first denotes the statistical and the second the systematical error.

The normalized over-all rates of 1μ events with $E_{tot} > 20$ GeV from the three experiments are listed in Table 3, and also shown in Fig. 6. For 1μ⁻ events, we note a fair agreement between the experiments. For 1μ⁺ events, the CDHS and CHARM rates differ somewhat, especially at density ⅓. The difference is most probably due to a statistical fluctuation.

The extrapolation and the subtraction methods yield consistent results for the prompt 1μ⁻ rate. For the prompt 1μ⁺ rate, the two methods give different answers in the CDHS experiment. This difference is most probably due to unknown systematic errors in one or both methods to determine the prompt 1μ rate.

From the point of view of systematic reliability, the extrapolation method is superior to the subtraction method, provided that the relative normalization of the runs at different density is precise. The systematic error of CDHS on the prompt 1μ rates from extrapolation arises from an estimated 5% normalization error, whereas CHARM estimates a 3% error.

From the point of view of statistical accuracy, the subtraction method is superior. The flux of conventional neutrinos originating from the interactions of 400 GeV/c protons in a solid copper dump has been calculated by Wachsmuth[17]. The calculation is made to fit the measurements of the over-all muon flux at four

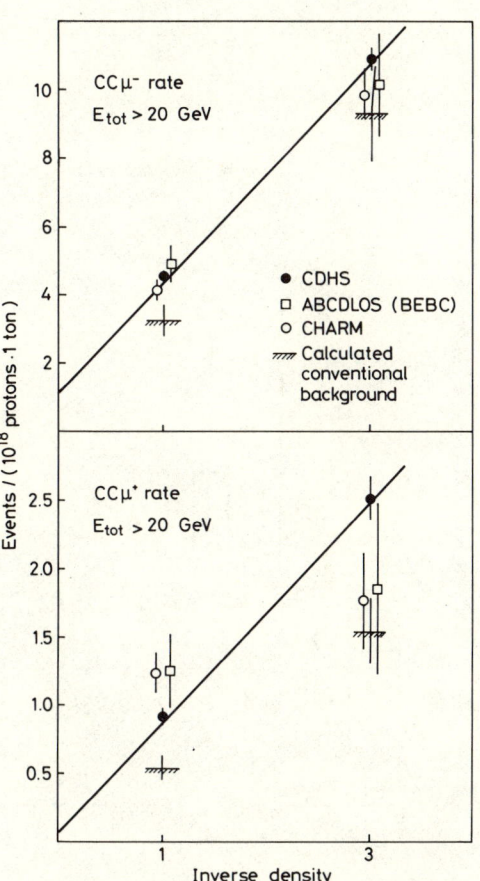

Fig. 6 Normalized over-all rates of 1μ events with $E_{tot} > 20$ GeV, as a function of the inverse dump density. The straight lines refer to the extrapolation of the CDHS data points, corrected for the increase of the conventional background due to some material in front of the dump.

depths of the iron shield, within a limited angle of 1 mrad with respect to the forward direction. The systematic uncertainty of the conventional neutrino flux is estimated to be at the level of 10%. An additional error arises from the neutrino total cross-section. CDHS assumes cross-section slopes $\sigma^\nu/E = 0.65 \pm 0.06$ and $\sigma^{\bar\nu}/E = 0.30 \pm 0.03$ in units of 10^{-38} cm^2/GeV, where the 10% error accounts for systematic uncertainties of the absolute normalization and of deviations from the linear energy dependence. The over-all systematic CDHS error on the conventional background is 15%. ABCDLOS and CHARM assume $\sigma^\nu/E = 0.62 \pm 0.03$ and $\sigma^{\bar\nu}/E = 0.30 \pm 0.02$, and quote an over-all uncertainty of 11%.

The ABCDLOS and CHARM results are consistent with equality of prompt ν_μ and $\bar\nu_\mu$ fluxes. The statistically more significant CDHS result indicates a smaller prompt $\bar\nu_\mu$ than ν_μ flux. Whereas the prompt ν_μ flux can be considered established, a prompt $\bar\nu_\mu$ flux still waits for experimental confirmation.

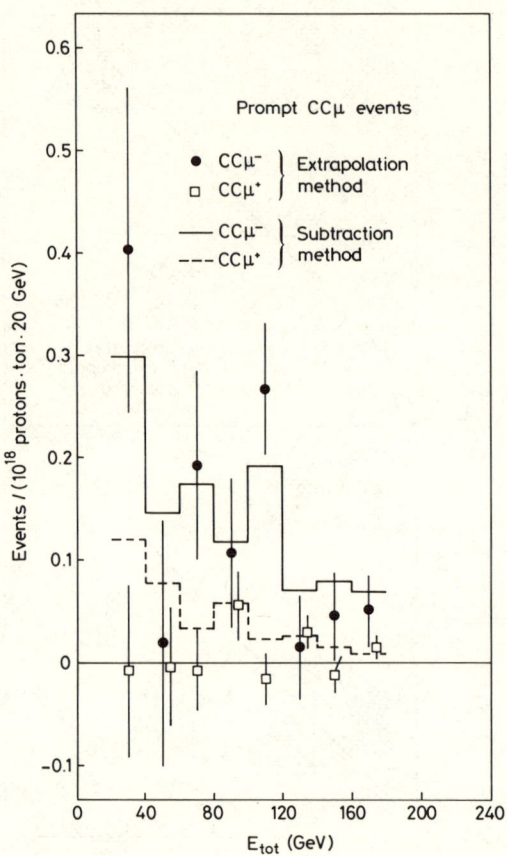

Fig. 7 Normalized prompt $1\mu^-$ and $1\mu^+$ rates as obtained from extrapolation and subtraction, as a function of E_{tot} (CDHS data).

The CDHS result for the prompt $1\mu^-$ and $1\mu^+$ rates as a function of E_{tot}, obtained with both the extrapolation and subtraction methods, is shown in Fig. 7.

The measurement of the prompt muon-neutrino flux is unsatisfactory as long as the results from extrapolation and subtraction do not agree, or the extrapolation result is not sufficiently precise. The 1979 CERN beam-dump experiments succeeded in demonstrating the existence of a prompt muon-neutrino flux, but failed to measure its magnitude with adequate precision.

4.3 The Prompt Electron-Neutrino Flux

Historically, the group working with BEBC was the first to announce an anomalous excess of events with a high-energy electron in the final state. The unambiguous identification of electron-neutrino charged-current (CC) events is still a domain of the ABCDLOS group, whereas CDHS and CHARM work with the concept of "muonless" (0μ) events: since the electromagnetic shower of the electron is hidden in the hadronic shower, electron-neutrino events are measured indirectly by a global excess of muonless events above the level expected from muon-neutrino neutral-current (NC) events. Since new types of neutrinos (e.g. ν_τ) yield predominantly muonless events, they are included in the excess of muonless events. Interpreting the observed excess as due to electron-neutrinos assumes that the contribution from, for example, ν_τ's is negligible.

In the case of electron-neutrino CC events, the measured shower energy (E_{sho}) is the total energy. In the case of genuine NC events, the shower energy is the hadronic energy. This ambiguity is to be kept in mind when interpreting a sample of muonless events.

The excess of muonless events is not directly accessible to measurement, but is the result after a number of subtractions. To give a feeling about the size of the various subtractions, Fig. 8 shows the data reduction of the CDHS experiment as a function of the shower energy. The steep increase of the subtractions towards lower energy is noted. For the fine-grain CHARM calorimeter, the cosmic-ray background and the background of muon-neutrino CC events with hidden muons is smaller, but the other subtractions are the same.

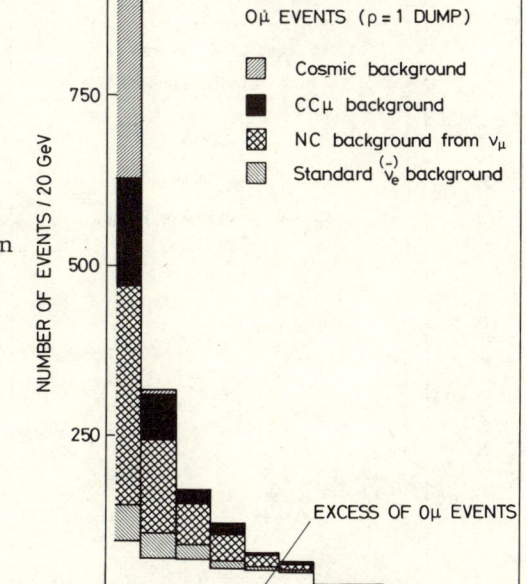

Fig. 8 Data reduction to obtain the excess of muonless events, as a function of shower energy (CDHS data).

Table 4

Muonless event rates and charged-current electron-neutrino rates of the CERN beam-dump experiments, normalized to 10^{18} protons on target and 1 t of detector at the CDHS position[a]

	ABCDLOS	CDHS	CHARM
Excess of 0μ events	–	1.16 ± 0.11 ± 0.11	1.18 ± 0.20 ± 0.09
Prompt e^- + e^+ events	1.18 ± 0.28	0.96 ± 0.09 ± 0.09	0.98 ± 0.17 ± 0.07

[a] If two errors are quoted, the first denotes the statistical and the second the systematical error.

Above E_{sho} = 20 GeV, there is good agreement between CDHS and CHARM on the excess of muonless events, as can be seen from Table 4 and from Fig. 9. To convert the rate of muonless events above E_{sho} = 20 GeV into the wanted rate of CC events with E_{tot} > 20 GeV, a factor 0.83 ± 0.02 is applied, which removes the contribution of genuine NC events.

The measured rate of prompt electron-neutrino CC events confirms the earlier measurement of the 1977 exposure, at a rate as reported by CDHS. The higher rates reported earlier by the BEBC[3] and Gargamelle[1] groups were the result of a statistical fluctuation.

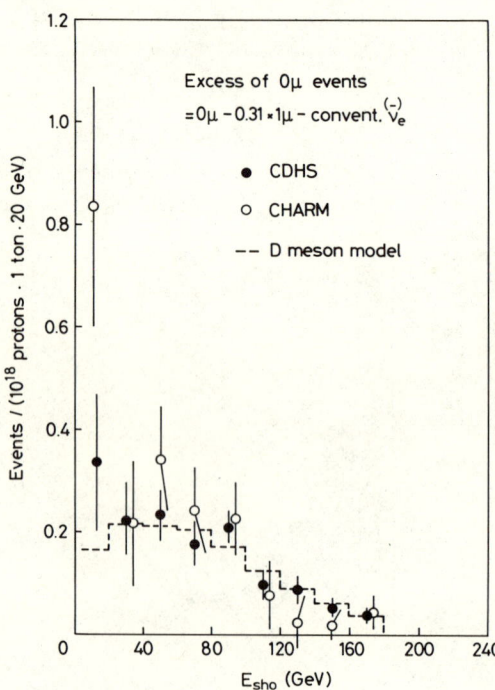

Fig. 9 Excess of muonless events, as a function of shower energy. The histogram shows the expectation of muonless events from the decay of D-mesons produced according to $(1 - |x_F|)^3 e^{-2PT}$.

The spectral shape of the excess of muonless events is qualitatively consistent with that expected from charmed meson production and decay. The smooth histogram in Fig. 9 shows the spectrum from D-mesons produced with an invariant inclusive cross-section

$$E \frac{d^3\sigma}{dp^3} \propto (1 - |x_F|)^3 e^{-2p_T} .$$

The ABCDLOS and the CHARM groups quote the following total cross-sections for $D\bar{D}$ production, assuming an 8% semileptonic branching ratio, linear A dependence, and equal ν_e and $\bar{\nu}_e$ fluxes:

$$\sigma(pN \to D\bar{D}X) = 15 \pm 4 \ \mu b \quad [18] \quad (\text{ABCDLOS})$$
$$\sigma(pN \to D\bar{D}X) = 19 \pm 6 \ \mu b \quad [19] \quad (\text{CHARM}) .$$

These results compare favourably with the result from the CFRS group (see Section 3). The CDHS group did not quote a cross-section, because the simple $D\bar{D}$ model appears incompatible with the possible inequality of the prompt $\bar{\nu}_\mu$ and ν_μ fluxes.

The CHARM group developed a method of identifying electron-neutrino CC events directly[14]. The method takes advantage of the small width of electromagnetic showers, the regular longitudinal profile, and the strong correlation between the total shower energy and the energy at the shower maximum, in contrast to the development of hadronic showers. In this way, they obtain a directly measured spectrum of electron-neutrino CC events, which is above E_{tot} = 20 GeV consistent with the one obtained from the excess of muonless events (Fig. 10).

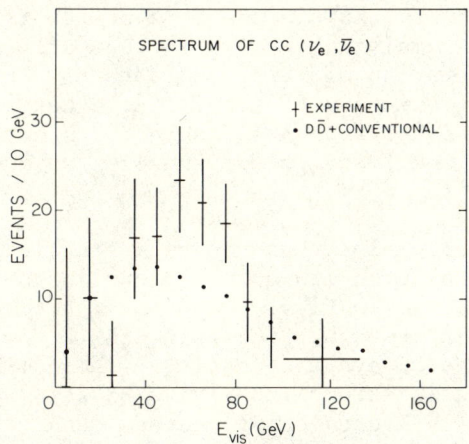

Fig. 10 Observed E_{tot} spectrum of charged-current electron-neutrino events of prompt and conventional origin (CHARM data). The expectation from a $D\bar{D}$ model and from conventional sources is also shown.

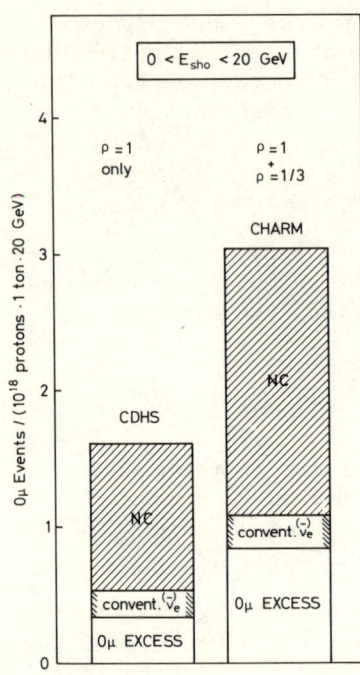

Fig. 11 Comparison of the subtractions to get the excess of prompt muonless events below E_{sho} = 20 GeV, between the CDHS and CHARM experiments. The data are scaled to a bin width of 20 GeV.

At low shower energies, $2 < E_{sho} < 20$ GeV, the CHARM Collaboration[19] observes 54 ± 19 (stat.) ± 9 (syst.) prompt muonless events, in excess of electron-neutrino interactions expected from $D\bar{D}$ production and decay. Since the time of the first announcement[20,6] of this anomaly, which was in the early stage interpreted as a manifestation of ν_τ's, the data have been carefully re-analysed. The previous 4.6σ significance has reduced to a 2.6σ significance. The CDHS group has examined its data in the bin $5 < E_{sho} < 20$ GeV, and found no evidence for an anomalous excess above charm production.

The CHARM group claims that the excess at low shower energies is due to muonless events with no electrons in the final state (compare the lowest bin in Figs. 9 and 10). From Fig. 11, where the subtractions performed by CDHS and CHARM are compared, it appears that genuine NC events constitute the dominant background. To minimize systematic errors from large background subtractions, CDHS uses the density-1 data only, whereas CHARM has chosen to combine the density-1 and density-$\frac{1}{3}$ data, thus accepting a larger background of NC events.

To summarize, the evidence for an anomaly at low shower energies is poor.

4.4 Interpretation of Results

By now there is general agreement that the bulk of the prompt leptons observed in high-energy proton beam-dump experiments is due to charm production with an inclusive cross-section of the order of 10 μb/nucleon, and subsequent semileptonic decay.

The CDHS experiment measures a ratio of a prompt $\bar{\nu}_\mu$ to ν_μ flux of $0.09 \pm ^{0.45}_{0.09}$. While this cannot be considered as a significant deviation from equality -- and is also not seen by other experiments -- it may point to a more complicated mechanism of charm production than assumed before. Significant Λ_c production, for instance, could alter the ratio from unity, as a consequence of a different x_F dependence.

The best evidence, if any, for a prompt antineutrino flux comes from BEBC: the ABCDLOS group observes nine events with an e^+, and expects 2.6 events from conventional sources. It appears unlikely that the nine events are all background.

The truly fundamental problem is the ratio of the prompt electron-neutrino to muon-neutrino flux, because this must be unity, irrespective of the detailed charm production mechanism. An inequality of fluxes, if true, would require new physics phenomena for its explanation.

Table 5 summarizes the results obtained by the three experiments on the ratio of the prompt electron-neutrino to muon-neutrino flux. All measured ratios are smaller than unity, a statistical fluctuation being excluded. Can systematic errors account for the deviation from unity? It is clearly hard to reconcile the results with unity where the prompt muons are obtained by subtraction. The results, where the extrapolation method has been employed, are compatible with unity.

Table 5

Ratio of prompt $(\nu_e + \bar{\nu}_e)$ and $(\nu_\mu + \bar{\nu}_\mu)$ CC events from the CERN beam-dump experiments[a]

	ABCDLOS[b]	CDHS[c]	CHARM[c]
Ratio with prompt 1μ events from extrapolation	-	0.78 ± 0.20 ± 0.24	0.49 ± 0.21
Ratio with prompt 1μ events from subtraction	0.59 + 0.35 − 0.21	0.56 ± 0.07 ± 0.19	0.48 ± 0.12 ± 0.10

a) If two errors are quoted, the first denotes the statistical and the second the systematical error. b) $E_{tot} > 10$ GeV. c) $E_{tot} > 20$ GeV.

There have been speculations about a possible interpretation of an apparently smaller electron-neutrino flux in terms of oscillations of ν_e's into another type of neutrino, possibly ν_τ's [21]. I believe that at present such far-reaching conclusions can hardly be justified in view of the large uncertainties of the experimental results.

Both CDHS and CHARM observe no significant energy dependence of the prompt $(\nu_e + \bar{\nu}_e)$ to $(\nu_\mu + \bar{\nu}_\mu)$ ratios. This observation implies $\Delta m^2 \gg 1$ eV^2, where Δm^2 denotes the difference of the squared masses of the neutrino mass eigenstates. This is at variance with two other relevant observations made in the same beam line: the ABCDLOS group finds in a 200 GeV narrow-band beam exposure a ratio of 1.04 ± 0.15 between the observed and the expected number of ν_e CC events from $K^\pm_{e_3}$ decay[22]. Similarly, the BEBC-TST Collaboration finds in a preliminary analysis no significant difference between the observed and expected number in a neutrino wide-band beam exposure, their ratio being 1.20 ± 0.25 [23]. We conclude that there is no compelling evidence for the existence of neutrino oscillations either.

5. SUMMARY AND OUTLOOK

The CFRS group has demonstrated prompt single muon events from high-energy proton-nucleus collisions. It is continuing its programme to explore further the mechanism of charm production, by using different projectiles and different energies.

The CERN beam-dump experiments have confirmed the earlier reported prompt electron-neutrino flux. They have established a prompt ν_μ flux. The prompt $\bar{\nu}_\mu$ flux is not established, and there is an indication from CDHS that it may be smaller than the prompt ν_μ flux.

The most embarrassing result is the apparent inequality of the electron-neutrino and muon-neutrino fluxes. Since systematic errors cannot be ruled out to be the source of the problem, it is planned to carry out a third beam-dump experiment at CERN, with greatly improved statistical and systematical accuracy. In parallel, beam-dump experiments are in progress or being planned at FNAL, so that the questions opened up by the recent experiments will hopefully be settled in one or two years time. Last, but not least: the beam dump offers a good chance to discover the ν_τ.

Acknowledgements

I would like to thank many colleagues from the various beam-dump groups, in particular Drs. P.O. Hulth, V. Khovansky, W. Kozanecki, K.H. Mess, R. Messner, F. Niebergall, J. Steinberger, L. Sulak and H. Wachsmuth, for a number of useful discussions.

REFERENCES

1. P. Alibran et al., Phys. Lett. 74B:134 (1978).
2. T. Hansl et al., Phys. Lett. 74B:139 (1978).
3. P.C. Bosetti et al., Phys. Lett. 74B:143 (1978).
4. K.W. Brown et al., Phys. Rev. Lett. 43:410 (1979).
5. H. Wachsmuth, Proc. Topical Conf. on Neutrino Physics at Accelerators, Oxford, 1978, Rutherford Laboratory, Chilton, Didcot (1978), p. 233.
6. H. Wachsmuth, Proc. Int. Symposium on Lepton and Photon Interactions at High Energies, Batavia, 1979, Fermilab, Batavia, Illinois (1979), p. 541.
7. P.F. Jacques et al., Phys. Rev. D 21:1206 (1980).
8. P. Coteus et al., Phys. Rev. Lett. 42:1438 (1979).
9. A. Soukas et al., Phys. Rev. Lett. 44:564 (1980).
10. R. Messner, talk given at the Third Warsaw Symposium on Elementary Particle Physics, Jodłowy Dwór, Poland, 1980; and private communication.
    K.W.B. Merritt, talk given at the Int. Conf. on High Energy Physics, Madison, USA, 1980.
11. See, for example, A. Bodek, Sources of prompt leptons in hadronic collisions, University of Rochester preprint UR-730 (1979).
12. J.L. Ritchie et al., Phys. Rev. Lett. 44:230 (1980).
13. M. Holder et al., Nucl. Instrum. Methods 151:69 (1978).
14. A.N. Diddens et al., A detector for neutral-current interactions of high-energy neutrinos, preprint CERN-EP/80-63 (1980), to be published in Nucl. Instrum. Methods;
    M. Jonker et al., Direct detection of charged-current electron-neutrino events in a fine-grain calorimeter, to be published in Nucl. Instrum. Methods.
15. H. Abramowicz et al., Prompt neutrino production in a proton beam-dump experiment, to be published in Z. Phys. C.
16. P. Monacelli, Seminar given at CERN (June 1980).
17. H. Wachsmuth, Neutrino and muon fluxes in the CERN 400 GeV proton beam-dump experiments, Internal report CERN-EP/79-125 (1979).
18. P. Fritze et al., Further study of the prompt neutrino flux from 400 GeV proton-nucleus collisions using BEBC, to be published in Phys. Lett. B.
19. M. Jonker et al., Experimental study of prompt neutrino production in 400 GeV proton-nucleus collisions, preprint CERN-EP/80-187 (1980), to be published in Phys. Lett. B.
20. K. Winter, Rapporteur's talk on Neutral Currents, given at the Symposium on Lepton and Photon Interactions at High Energies, Batavia, USA, 1979.
21. See, for example, A. de Rújula et al., Nucl. Phys. B168:54 (1980).
22. H. Deden et al., A test of ν stability using a 200 GeV narrow-band neutrino beam at BEBC, preprint CERN-EP/80-164 (1980), to be published in Phys. Lett. B.
23. W. Venus, private communication.

DISCUSSION   (Chairman: A. Mann)

J. LOSECCO, Michigan: I would like to comment on the 28 GeV/c beam-dump experiments at BNL. All three experiments have a comparable event rate. The Brookhaven-Columbia-Illinois group has calculated the conventional rate in two ways. With one they get no prompt signal, but with the other calculation they get an excess comparable to that observed by the Brookhaven-Harvard-Oak Ridge-Pennsylvania group. This excess corresponds to a 180 nb production cross-section into the 10 mrad acceptance of the detector, which is much smaller than cross-sections observed at higher energies.

A. BODEK, Rochester: Do the conventional neutrino and antineutrino rates obtained from the extrapolation agree with the fact that the π^- flux is smaller than the π^+ flux? What are those rates?

F. DYDAK: The CDHS results for the conventional neutrino rate are 3.21 ± 0.18 events per 10^{18} protons and 1 t of detector, and for the conventional antineutrino rate 0.81 ± 0.09 events, both for $E_{tot} > 20$ GeV. These neutrinos come mostly from π's with average energies above 100 GeV. H. Wachsmuth calculates above 100 GeV a π^+/π^- ratio of about 3.5, consistent with the ratio of the conventional neutrino rates.

G. MYATT, Oxford: You have implied that the different conclusions as to a prompt $1\mu^+$ signal between the different CERN experiments is a matter of interpretation rather than differences in the data. However, the origin of this discrepancy is the experimentally higher rate of $1\mu^+$ events observed at density $\frac{1}{3}$ by the CDHS experiment compared to that of CHARM and BEBC. This can be seen in your Fig. 6.

F. DYDAK: My statement referred to the different results from the CDHS experiment for the prompt $1\mu^+$ rate from extrapolation and subtraction. This difference is not understood. It may be due to a statistical fluctuation, as well as the differences from the other two experiments. The weighted average of all experiments may well be the right answer.

G. MYATT, Oxford: Your statement that the ABCDLOS group has claimed to observe a ratio of the prompt ν_e to ν_μ flux different from unity is not quite correct. Given the size of the errors our result is only 2σ from unity, and we would not make any claim on this basis; we merely presented our data.

FUTURE ACCELERATORS

Klaus Winter

CERN, Geneva, Switzerland

ABSTRACT

What will future accelerators, proposed or at present under construction be able to teach us about weak interactions? Three accelerators are briefly described: the $p\bar{p}$ collider at CERN; the large e^+e^- collider project LEP; and the electron-proton collider project HERA at DESY. The first two aim at discovering the vector bosons of the weak interaction and at studying their properties; LEP can be used to search for the Higgs meson and to study the fermion spectrum. HERA aims at a study of structure and a search for the substructure of ordinary matter; it is sensitive to effects due to the exchange of heavier vector bosons with masses up to 300 GeV.

* * *

1. INTRODUCTION

Lepton and quarks are thought to be the elementary constituents of matter. Experimental high-energy physics now aims at studying the interactions between these elementary constituents in quark-quark, lepton-lepton, and lepton-quark collisions. Proposals have been put forward and there are machines under construction for studying these three different types of fundamental collisions:

i) the $p\bar{p}$ collider at CERN, the Tevatron at FNAL, and the pp collider ISABELLE at BNL aim at studying quark-quark collisions in the energy range $\sqrt{s}$ = 300 - 2000 GeV;

ii) large e^+e^- collider projects -- LEP at CERN, the single-pass collider at SLAC, and a Z^0 factory at Cornell -- aim at studying lepton-lepton interactions at energies of the conjectured Z^0 pole and beyond;

iii) electron-proton collider projects are proposed at DESY and at FNAL, for the study of lepton-quark collisions.

The conception and the parameters of these accelerators are strongly motivated by the physics of weak interactions. In particular, the $p\bar{p}$ colliders are being built with the main aim of discovering the vector bosons of the weak interaction[1]. The physics that can be studied with very large e^+e^- colliders has been investigated by many people[2] since 1976, following the publication of a paper by B. Richter[3]. These colliders aim at studying the properties of the Z^0 and the W, at investigating whether the Higgs meson is a physical particle, and at studying the spectrum of fermions. The aim of electron-proton colliders is to push the study of the structure of matter to smaller dimensions and investigate leptons and quarks for internal structure[4]; they will also search for effects due to the exchange of heavier vector bosons with masses up to 300 GeV [4].

In the following sections, three projects will be described: the CERN $p\bar{p}$ collider; LEP, an e^+e^- collider project at CERN; and HERA, an ep collider project at DESY; and their scope for studying these questions will be discussed.

2. THE $p\bar{p}$ COLLIDER PROGRAM

Figure 1 shows a layout of the $p\bar{p}$ collider complex at CERN[5]. Protons are accelerated to 26 GeV/c in the Proton Synchrotron (PS), ejected, and directed onto a target. Antiprotons of 3.5 GeV/c produced at 0° are selected, focused, and injected into the antiproton accumulator (AA) ring. Their phase-space density is increased by stochastic cooling[6] in two steps; first on the injection orbit during 2.4 s; then for 12 hours in a stack screened by a shutter. Every 12 hours, about 10^{12} antiprotons would be accumulated and ejected to the PS, accelerated to 26 GeV, and injected into the Super Proton Synchrotron (SPS) to fill three bunches. Another three bunches circulating in the opposite direction will be filled with 10^{12} protons; the beams will then be accelerated to 270 GeV.

Collisions can be observed in two interaction regions where two large detectors, UA1 and UA2, will be installed in 1981. A first physics run is foreseen for the end of that year. The luminosity of the complex is estimated at 10^{30} cm^{-2} s^{-1}. For the purpose of estimating counting rates, we assume that an integrated luminosity of 10^{36} cm^{-2} can be obtained in a physics run of 1000 hours.

The main aim of the physics program is a search for the vector bosons of the weak interaction, Z^0 and $W^\pm$. Their masses are predicted

FUTURE ACCELERATORS   363

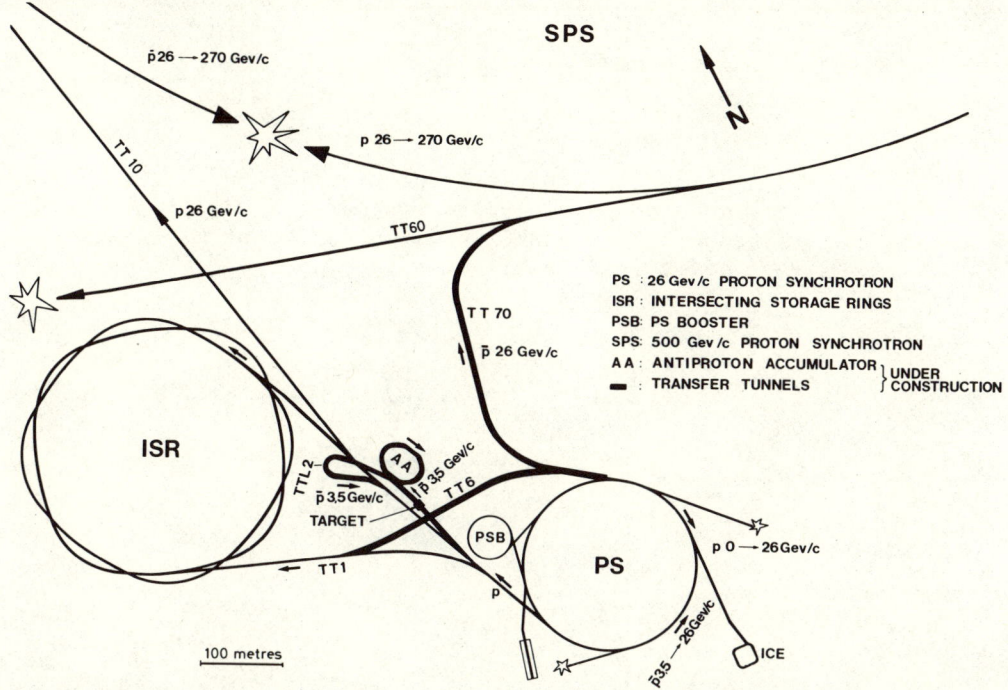

Fig. 1 The CERN p$\bar{\text{p}}$ collider system showing the antiproton accumulator (AA) ring, the 26 GeV $\bar{\text{p}}$ ejection line from the PS to the SPS and to the ISR

by the Glashow-Salam-Weinberg model on the basis of measurements of the coupling constants at low Q^2; using the most recent data, and taking into account radiative corrections, these predictions are[7]

$$m_W \cong 81 \text{ GeV}, \quad m_Z \cong 92 \text{ GeV}. \tag{1}$$

Proton-antiproton collisions produce unseparated colliding wide-band beams of gluons and of quarks and antiquarks; their collision produces heavy vector states (see Fig. 2) decaying into fermion-antifermion pairs. Figure 3 shows the cross-section for pp → W^+X and for $\bar{\text{p}}$p → $W^{\pm}X$ as a function of the scaling variable m_W^2/s as calculated by Peierls and Quigg[8]. Effects due to the emission of gluons, responsible for scaling violations, are expected to be small at small values of the scaling variable, $\tau = m_W^2/s < 0.1$ [8]. It is therefore of great advantage to work at a high energy so as to reduce the uncertainty of the cross-section prediction. At $\sqrt{s}$ = 540 GeV the following values are obtained:

$$\sigma(p\bar{p} \to WX) \simeq 3.2 \times 10^{-33} \text{ cm}^2$$

$$\sigma(p\bar{p} \to Z^0X) \simeq 1.6 \times 10^{-33} \text{ cm}^2 .$$

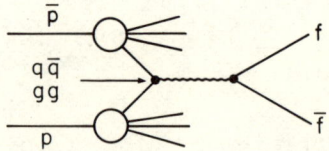

Fig. 2 Heavy vector particle production by wide-band unseparated quark-antiquark and gluon colliding beams in pp̄ collisions

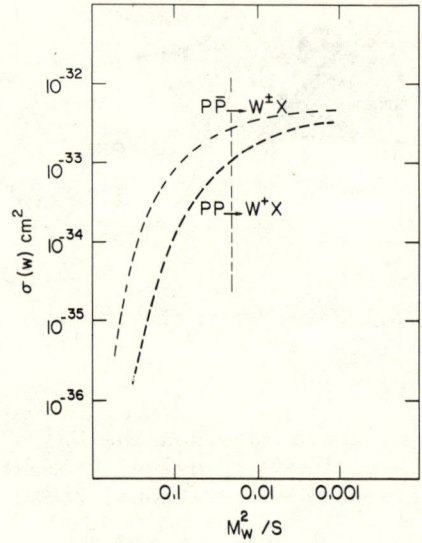

Fig. 3 Total cross-section of the reactions pp → WX and pp̄ → WX as a function of the scaling variable $\tau = m_W^2/s$, as calculated by Quigg and Peierls (Ref. 8)

Detection of hadronic decay modes of the vector bosons is difficult because of the large cross-section for hadron production at large p_T. The experiments therefore aim at detecting the leptonic decay modes occurring with the following branching ratios:

$$BR(W \to e\nu) = \frac{1}{4N} = 8.3\% \quad \text{(for N = 3 fermion families)}$$

$$BR(Z \to e^+e^-) = \frac{1}{9.4N} = 3.5\% \ .$$

The rates expected in a physics run of 1000 hours for an integrated luminosity of 10^{36} cm^{-2} are summarized in Table 1. Assuming a detection efficiency of $\sim$ 50%, a rate of about 30 events of $Z^0 \to e^+e^-$ can be expected, enough to form a significant peak if the mass resolution is $\sim$ 1%, as is foreseen in the design of the experiments[9].

Table 1

Rates of W and Z^0 production in $\bar{p}p$ collisions at $\sqrt{s}$ = 540 GeV for $\int L\, dt = 10^{36}$ cm^{-2}

Process	BR·σ (cm^2)	Events/10^{36} cm^{-2}
$W^+ \to$ all	3.2×10^{-33}	3200
$W^+ \to e^+\nu$ a)	2.6×10^{-34}	260
$Z^0 \to$ all	1.6×10^{-33}	1600
$Z^0 \to e^+e^-$ b)	5.7×10^{-33}	57

a) BR($W \to e\nu$) = 8.3% for three fermion families.
b) BR($Z^0 \to e^+e^-$) = 3.5% for three fermion families.

Recently, it has been found experimentally[10] that the cross-section of lepton pair production in hadron-hadron collisions is enhanced by a factor of approximately 2 with respect to calculations such as those of Peierls and Quigg starting from structure functions measured in deep inelastic lepton-hadron scattering. It is therefore to be expected that the cross-sections for W and Z production will be enhanced by the same factor.

It may be noted for comparison that ISABELLE, a 400 GeV + + 400 GeV proton-proton collider under construction at Brookhaven, may reach a luminosity of 10^{32} cm^{-2} s^{-1} and a rate of $N(Z^0 \to e^+e^-)$ $\sim$ 5000 in 1000 hours. LEP, with a design luminosity of 5×10^{31} cm^{-2} s^{-1}, would give a rate of $N(Z^0 \to e^+e^-) \sim 3000$ per hour.

3. THE LARGE e^+e^- COLLIDER PROJECT LEP

Large e^+e^- colliders are being designed to study weak interactions at high energy, in the energy range in which the Z^0 pole is expected, and their unification with the electromagnetic interaction. The weak neutral-current coupling of all leptons can be measured, the vector bosons can be produced, and their self-coupling can be studied. The Higgs meson can be produced if its mass is less than 90 GeV and if it is a physical particle. The fermion spectrum can be studied.

At CERN a project called LEP[11] is being considered, with total energy up to 170 GeV using copper cavities and up to 260 GeV using superconducting cavities. The design is based on a cost optimization at $\sqrt{s}$ = 170 GeV, with a peak luminosity of 10^{32} cm^{-2} s^{-1}. The main design parameters are summarized in Table 2. The energy and

Table 2

Main design parameters of the large e^+e^- collider LEP

Circumference	30 km
Magnetic field at 85 GeV beam energy	800 G
Energy loss per revolution at 85 GeV	1.37 GeV
Length of RF cavities	1628 m
Number of interaction regions	8

the luminosity that can be reached in different phases of the LEP project are shown in Fig. 4. In phase 1, an energy up to 100 GeV can be reached, covering the Z^0 pole, at a luminosity of 5×10^{31} cm^{-2} s^{-1}, using 16 MW of radiofrequency power. A layout of LEP is shown in Fig. 5. To save on cost and on construction time,

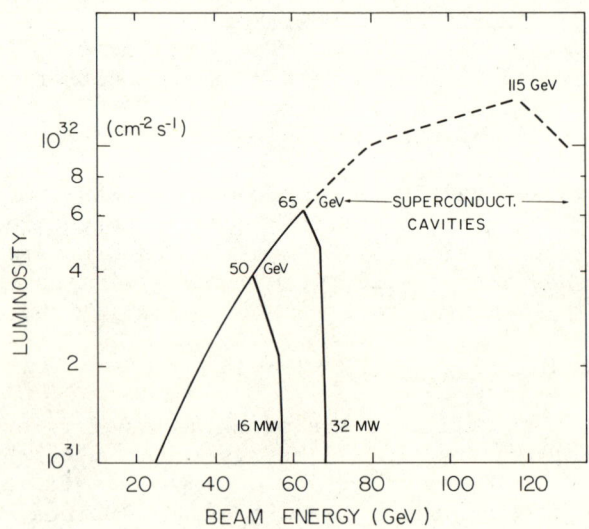

Fig. 4 Design luminosity of the different phases of LEP (Ref. 11). Also shown is the RF power required for phase I and phase II.

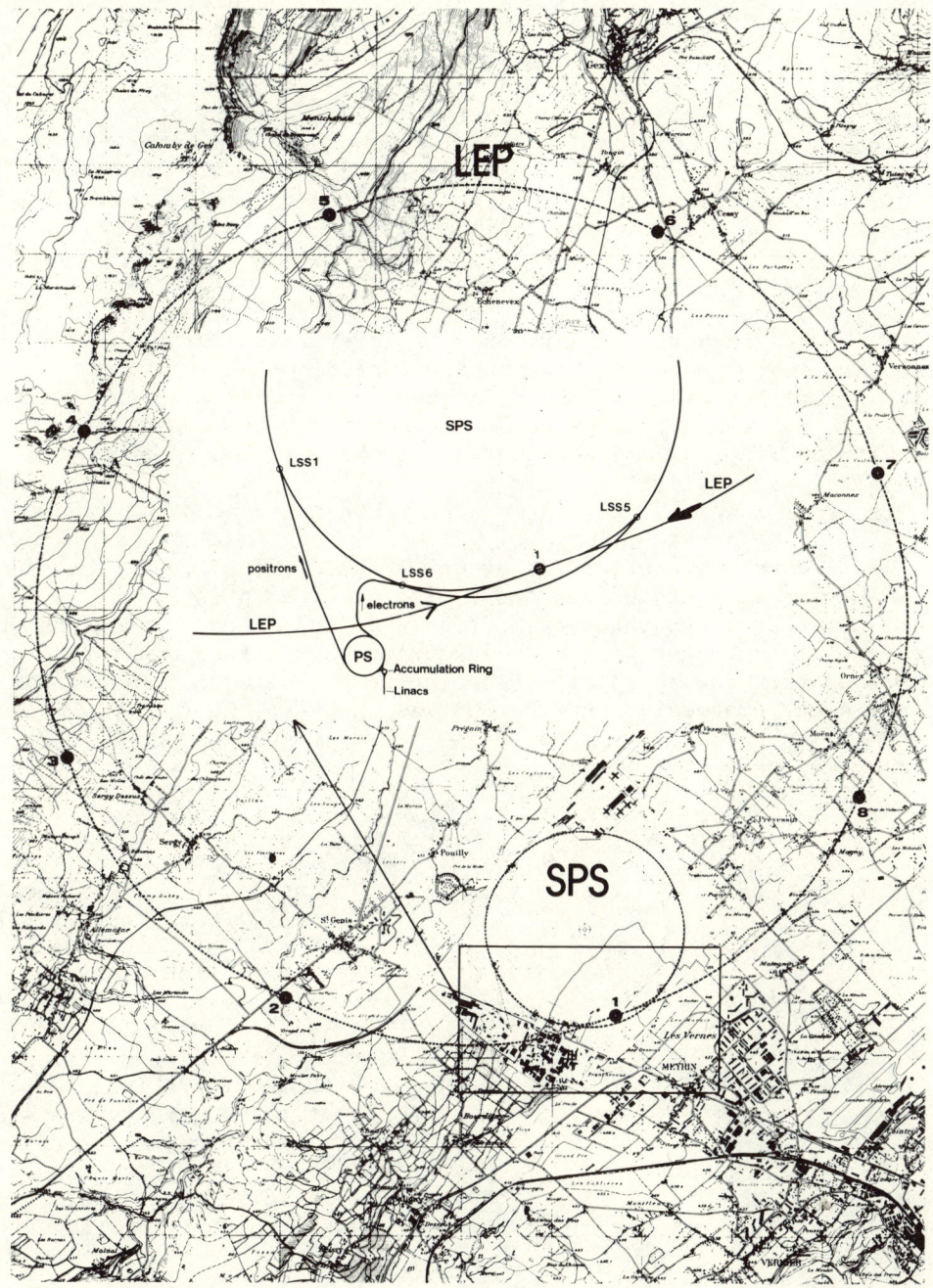

Fig. 5 Layout of the LEP project, of the injection system using the PS-SPS complex, and of the eight beam crossing-points

the CERN PS will be used as a pre-injector to accelerate beams of electrons and positrons to 2.5 GeV, and the CERN SPS as an injector synchrotron of 22 GeV. These beams will then be transferred to the LEP ring, which will be constructed at a level below the SPS. The beams of LEP will cross each other at eight points; at position 1, located near to the transfer tunnel from the SPS injector, a by-pass for protons can be built; at a later stage this will allow the study of electron-proton collisions. During phase 1 of the LEP project, only four interaction regions may be equipped for experiments.

At the Z^0 pole, which is expected to occur at an energy of 91 GeV [12], a rapid rise of the total cross-section is to be expected, by a factor of approximately 1000, the peak value going up to R = 4316 (see Fig. 6). The width of the pole depends on the number of fermion generations into which the Z^0 can decay. Assuming three generations, the width is expected to be Γ_{Z^0} = 2.2 GeV. Another spectacular prediction based on the Glashow-Salam-Weinberg theory is the disappearance of this peak if the colliding beams are polarized to the same longitudinal state, e.g. if they are both left-handed.

The event rate predicted at the Z^0 pole, of 172,000 events/day for a luminosity of 10^{32} cm^{-2} s^{-1}, is larger by a factor of several thousand than event rates at present e^+e^- colliders. Should the Z^0 pole not exist, then we would expect to observe a constant rise in cross-section due to the four-fermion weak interaction, and a constant electromagnetic contribution, as shown in Fig. 6; the sum of

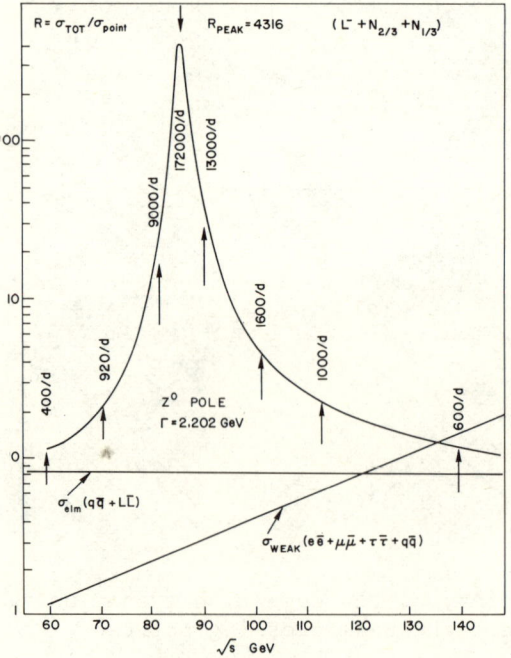

Fig. 6 Rise of the ratio R of the total hadronic cross-section and the electromagnetic point-like cross-section as expected at the Z^0 pole in the Glashow-Salam-Weinberg theory ($\sin^2 \theta = \frac{1}{4}$). Event rates per day are shown for a peak luminosity of 10^{32} cm^{-2} s^{-1}. The curves labelled σ_{weak} and σ_{elm} give the ratio R in the case where no Z^0 pole exists.

the two and of their interference term would give an event rate of a few hundred per day at $\sqrt{s}$ = 200 GeV.

At the Z^0 peak, measurements of the rate to the different fermion-antifermion pairs (e^+e^-, $\mu^+\mu^-$, $\tau^+\tau^-$, $u\bar{u}$, $d\bar{d}$, $s\bar{s}$, etc.) will determine the relative decay widths of the Z^0 and thus the relative neutral-current coupling constants in the combination[2]

$$v_f^2 + a_f^2 . \qquad (2)$$

Measurements of the forward-backward asymmetry at the Z^0 pole can be used to determine the product of the coupling constants[2]

$$v_f \cdot a_f . \qquad (3)$$

This information can also be extracted from measurements of the polarization of final-state fermions produced in the decay of polarized Z^0. These measurements give values for the two combinations in Eqs. (2) and (3), and leave a fourfold ambiguity in the determination of a_f and v_f [13]. To solve this ambiguity, measurements of the asymmetry off-resonance (see Fig. 7) or of helicity off-resonance are required.

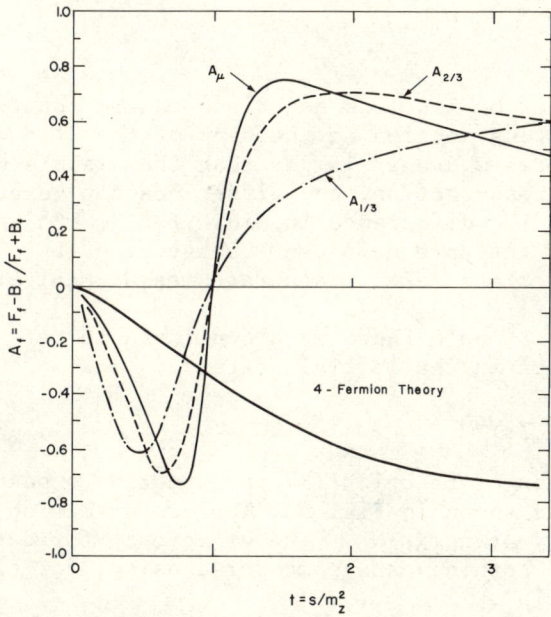

Fig. 7 Energy dependence of the forward-backward asymmetry of the reaction $e^+e^- \to f^+f^-$ for fermions of different charge as predicted for $\sin^2 \theta = \frac{1}{4}$. Also shown is the asymmetry of $\mu^+\mu^-$ in a four-fermion theory without a Z^0 pole.

The large event rate at the Z^0 pole will allow sensitive searches for rare phenomena. For instance, the cross-section for $e^+e^- \to b\bar{b}$ is expected to be 30 times larger at the Z^0 pole than at threshold. The production of Higgs mesons can proceed via the Z^0 if the mass is smaller than the Z^0 mass. For m_H = 50 GeV the branching ratio BR ($Z^0 \to H\mu^+\mu^-$) is expected[2] to be 10^{-5}. A convincing result can be obtained from $\sim 5 \times 10^6$ events, which can be obtained in about 100 days.

There are several ways of determining by experiment the number of fermion generations. In an anomaly-free world the sum of all fermion charges, accounting for three colours of the quarks, is zero. Hence the number of fermion generations is equal to the number of neutrino generations. From the experimental limit on the branching ratio of $K \to \pi\nu\bar{\nu}$, we deduce $N_\nu < 6000$. In the decay of a heavy onium state, e.g. of toponium if found,

$$O' \to O + \pi^+\pi^- \to \pi^+\pi^- + \text{nothing} ,$$
$$\!\!\!\downarrow \nu\bar{\nu}$$

the branching ratio of $O \to \nu\bar{\nu}$ is related to the number of neutrino generations,

$$\text{BR } (O \to \nu\bar{\nu}) \propto N_\nu \left(\frac{m_O}{m_{Z^0}}\right)^2 . \tag{4}$$

At LEP the number of massless neutrinos can be counted by carefully measuring the cross-section at the peak of the Z^0 and the total width Γ_{Z^0}. If the total width is larger than the visible width deduced from the peak cross-section, the difference can be attributed to the decay $Z^0 \to \nu\bar{\nu}$. The difference in width for N_ν = 3 and 4 is 140 MeV, comparable with the spread in beam energy of $\Delta\Gamma$(LEP) $\sim$ 100 MeV. Also, radiative corrections make an accurate measurement of N_ν difficult.

Beyond the Z^0 pole there is a considerable rate of Z^0 production by radiation in the initial state,

$$e^+e^- \to \gamma Z^0 \to \gamma\nu\bar{\nu} . \tag{5}$$

The cross-section of reaction (5) is expected to peak at E_γ = = $E_{LEP} - m_{Z^0}$, as shown in Fig. 8. The cross-section at the pole[14] is approximately $3 \cdot \sigma_{point} \cdot N_\nu$. The efficiency for detecting the final state γ + nothing can be measured using, for example, the decay mode $Z^0 \to \mu^+\mu^-$.

A study of reaction (5) can also provide evidence for right-handed neutrinos if they exist in nature. In a left-right symmetric theory $SU(2)_L \times SU(2)_R \times U(1)$, the ratio of the cross-section of reaction (5) for right-handed and for left-handed neutrino is[15]

FUTURE ACCELERATORS

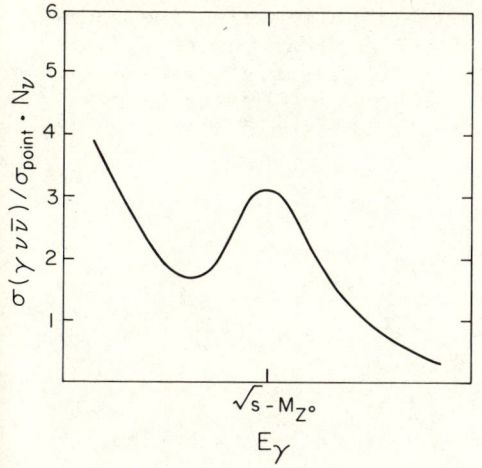

Fig. 8 Cross-section of the reaction $e^+e^- \to \gamma Z^0 \to \gamma + \nu\bar{\nu}$ as a function of the E_γ above the Z^0 pole in units of $\sigma(\gamma\nu\bar{\nu})/\sigma_{point} \cdot N_\nu$, N_ν being the number of neutrino families

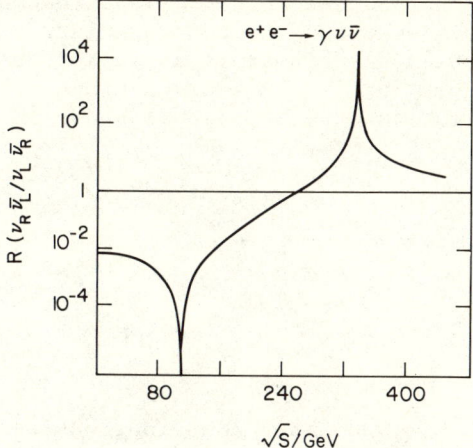

Fig. 9 Ratio of the cross-section of the reaction $e^+e^- \to \gamma\nu\bar{\nu}$ with $\nu_R\bar{\nu}_L$ and $\nu_L\bar{\nu}_R$ as a function of $\sqrt{s}$, assuming a left-right symmetric $SU_L(2) \times SU_R(2) \times U(1)$ theory with $m_{Z^0(R)} = 320$ GeV (Ref. 15)

$$R = \frac{\sigma(\gamma\nu_R\bar{\nu}_L)}{\sigma(\gamma\nu_L\bar{\nu}_R)} \propto \left(\frac{m_{Z,L}}{m_{Z,R}}\right)^4 .$$

For $m_{Z,L}/m_{Z,R} = 0.3$, R is expected to vary with energy as shown in Fig. 9.

The ratio of cross-sections for scattering of right-handed and left-handed electrons on protons,

$$\frac{\sigma(e_R^- p \to \nu_R X)}{\sigma(e_L^- p \to \nu_L X)} ,$$

is expected[16] to reach a value of 10% at $Q^2 = (320 \text{ GeV})^2$ in the context of the same assumptions.

W pairs are expected to be produced in the reaction

$$e^+e^- \to W^+W^- \tag{6}$$

with a rate which can be calculated following the assumed gauge couplings. The self-coupling of the bosons depends on the gauge group, and large cancellation effects between the three dominant graphs shown in Fig. 10 lead to characteristic effects in the energy dependence and the angular distribution. The calculated cross-section rises sharply above threshold (172 GeV for $\sin^2\theta = 0.2$ when accounting for radiative corrections); its maximum, owing to cancellation effects, occurs about 30 GeV above threshold (see Fig. 11).

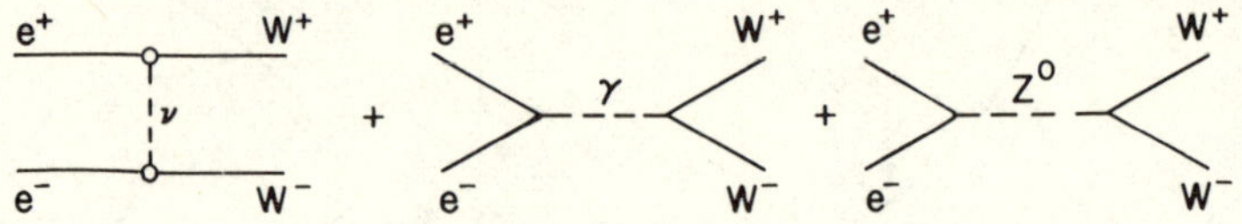

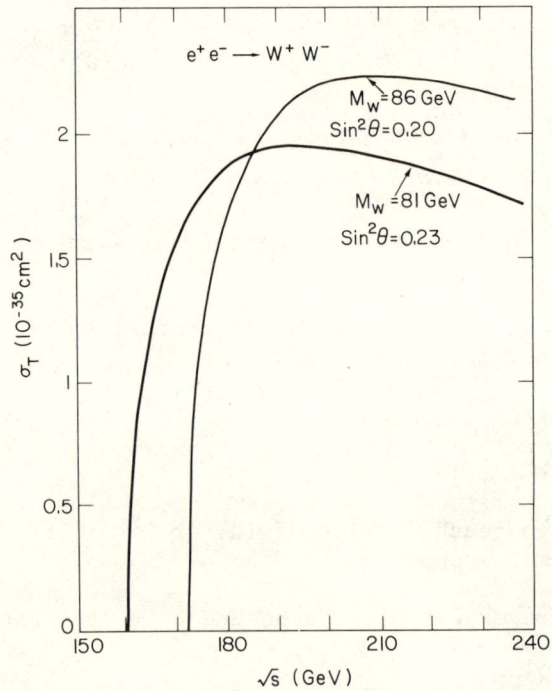

Fig. 10 Dominant contributions to the reaction $e^+e^- \to W^+W^-$ involving the boson self-coupling vertices ZWW and γWW

Fig. 11 Energy dependence of the reaction $e^+e^- \to W^+W^-$ for $\sin^2\theta = 0.23$ and 0.20 showing large cancellation effects owing to boson self-coupling graphs (Ref. 2)

FUTURE ACCELERATORS

A measurement of the total cross-section of reaction (6) is a fundamental test of the gauge nature of the weak bosons. The required maximum energy of LEP is largely determined by this reaction. However, it may take a long time to build LEP up to this final stage.

More details about the physics that can be studied at LEP can be found in Ref. 2.

4. THE LARGE ep COLLIDER PROJECT HERA

Large $e^{\pm}p$ colliders have been designed since 1966 [17]; interest in their physics potential was revived in 1971 [18] after the discovery of point-like structures in deep inelastic lepton scattering.

More recent studies[19] of the physics potential of large electron-proton colliding rings have concentrated on neutral and charged weak currents as one of the main motivations for constructing a large facility. The properties of charged weak currents at short distances can be further explored, extending present studies in neutrino beams, by electron-proton colliders. It was found crucial to explore the region above 100 GeV in the centre-of-mass system, the presumed mass scale of the vector bosons of weak interactions. DESY has studied a project of an ep machine[4] in a 6.5 km long tunnel adjoining the DESY site, called HERA, designed to collide 820 GeV protons with 30 GeV electrons. The layout of HERA is shown in Fig. 12. HERA has two

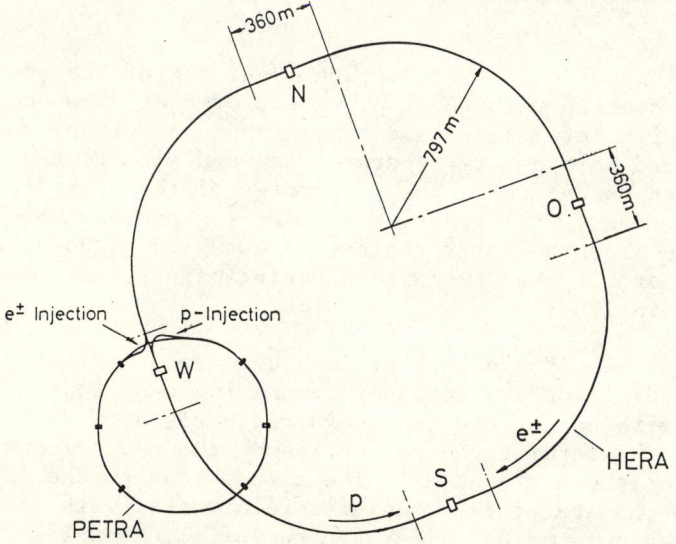

Fig. 12 Layout of the HERA electron-proton collider project at DESY (Hamburg) and its injector PETRA. Four beam crossing-points for experiments are planned (Ref. 4).

rings, one for electrons and positrons, the other for protons. They cross each other in four long straight sections enlarged to accommodate experimental areas. PETRA is used as an injector synchrotron for electrons and protons.

The main parameters of HERA are summarized in Table 3. The proton ring will be constructed from superconducting magnets with a

Table 3

Main design parameters of HERA

Parameter	Proton ring	Electron ring
Energy range	100–820 GeV	10–33 GeV
Peak luminosity	3.5×10^{31} cm^{-2} s^{-1}	
Polarization time	19.5 min	
Number of interaction regions	4	
Circumference	6451 m	
Magnetic field at E_{max}	47.25 kG	1.819 kG
Circulating current	500 mA	58 mA
Injection energy	40 GeV	14 GeV

maximum field of 4.725 tesla. The lower end of the energy range is chosen to overlap with the top end explored at CERN and FNAL. Measurements with left-handed and right-handed electrons and positrons are required in order to separate weak and electromagnetic contributions and to determine the properties of the neutral and of the charged current. All four interaction regions have been designed to produce these helicity states. A detailed proposal will be made in 1981; prototypes of superconducting magnets are expected to be available in 1982.

It has been found in experiments on deep inelastic neutrino, muon, and electron scattering that the incident leptons interact directly with one of the quarks in the nucleon. The data can be analysed to determine the properties of the neutral and charged weak currents at short distances. The effects due to the exchange of the vector bosons and of their heavier recurrences with masses up to 500 GeV can be studied. The spectrum of electron-like leptons or the spectrum of heavy quarks coupling to u or d quarks in the proton can be investigated up to 200 GeV mass. New currents, if they are mediated by vector bosons with masses up to 300 GeV and coupled to new heavy leptons and quarks, can be detected. Once the structure of the currents is determined, measurements of deep inelastic scattering can explore the structure of the proton and of its constituent quarks down to distances of 10^{-17} cm. The scale of masses which can thus be explored is shown in Fig. 13 as a function of the available centre-of-mass energy.

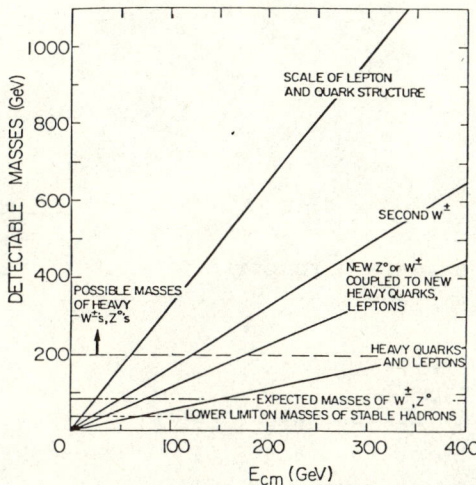

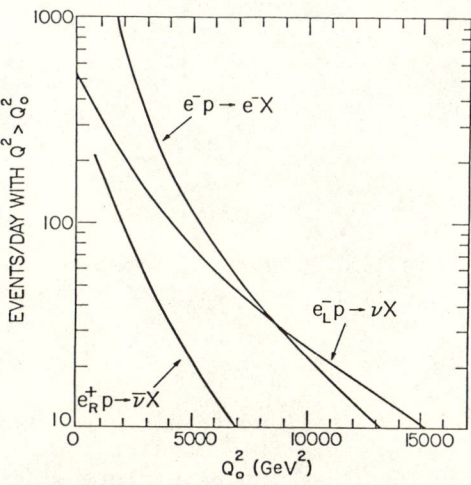

Fig. 13 The masses of heavy particles produced directly (leptons and quarks) or observed indirectly by their propagator effects (Z^0, W) as a function of available energy in the centre of mass (E_{cm}). The substructure of leptons and quarks can be probed far beyond E_{cm} in ep collisions (Ref. 4).

Fig. 14 Event rates per day at momentum transfers larger than a given value Q_0^2 for a peak luminosity of 10^{32} cm^{-2} s^{-1}. The rate of neutral-current events shown is due to one-photon exchange only (Ref. 4).

Present data on charged-current interactions at short distances are consistent with left-handed currents mediated by a single vector boson of 80 GeV mass. At HERA energies, many vector bosons, some mediating right-handed currents, may contribute to deep inelastic scattering. Figure 14 shows that the event rate beyond the mass scale of 80 GeV ($Q^2 > 6400$ GeV^2) is adequate, and that its variation with Q^2 can be used to determine the mass of the propagator as long as it is below 500 GeV (see Fig. 15).

The asymmetry of the cross-section of left-handed and right-handed electrons measures parity violation. In left-right symmetric theories of neutral currents of the type $SU(2)_L \times SU(2)_R \times U(1)$,

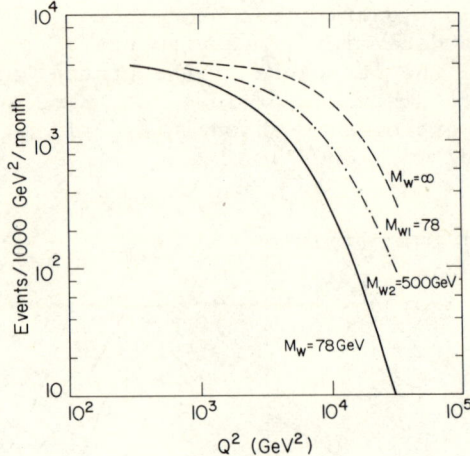

Fig. 15 Rate of charged-current events as a function of Q^2 for different mass values of the charged boson; $s = 98400$ GeV^2 and $L = 10^{32}$ cm^{-2} s^{-1} (Ref. 4)

the variation of this asymmetry with the inelasticity y is characteristic of the mass of the second boson (see Fig. 16); masses up to 300 GeV can be explored.

The existence of right-handed charged currents can be deduced from a comparison of $\sigma(e_R^- p \to \nu X)$ or $\sigma(e_L^+ p \to \bar\nu X)$ with $\sigma(e_L^- p \to \nu X)$ or $\sigma(e_R^+ p \to \bar\nu X)$ [16]. As already remarked in Section 3, a sensitive search for right-handed currents can be made as well by measuring $\sigma(e^+e^- \to \gamma \nu \bar\nu)$ as a function of energy[15].

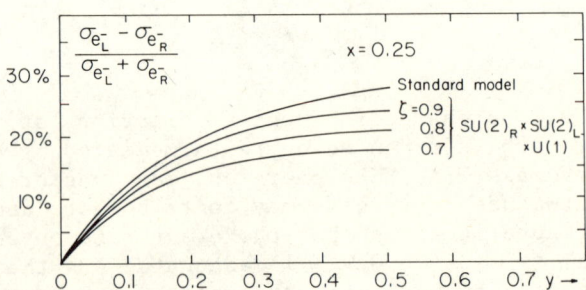

Fig. 16 Parity violating asymmetry at $x = 0.25$ for SU(2)$_L \times$ SU(2)$_R \times$ U(1) models with different values of ζ; $\zeta = 0.7$ corresponds to the mass of a second Z^0 of 224 GeV; $\zeta = 0.8$ to $m_{Z_2} = 280$ GeV; $\zeta = 0.9$ to $m_{Z_2} = 407$ GeV; and $\zeta = 1$ to $m_{Z_2} = \infty$ (Ref. 4)

Electron-like charged or neutral leptons, and new heavy quarks which couple to u or d quarks in the proton, can be produced with large rates (see Fig. 17). This is a domain that cannot be explored by e^+e^- colliders.

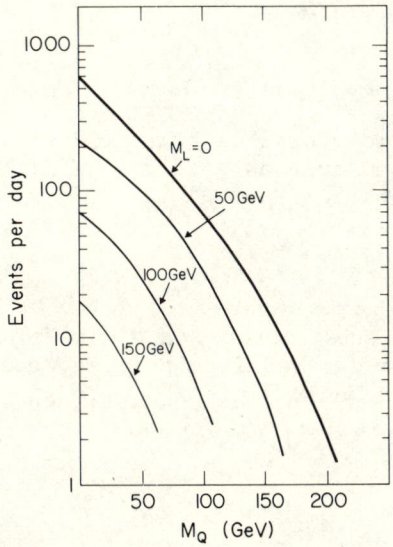

Fig. 17 Rate of events due to new currents involving new heavy quarks of mass m_Q and new heavy leptons of mass m_L (Ref. 4)

Three generations of quarks and leptons have been discovered, and with this proliferation we may no longer be justified in considering leptons and quarks as the elementary constituents of matter. They may instead be composed of common elementary subconstituents. HERA can probe the structure of leptons and quarks in deep inelastic scattering to distances of 10^{-17} cm or to mass scales of 1 TeV (see Fig. 13). The cross-section would be modified in the case of quark substructure, e.g. if the quarks would exchange gluon-like objects which do not interact weakly or electromagnetically. Their exchange would, for example, be visible as a step in the momentum fraction of protons carried by quarks, $\int_0^1 F_2(x) \, dx$, as a threshold at $Q^2 = 4 m_g^2$ is crossed (see Fig. 18). Brodsky and Drell[20] have shown that in some theories of lepton and quark substructure the anomalous magnetic momentum, in excess of the point-like value calculated by QED, i.e. δa, is related linearly to the mass characterizing the

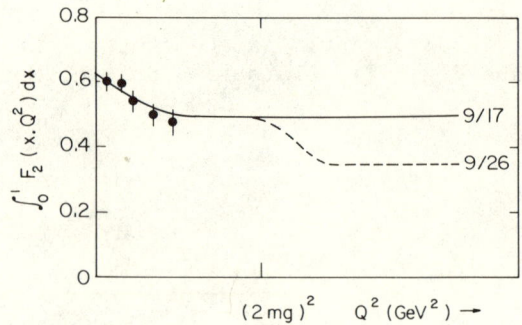

Fig. 18 Possible behaviour of the momentum sum rule as a threshold at $Q^2 = 4 m_g^2$ is crossed (Ref. 4)

form factor due to substructure, $\delta a \simeq \mathcal{O}(m_q/m^*_{const})$. Present experimental limits on δa derived from precise measurements of $(g-2)$ for electrons and muons would limit $m^*_{constituent}$ to be larger than 10^3 TeV.

5. SUMMARY

The results obtained in neutrino physics over the past 20 years have contributed in a decisive way to our present understanding of the constituents of matter and of their interactions. Many new questions can now be asked, guided by the successful theory of electroweak interactions; these questions will be explored using the new accelerators now proposed. Future neutrino conferences therefore promise to be exciting.

ACKNOWLEDGEMENTS

The stimulating atmosphere created by the organizers, in particular by Prof. E. Fiorini, has made Neutrino '80 a great success. My most sincere gratitude goes to them and to my colleagues who have patiently helped me during the preparation of this report.

REFERENCES

1. C. Rubbia, P. McIntyre and D. Cline, Proc. Int. Neutrino Conf., Aachen, 1976 (Vieweg, Braunschweig, 1977) p. 683.
   C. Rubbia, Proposal for a $\bar{p}p$ collider using the CERN SPS, CERN/NP/77-1 (1977).

2. P. Darriulat (ed.), Physics with very high energy e^+e^- colliders, CERN 76-18 (1976).
   M. Jacob (ed.), Proc. LEP Summer Study, Les Houches and CERN, CERN 79-01 (1979).
3. B. Richter, Nucl. Instrum. Methods 136, 47 (1976).
4. U. Amaldi (ed.), Report of the Electron-Proton Working Group of ECFA: Study on the electron-proton storage ring project HERA, ECFA 80/42 (1980).
   B. Wiik, Future accelerator plans at DESY, DESY 80/56 (1980).
5. Design study of a proton-antiproton colliding beam facility, CERN/PS/AA 78-3 (1978).
6. S. van der Meer, Stochastic damping of betatron oscillations, CERN/ISR-PO/72-31 (1972).
7. See, for example, B. Borgia, rapporteur's talk, these proceedings.
8. C. Quigg, Rev. Mod. Phys. 49, 297 (1977).
9. Aachen-Annecy-Birmingham-CERN-London (QMC)-Paris (Collège de France)-Riverside-Rome-Rutherford-Saclay-Vienna Collaboration (A. Astbury et al.), A 4π solid angle detector for the CERN SPS $p\bar{p}$ collider, UA1 proposal SPSC/78-6/P92 (1978).
   Bern-CERN-Copenhagen-Orsay (LAL)-Pavia-Saclay Collaboration (M. Banner et al.), Study of $\bar{p}p$ interactions at 540 GeV c.m. energy, UA2 proposal SPSC/78-8/P93 (1978).
10. J. Badier et al., Phys. Lett. 89B, 145 (1979).
11. Design study of a 22 to 130 GeV e^+e^- colliding beam machine (LEP), CERN/ISR-LEP/79-33 (1979).
12. See, for example, T.J. Goldmann and D.A. Ross, Phys. Lett. 84B, 208 (1979).
13. T.G. Rizzo, Determining the neutral-current couplings of second and third generations of quarks and leptons, BNL 27751 (1980) (unpublished).
14. K.J.F. Gaemers, R. Gastmans and F.M. Renard, ECFA-LEP study group, ECFA/LEP 45 (1979).
15. Q. Shafi and Ch. Wetterich, Phys. Lett. 81B, 346 (1978).
16. K. Winter, Phys. Lett. 67B, 236 (1977).
17. L. Goldzahl and E.G. Michaelis, CERN 66-12 (1966).
    H.G. Hereward, K. Johnson, A. Schoch and C.J. Zilverschoon, Proc. 3rd Int. Conf. on High-Energy Accelerators, Brookhaven, 1961 (BNL, Upton, 1961) p. 265.
18. C. Pellegrini, J. Rees, B. Richter, M. Schwartz, D. Möhl and A. Sessler, Proc. 8th Int. Conf. on High-Energy Accelerators, CERN, 1971 (CERN, Geneva, 1971) p. 153.
19. See Ref. 4 and C.H. Llewellyn Smith and B. Wiik, DESY 77/38 (1977).
20. S.J. Brodsky and S.D. Drell, Fermion substructure and anomalous magnetic moment, SLAC-Pub-2534 (1980), submitted to Phys. Rev. D.

NEW NEUTRINO DETECTION TECHNOLOGY:

APPLICATION OF MASSIVE WATER DETECTORS TO ACCELERATOR NEUTRINO PHYSICS

L. Sulak

University of Michigan
Ann Arbor, MI 48109

ABSTRACT

In surveying the field of new detector technology, it appears that the advent of massive, inexpensive water Cerenkov detectors may have a significant impact on future neutrino physics. These detectors offer the volumes necessary to perform experiments at very low fluxes, for example with long neutrino flight paths or with rare neutrino species (e.g. ν_e). As an illustration of the potential on the new techniques, we consider in detail an experiment dedicated to the study of the time evolution of a neutrino beam enriched with ν_e's. The highest fluxes of ν_e appear to be achieved with current beam lines at the Brookhaven AGS or the CERN PS. An array of massive, inexpensive detectors allows a configuration optimized for good sensitivity to neutrino eigenmass differences from 0.6 eV to 20 eV and mixing angles down to 15° (comparable to the Cabibbo angle). The ν_e beam is formed using $K^°_{e3}$ decays. A simultaneously produced ν_μ beam from $K^°_{e3}$ decay serves as the normalizer. Pion generated ν_μ's are suppressed to limit background. The detector consists of a series of seven water Cerenkov modules (each with 175T fiducial mass), judiciously spaced along the ν line to provide flight paths from 40m to 1000m. Simulation and reconstruction of neutrino events in a detector similar to the one considered show sufficient resolution in angle, energy, position and event timing relative to the beam.

INTRODUCTION

The possibilities of neutrino mixing and non-zero neutrino

mass continue to intrigue physicists due to the parallels between the quark and lepton sectors in the unification schemes. However, little is known experimentally of the conjectured three masses, three mixing angles ξ, and one CP violating phase involved in the mass matrix of the three known neutrino states. Ancilliary to other neutrino experiments, attempts have been made to study the time evolution of ν_μ beams, some to reasonably high sensitivity. However, few accelerator experiments have obtained information involving evolution in the ν_e or ν_τ sectors.

Recall that for oscillations[1] to occur there must be both a violation of the specific lepton numbers involved and at least one non-zero neutrino mass. However, accepted theory[2] places no constraints on neutrino masses or mixing angles (which we choose to call Pontecorvo angles), although angles on the order of the Cabibbo angle are perhaps to be expected.

From a completely different point of view, astrophysicists support the idea of a massive neutrino[3] in the range of 4 eV to 80 eV to provide the 80% of the mass of clusters of galaxies that is evident from their rotation (but otherwise is invisible). However, evidence from $\nu_\mu \to \nu_\mu$ experiments[4] argues against oscillation masses in this region for sizeable mixing angles. On the other hand, indication of oscillations from the ν_e sector into (possibly) the ν_τ sector exist. The latest round of CERN beam dump experiments[5] finds a prompt ν_e/ν_μ ratio of 1:2. If charm is the source of the beam dump neutrinos, one would expect a ratio of 1:1; i.e., the ν_e flux seems to have been attenuated in traveling to the detectors. These experiments are done with a neutrino flight path ℓ of $\sim$ 900 m and $E_\nu \sim$ 80 GeV ($\ell/E = 10^{-2}$ m/MeV). The results have been interpreted[6] as evidence for a neutrino mass difference of 7 eV and a mixing angle of $\sin^2 2\xi = 0.5$ for oscillations $\nu_e \to \nu_\tau$.

Reines[7] has recently presented data indicating that the rate of ν_e charged-current interactions near a reactor is about half of what would be expected when normalized by the rate of neutral-current interactions. This suggests that the evolved neutrino state has produced neutral-current interactions but no charged-current ones. The evolved state presumably includes ν_μ, ν_τ, or ν_x's below charged current threshold. The detector is located at a distance of 11.2 m from the ν source, with an energy threshold of 4 MeV ($\ell/E \sim$ 3 m/MeV). The data are consistent with a neutrino mass difference of $\sim$ 0.8 eV or $\gtrsim$ 1.5 eV with $\sin^2 2\xi \sim 0.8$.

The ITEP group has improved their analysis of the end point energy in H^3 beta decay. Preliminary indications[8] are that the effective mass (which could be a superposition of two or more eigenmasses) is 14 eV < M < 46 eV with 99% CL.

It is clear that a dedicated experiment, capable of exploring a large range of ℓ and E so as to search for an oscillatory behavior, is necessary. Further, the experiment should concentrate on the ν_e sector, and optimize on the region of phase space delineated by the recent indications of positive effects -- 0.5 eV to 30 eV, and non-maximal mixing angles.

For such an experiment, the 30 GeV proton machines have clear advantages over alternative neutrino sources. They accelerate the highest proton flux[9] above kaon production threshold, a particularly attractive source of ν_e's. Low energy sources of ν_e's, muon decay at LAMPF, for example, are limited in two ways. The upper limit on sensitivity to neutrino mass differences is limited by the maximum neutrino energy, 53 MeV. A lower limit comes from the 6% duty cycle which leads to an unavoidable cosmic ray background at large flight paths (where event rate is necessarily low).

Fermilab and SPS energies are too high to achieve adequate sensitivity to small oscillation masses. In addition, production processes above charm threshold are not well understood thus preventing unambiguous interpretations of deviations from nominal ν event rates.

From the study of ν_e interactions at Gargamelle, a good deal is known about the generation of electron neutrino beams at AGS and PS energies. Gargamelle[4] observed 200 ν_e interactions in their 5 ton detector at a distance of 90m from a neutrino source in which positive secondaries were focussed by a magnetic horn. The detector was exposed to 1.1×10^{18} protons. This allows us to estimate the ν_e event rate in a dedicated experiment. For concreteness consider a sensitive fiducial mass of 175 tons, exposed to 2×10^{19} protons ($\sim$ 6 weeks at the AGS), located at the position of the bubble chamber $\ell \approx 90$ m from the target. Scaling the Gargemelle event rates yields 1×10^4 ν_e events from K° decay. At most, this rate falls as $1/\ell^2$ (if ν_e oscillations do not occur). At a distance of 1 km, 200 events are expected in a detector consisting of four modules of 175T.

For typical ν_e energies ranging from 500 MeV to 3 GeV we get the range of sensitivity for PS/BNL experiments. At 40 meters the target (the minimum flight path imposed by the muon shield) one could be sensitive up to mass differences as high as 20 eV. At large distances (1 km) and low momenta, one is sensitive to about 0.6 eV. The effects of even smaller mass differences would manifest themselves as deviations from $1/\ell^2$ and a neutrino event rate that is dependent upon the position in the detector.

We consider searching for ν_e oscillations using seven massive detectors of novel technology. They are located from 40m to 1 km from the target. Each module contains 175 fiducial tons of water and an active veto mass of 450T. The detector technology has been developed for use in the IMB proton decay experiment.[10] The technique should yield and energy resolution of 15% at 0.5 GeV, and directional resolution of $\sim$ 15°. This value of energy resolution has been achieved in previous similar detectors.[11]

Using a water medium and a detector with a minimum of mechanical structure enables one to achieve a large mass at low cost. Custom electronics to handle the large number of phototubes (1000 are needed) inexpensively is also advocated. Since designs, vendors and construction techniques now exist, application of this detector technique to accelerator neutrino experiments is imminent.

Production of a ν_e Beam

The sources of ν_e's in the normal positive-focussed PS/BNL beam ($\sim$ 0.5% of ν_μ) are the following:

1) muon decay $\quad\quad\quad\quad \mu^+ \to \nu_e\, e^+\, \bar{\nu}_\mu$
2) K^+_{e3} decay $\quad\quad\quad\quad K^+ \to \pi^\circ\, e^+\, \nu_e$ (5% B.R.)
3) K°_{e3} decay $\quad\quad\quad\quad K^\circ_L \to \pi^\pm\, e^\pm\, \bar{\nu}_e$ (38% B.R.)

Figure 1 shows H. Wachmuth's calculation of the flux for the Gargamelle experiment.[4] Dominating the processes are ν_μ's produced by π^+ decay (and $K^+_{\mu 2}$ and $K^\circ_{\mu 3}$ decay) in the beam tunnel. Only for K° decay are the ν_μ and ν_e branching ratios comparable. Since background rejection and μ/e separation require comparable ν_μ and ν_e fluxes, we advocate a beam enriched in neutral kaons. We suggest a sweeping magnet surrounding the target to bend all charged particles out of the beam line before they can decay. With typical field integrals, one can achieve > 99% reduction in ν_μ flux. The only remaining particles are the K°_L's ($\pi^+\pi^-$ from K°_S are also swept out). The beam will therefore consist of neutrinos from K°_L decay: 30% ν_e, 30% $\bar{\nu}_e$, 20% ν_μ and 20% $\bar{\nu}_\mu$ plus whatever remains from early decays before sufficient bending occurred. The total flux is down by $\sim$ 2000 from normal running, and the event rate is peaked at $\sim$ 1 GeV.

Event Rates

To calculate expected event rates, we scale the data of the CERN Gargamelle collaboration.[4] They report 200 charged-current ν_e events from an exposure of 1.1×10^{18} protons on target (pot) at an energy (24 GeV) and decay tunnel configuration similar to BNL. Scaling this event rate for distance from target, fiducial mass, typical BNL proton exposure, and fraction of ν_e events due to K°_L decay ($\sim$ 1/9 of all potential ν_e sources), we find

$$200 \text{ events} \times \left[\frac{60\text{m GGM}}{500\text{m}}\right]^2 \times \frac{175 \text{ tons}}{5 \text{ tons}} \times \frac{2 \times 10^{19}}{1.1 \times 10^{18}} \times \left[\begin{array}{l}\nu_e \text{ fraction} \\ \text{from } K^\circ_L \\ \text{decay}\end{array}\right] \frac{1}{9}$$

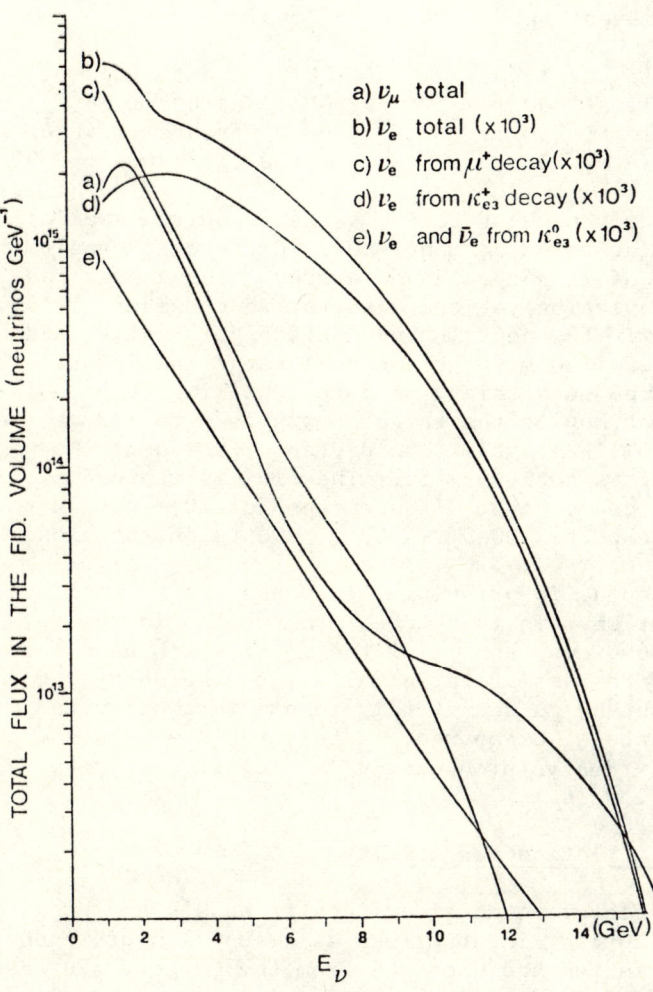

Fig. 1

= 200 ν_e events in a single 175T (fiducial) module at a distance of 0.5 km.

The detector we consider consists of an array of seven 175T modules at various distances along the ν beam line. The approximate spacing of these modules is illustrated in Figures 2a & b. At BNL/PS energies, the events are mainly elastic and N^* production. The anticipated event rates are as follows:

4 modules at 1 km	200 ν_e events + 66 $\bar{\nu}_e$ events
1 module at 500 m	200 ν_e events + 66 $\bar{\nu}_e$ events
1 module at 250 m	800 ν_e events + 240 $\bar{\nu}_e$ events
1 module at 70 m	10,400 ν_e events + 3500 $\bar{\nu}_e$ events

The module closest to the target requires special treatment in an oscillation experiment. Mass difference sensitivity requires a knowledge of flight path and energy. The energy resolution (standard deviation) of the detector we consider is 15%, and the uncertainty in the neutrino production point gives a distance resolution of $\pm$ 30 m for the normal target position. To reduce the fractional uncertainty in ℓ to something comparable to that in p, we suggest moving the target downstream to reduce the decay region to 10 m in length. The distance from neutrino production to the detector is then 45 $\pm$ 5m. The flux is reduced by 1/6 due to the shorter decay path. If one runs for 10^{19} pot in this configuration, we expect 2100 ν_e + 700 $\bar{\nu}_e$ events in the closest module.

(One could, in principle, increase the ν_e flux by using the K^+ contribution. In this mode, a beam stop in the forward direction would absorb pions, and allow the K^+'s, which have a larger angular divergence, to pass preferentially into the decay region. A factor of five might be gained if this scheme were optimized properly. The disadvantage, of course, is that the K^+_{e3} decays (5% BR) are swamped by K decays producing ν_μ's, leading to probably unacceptable background levels.)

Neutrino Oscillation Sensitivity

The detector we consider identifies e's and μ's from ν_e and ν_μ charged-current interactions as a function of reconstructed neutrino momentum and distance from the target. For non-evolving neutrinos from K^0_L decay, the ratio of event rates should be independent of p and ℓ. If only the ν_e component of the beam oscillates and two states participate (between ν_e and ν_τ for example), the ν_e fraction will vary as

$$\frac{\nu_e}{\nu_\mu} = \frac{\nu_e}{\nu_\mu}\bigg|_o [1 - (\frac{\sin^2 2\xi}{2})(1 - \cos\frac{(m_1^2 - m_2^2)}{2p}\ell)],$$

where ξ is the mixing angle and m_1 and m_2 are the masses of the

NEW NEUTRINO DETECTION TECHNOLOGY

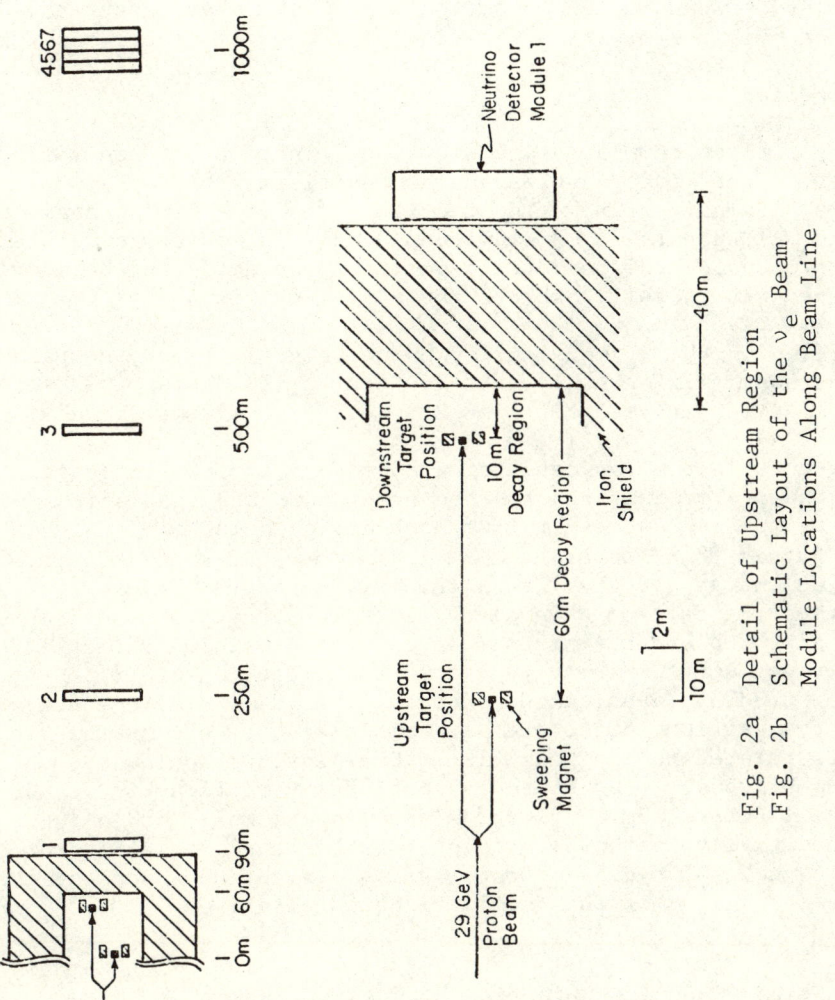

Fig. 2a Detail of Upstream Region
Fig. 2b Schematic Layout of the ν_e Beam Module Locations Along Beam Line

neutrino eigenstates. (For the mass difference range of the suggested experiment, Gargamelle has shown[4] that ν_e and ν_μ mixing is small). Thus ν_e will vary as a function of ℓ and p, while the ν_μ event rate will serve to normalize the flux. The dynamic range of the experiment is from 500 to 2000 MeV/c in p and from 50 to 1000 m in ℓ. Therefore ℓ/p ranges from 0.025m/MeV to $\sim$ 2 m/MeV, corresponding to $\Delta M = \sqrt{m_1^2 - m_2^2} \sim 0.6$ eV. At the other end of the dynamic range, oscillations can be sensed up to $\Delta M = 20$ eV. For larger ΔM, there will be complete mixing at all energies and distances, and ν_e/ν_μ will be constant, but down by a factor of $(1 - \frac{\sin^2 2\xi}{2})$ from what is expected.

For three communicating neutrinos, there are three mass eigenvalues, three mixing angles and a CP violating phase for the most general mixing matrix.[6] The fact that ξ_{12} (the mixing angle between ν_e and ν_μ) appears to be small could make it easier to unfold the others. Undoubtedly several experiments in different beams will be necessary to fully explore the mixing matrix.

Even if the $\nu_e - \nu_\mu$ mixing angle is small (i.e., comparable to the quark mixing angle, the Cabibbo angle), this experiment should be sensitive. For $\frac{\sin^2 2\xi}{2} = 0.3$ or $\xi = 14°$, the nominal 200 events in the far detector are reduced to 160 events (a 3σ decrease). For the nearer detectors, increased statistical power permits measurement of smaller Pontecorvo angles.

Rather than comparing a ratio (at one distance integrated over an energy spectrum) to a theoretical value, this experiment observes the rate of several distances. Energy dependences of the backgound and the signal cancel out, providing a flux independent measurement. The ν_e/ν_μ ratio as a function of ℓ/p is measured by identical detectors. Thus, systematic effects, detection efficiencies, etc. are minimized, as are systematic errors related to the beam flux. Oscillations would show up in an unambiguous way as damped sinusiodal variations in ν_e, ν_μ or ν_e/ν_μ. For values of Δm and mixing angles well within the limits of sensitivity of the experiment, one has the possibility of observing several cycles of oscillation. Figure 3 shows the eventual smearing out of the oscillations which arises if the resolution in ℓ/p of the detector is 15%.

Accelerator Neutrino Detection by Water Cerenkov Techniques

The new technology in water Cerenkov detection can provide the high mass, low cost detectors necessary to perform this experiment.

A detector module might schematically consist of a cylindrical reservoir or water, coaxial with the beam, viewed by many submerged PM tubes. Charged particles are identified by Cerenkov

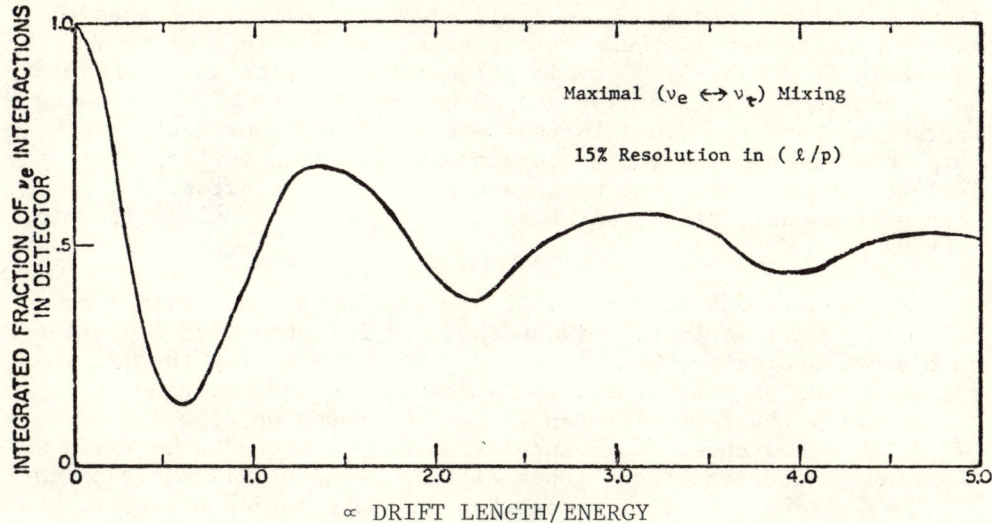

Fig. 3

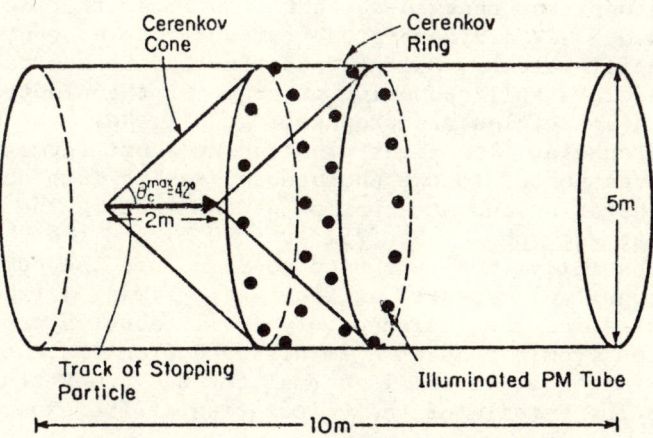

Fig. 4 Cerenkov Pattern from 0.5 GeV Idealized Stopping Track in Schematic Detector Module

light which is produced at an angle of $\lesssim 42°$ with respect to the particle direction if the particle speed exceeds the Cerenkov threshold of $\beta = 0.75$. The directionality of light allows angular and position resolution (15° and 30 cm) consistent with the needs of the experiment. Since in past BNL neutrino experiments[12] simple cuts on rf timing and event containment have been sufficient to isolate the ν signal, the main function of the detector is to obtain reasonable e-μ separation in the largest possible fiducial mass at low cost.

Figure 4 shows schematically how a forward track with a range of 2m (0.5 GeV) would illuminate about 25 PM tubes with approximately 125 photoelectrons. (A total of 128 PM's view the 175 fiducial tons of the figure; only the lit tubes are drawn in.) By utilizing the time differences between tubes on opposite sides of the Cerenkov cone, small angular differences can be measured. The time resolution of the tubes at this level of illumination is $\lesssim$ 2.5 nsec (see Ref. 10 for measured characteristics of the PM tubes) which is adequate to provide the position and angular resolution necessary to reject background.

At 0.5 GeV, the radiating track lengths for both muons and electrons are about 2m. Thus events will be efficiently contained in a 10m long module. Good containment occurs for most events since the typical neutrino energy is only 1.5 GeV. The transverse size of the detector (5m) provides containment of events up to the largest angles (about 30°) characteristic of the charged lepton from elastic neutrino interactions at this energy.

Detailed Monte Carlo calculations to simulate the detector response have been carried out for this type of water detector.[10] Charged leptons from charged-current elastic scattering with energies around 0.5 GeV are generated from nucleons at random positions in the fiducial volume. Multiple scattering, bremsstrahlung, etc. are included in a full showering program for the electrons down to 0.5 MeV. The particles are propagated until they pass below Cerenkov threshold. The light produced by short segments of tracks is transported to the phototubes, taking into account Rayleigh scattering and attenuation of the water. The quantum efficiency at each phototube (25% at 4000 Å) and the photoelectron capture probability (50%) are used to calculate the number of photoelectrons (pe) measured at the tube. Timing jitter and Poisson statistics are included. Typically about 5 photoelectrons are collected from a tube at a 3m distance from track to phototube —a yield of 125 pe overall is normal for a 0.5 GeV track. See Ref. 10 for the details of the calculation of the expected Cerenkov light.

These light yield calculations have been independently verified in an actual 10 meter tank of water simulating the detector.

See Figure 5 and Ref. 10. Two scintillation counters define the light source, muon tracks ~ 10m from the phototubes. At the full 10m distance, an average of 1.5 pe are observed in 5" diameter hemispherical PM's. Since Cerenkov light intensity scales as $1/r$, this verifies the above predictions of 5 pe/tube for an $<r>$ of 2.5m. (The tank is outfitted with a reservoir bladder and detector materials in a surface to volume ratio mimicing that of the suggested detector; water purified by reverse osmosis provides a long transmission length, > 40 m at 4400Å.)

Tapes of events generated by the simulation described above are read into a reconstruction program which uses the time, pulse height and position of each struck tube. A χ^2 minimization performs track fitting. Although the program is rather unsophisticated, it achieves sufficient reconstruction resolution. Figure 6 shows the position resolution. For nonshowering tracks (μ) the vertex resolution is 20 cm. For showering tracks, where more fluctuations occur, this number is 30 cm. The angular resolutions are about $5°$ for nonshowering tracks and $15°$ for showering tracks, which are essentially the physical limits imposed by multiple scattering and shower fluctuations. The total energy desposition is measured by adding up the number of photoelectrons from all struck tubes. At 0.5 GeV the resolution is 15%, as shown in Figure 7; it scales as $\sqrt{E}$. Again, this resolution is dominated by smearing processes; more tubes, mirrors, waveshifters, etc., would not improve any of these values.

The detector is instrumented at low cost using custom readout electronics developed for the proton decay experiment. The detector is enabled during each fast beam spill. Any trigger firing > 3 PM tubes in a 50 nsec coincidence window will cause the time of firing and the charge deposited on each tube to be recorded. The electronics then wait 10 μsec for a possible second pulse which could be the result of an associated muon decay. All digitized information is transmitted serially from each of the remote detectors to a computer for recording on magnetic tape and subsequent analysis.

In the absence of oscillations, the neutrino beam reaching the detector will be essentially 50% ν_μ and 50% ν_e, peaked at about 1 GeV. The majority of interactions will be charged-current elastic, with a smaller fraction of single pion production. Although e, μ and π are measured, the nucleon is invisible since it does not exceed the Cerenkov velocity threshold. Thus elastic neutral current events are invisible, although neutral current single pion production is detectable.

Electrons and muons are distinguished in two ways. First, the muon decay signature will be detected about 65% of the time.

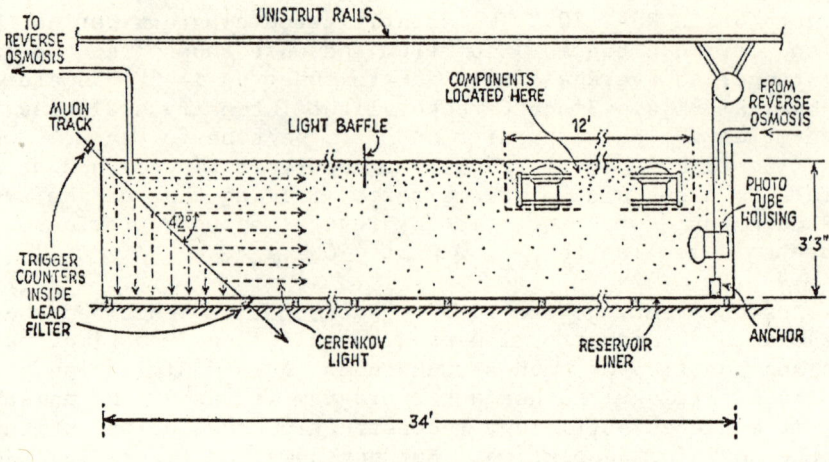

Fig. 5 Side View of Long Baseline Test Facility

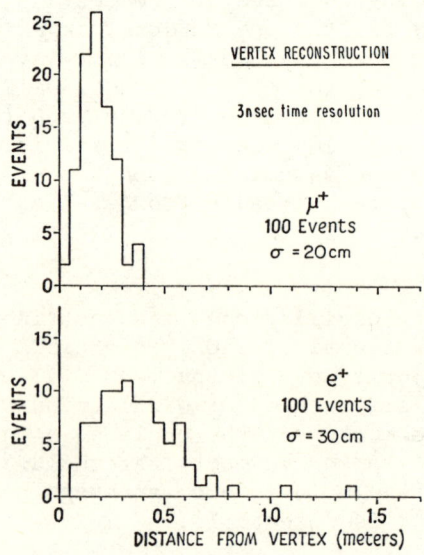

Fig. 6 Vertex Reconstruction
3nsec time resolution

Figure 8 shows the response of the PM tube array to a 40 MeV electron from muon decay. The electron light fires 5-10 phototubes. The firing times must be consistent with electron production at a point coincident with the stopping point of the initial muon. Secondly, the patterns produced by showering and nonshowering tracks are significantly different. Light is scattered more broadly by showering tracks, filling in the center of the rings characteristically produced by nonshowering tracks. Also, for a given range, showering tracks have a larger density of radiating particles than a muon track. Thus, by comparing the density of light to the track length, one can separate e and μ events. These topological and energy factors appear to admit an 80% positive identification of μ and e. Coupled with the muon decay signature, we expect > 90% identification confidence.

Background Rejection

The backgrounds can be divided into two categories:

1. Cosmic Rays

Due to the short duty cycle of the beam (3 μsec at about 1 Hz repetition rate) and the ability to use the RF_o timing structure of the beam, the total live time for a 2×10^{19} pot exposure is about 1 sec (assuming 10^{13} pot/pulse). The raw background rate from cosmic rays is $1/cm^2$ min sr = $\frac{50 \times 10^4}{60}$ $2\pi = 5 \times 10^5$ per module. This rate can be eliminated by an active veto above the detector, or by the topology--a long minimum ionizing downward going track--for events which sneak through the veto. On the other hand these muons, and the electrons from decays of stopping muons, are useful for calibrating and monitoring the detector. In addition, the detector would be enabled halfway between each beam spill to independently monitor the cosmic ray background.

2. Beam-Associated Background

Neutrons have been problematic for sensitive neutrino experiments. Here, however, the Cerenkov detection technique means that protons from neutron interactions will be invisible below a kinematic threshold of E $\sim$ 1.5 GeV. In addition, well known cuts (see Reference 12) from previous experiments appear to be sufficient to eliminate any remaining background involving pion production. Primarily this involves using the RF timing to reject slow neutrons by time of flight difference. Additionally, the large size of the detector (compared to a 60 cm nuclear collision length) could be used to search for spatial asymmetries characteristic of any remaining neutrons.

Although neutral-current elastic scattering is completely invisible in this detector, single pion production could present

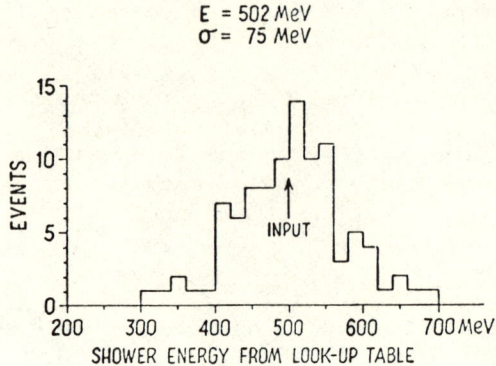

Fig. 7 Shower Energy Resolution Function
$\overline{E}$ = 502 MeV
σ = 75 MeV

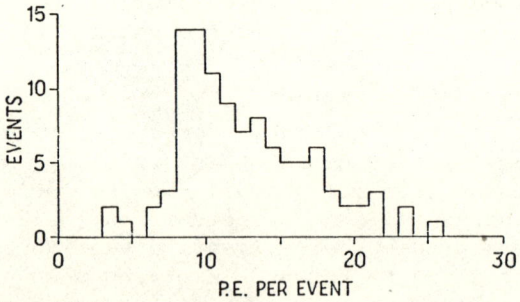

Fig. 8 P.E. From 40 MeV $e^{\pm}$

a background. Asymmetric π^o decay might produce a shower indistinguishable from a single electron. If induced by ν_μ, this would appear as a charged current ν_e event. However, neutral-current events occur at a rate of a factor of 5 lower than charged-current events in cross section. Additionally, Δ's produce π's with low energy, and the decay distribution is essentially isotropic. Further, typical π^o decays produce showers that are well separated. We expect an overall rejection factor of at least 25 by applying cuts based on these characteristics. If the ν_μ flux equals the ν_e flux, neutral-current backgrounds should be less than 5%.

Detector Construction Details

The basic detector configuration (see Figures 4 and 9) relies heavily on technology and hardware in production for the IBM proton decay detector. The sensitive volume of the detector module suggested here consists of about 780 tons of reverse-osmosis water contained inside of a commercially available reservoir liner. It is optically separated into an upper veto region of about 450 tons sheltering an approximate 330-ton interior volume in which an array of 5" hemispherical PM tubes view the Cerenkov light from charged tracks and showers. The interior volume contains a fiducial region of about 175 tons; event reconstruction relies on the timing and directionality properties of the Cerenkov cone.

The containment vessel might consist of an excavated trough with surrounding sand embankments orientated along the beam axis. The trough is lined with a factory-assembled "bladder" of DuPont "hypalon" reservoir lining material. Reservoir liners of this size and type are a standard product of the manufacturer. See Ref. 10 for details. The roof of the bladder floats on the water. It also contains grommets which permit it to be stretched taut, allowing access to the interior of the bladder with the water drained. This eliminates the expense of a mechanical roof over the detector. The liner is light tight and has no deleterious effects on the water clarity with regular refiltering.

Each trough is 22m long with a trapezoidal cross section. The depth of the water is six meters, with a 3m flat region on the bottom, a 15 meter top surface, and 45° sloping walls.

Immersed in the water would be 128 phototubes per module. Ninety-six of these would be arrayed facing inward along the surface of the cylindrical interior volume. Sixteen tubes might serve to cover the end caps of the cylinder, eight on each end. The remaining 16 could be large 8" hemispherical PM tubes which populate the optically separate veto region.

The rear of each tube is sealed in a watertight, neutrally-

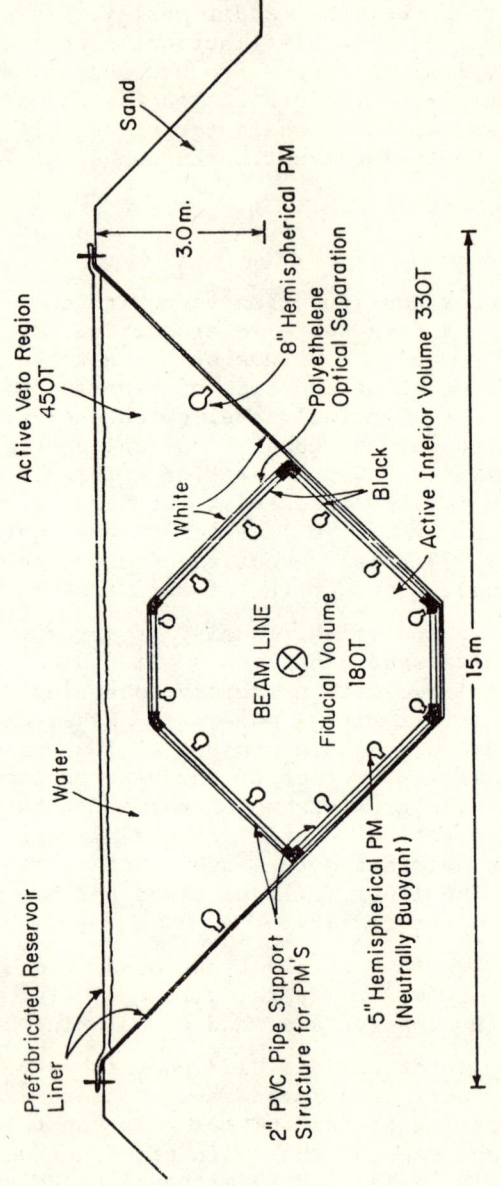

Fig. 9 Cross-Section of Water Cerenkov Module (PM Pattern Repeated 8 Times/Module)

buoyant PVC housing. The hemispherical bulb is in contact with the water. One cable brings high voltage in and takes signals out through a nylon compression fitting. The PM base printed circuit board contains no active components. Such housings have been operated underwater for periods up to 6 months in tests for the IMB detector.

The phototubes which view the interior volume are supported by a hexagonal frame made from conventional 2" PVC pipe, again as employed in the IMB detector. A polyethylene sheet draped over the frame forms the optical divider between the veto region and the interior. Hypalon sheets in the veto region will have reflective white polyethylene bonded to them to maximize light collection, whereas sheets in the interior volume will be black to minimize reflections that could confuse reconstruction.

The efficiency of the veto region in rejecting entering charged particles can be estimated from the measured 95% trigger efficiency of a single 8" PMT viewing a 1m track at a distance of 10m in a non-reflecting test tank (see Ref. 10). In the detector configuration we consider, 5-10 veto tubes are within several meters of any top entering charged track. Veto efficiencies of 10^4 could be reasonably expected. In view of the stringent timing and topology cuts which are possible using the interior PM tubes, this should be adequate to reject non-neutrino induced backgrounds.

The water purification system proposed for this detector is a scaled-down version of the reverse-osmosis (RO) filtering system used in the IMB proton decay experiment. An RO filtering system of the size needed in this experiment has been in use for several months at a 70-foot vertical test tank at Ann Arbor and also at the horizontal test tank (Figure 5) used for the light yield tests. It has maintained water clarity at a > 40m attenuation length level for months. Note that typical optical paths lengths are only 3-5m in the detector we consider, although 10m paths can occur.

The electronics readout system could be identical to the system in production for the IMB experiment. The system for each module consists of one crate of custom readout and digitization logic (16 such crates are used in the proton decay detector).

In summary, it appears that a dedicated electron neutrino oscillation experiment is currently feasible, and that the technology for the large targets necessary to be deployed at several distances to actually plot out an oscillatory curve (should oscillations exist) are available at economical cost.

Acknowledgements

The inspiration for the detector described in this paper derives primarily from the fruitful relationship of the author with the following fellow collaborators on the Irvine-Michigan-Brookhaven proton decay project: W. Bratton, M. Goldhaber, W. Kropp, J. Learned, F. Reines, J. Schultz, E. Shumard, D. Sinclair, D. Smith, H. Sobel, J. VanderVelde and C. Wuest. Other IMB collaborators, B. Cortez, G.W. Foster, J. LoSecco, and J. Stone are due special acknowledgment; we worked out together the details of the detector presented as an illustration here. They are a constant source of inspiration and inventiveness.

I express my sincere gratitude to the tens of neutrino physicists who contributed voluminously to my review talk at this conference. Regretfully, inclusion of each of their work proved to be piecemeal, fragmented, and overlength; in the end it was abandoned.

References

1. For a review see S. M. Bilenky and B. Pontecorvo, Phys. Reports 41 (1978) 225.
2. H. Georgi and S. Glashow, Phys. Rev. Lett. 32, 438 (1974); A. J. Buras, J. Ellis, M. K. Gaillard and D. V. Nanopoulos, Nuc. Phys. B135 (1978) 66-92.
3. G. Steigman, First Workshop on Grand Unification, P. Frampton Ed. (to be published); S. Tremaine and J. Gunn, PRL 42 (1979) 407.
4. J. Blietschau, et al. Nucl. Phys. B133 (1978) 205-219.
5. H. Wachsmuth, CERN-EP/79-115, 1979, to appear in Proceedings of International Symposium on Lepton and Photon Interactions at High Energies, FNAL 1979; P. Alibran et al., Gargamelle Collaboration, Phys. Lett. 74B (1978) 134; T. Hansl et al., CDHS Collaboration, Phys. Lett. 74B (1978) 139; P.C. Bosetti et al., BEBC-ABCLOS Collaboration, Phy. Lett. 74B (1978) 143.
6. "A Fresh Look at Neutrino Oscillations", A. DeRujula, et al Ref. TH. 2788-CERN; "Mass and Mixing Scales of Neutrino Oscillations", V. Barger et al. coo-881-135 (U. Wisconsin preprint).
7. "Evidence for Neutrino Instability", F. Reines, H. Sobel and E. Pasierb (U. Irvine Preprint) submitted to PRL.
8. V. A. Lubimov et al., ITEP-62 print and E. T. Tretyakov, et al., Proceedings of the International Neutrino Conference, Aachen 1976, p. 663 for a description of the experiment and a preliminary analysis of the data.
9. L. C. Teng, 1976 Summer Study on Kaon Factories, BNL 50579, p. 189. The FNAL Booster could produce somewhat fewer K's/sec than the AGS.
10. M. Goldhaber, et al., Proceedings of International Neutrino Conference, Bergen 1979, p. 121, C. Jarlskog, ed.
11. S. E. Willis, et al. PRL. 44, 522 (1980).
12. "Observation of Elastic Neutrino-Proton Scattering", W. Kozanecki, Harvard University Thesis, May 1978.

CONCLUDING REMARKS

D. H. Perkins

Department of Nuclear Physics
University of Oxford
Keble Road
Oxford, OX1 3RH

This is the tenth in a series of conferences on neutrino physics extending over more than a decade. Neutrino conferences have acquired a reputation for exuding an air of expectation, excitement and even of incredibility. Neutrino '80, in Erice, has been no exception. At what other conference in the last years could you, in the course of a few days, have heard such wonderful things as the evidence for a finite neutrino mass and for the instability of the proton?

We have heard presentations of contributed papers as well as excellent reviews and invited papers from a broad field. It would be both improper and impossible for me to try to summarize all this material. You can make your own judgements. All I can do is to give my personal impressions, mentioning a few of those highlights which I thought particularly significant, as well as some of the less dramatic results which I think are nevertheless important for the future.

What have we learned? Consolidation of established results is always welcome, and in this connection it is comforting to know that solar neutrinos are there, at the same rate as last year. The Brookhaven experiment at the Homestake Mine finds a time averaged intensity of 2.2 ± 0.4 SNU, the same as last year, to be compared with the expected value of 7.8 ± 1.5 SNU from the standard solar model (7.5 ± 1.5 SNU at Bergen '79). The neutrinos are detected by the transition $Cl^{37} \rightarrow Ar^{37}$, which can be produced only by the high energy neutrinos from Be^7 and B^8 decay created in side reactions of the solar cycle. This high-energy neutrino flux depends critically on the core temperature and the cross-section for $He^3 + He^4 \rightarrow Be^7 + \gamma$ in the KeV region. The energy dependence of this cross-section has been re-determined by Rolfs[1]; if it is normalized to the previous cross-section measurements in the MeV region it implies a 40%

reduction in the KeV cross-sections. The consequent reduction in the standard model prediction brings the expected rate to 5.5 SNU (4.8 SNU at Bergen). As is well known, turbulent mixing in the solar core can also lead to a reduction in the flux of Be^7 and B^8 neutrinos.

In summary, the observed neutrino flux is healthily above background and in conflict with the standard model. One's only comment can be - so much the worse for the standard model.

New experiments, sensitive to the lower energy but more numerous pp and pe p neutrinos from the first stage of the hydrogen cycle ($p + p \to d + e^+ + \nu$) have been proposed. These use targets of indium or gallium:-

$$\nu + In^{115} \to e^- + Sn^{115m}$$
$$\hookrightarrow 2\gamma + Sn^{115}$$

$$\nu + Ga^{71} \to e^- + Ge^{71}$$

We heard that a pilot device (1.3 tons gallium) has been successfully operated and that a 50-ton detector is under construction. Thus, in a few years, a more definitive and reliable comparison between expected and observed neutrino counting rates should be possible.

Without doubt, the experiment of this conference must be that of the ITEP group on the neutrino mass measurement from the Kurie plot of tritium β-decay. Fig.1 shows results from two precision experiments carried out over the last 8 years, both using a special (8π rotation) spectrometer to determine the momentum of electrons from a thin H^3 source adsorbed on a target of aluminium (Berkgvist [2]) or contained in valine (ITEP[3]). The Bergkvist results (1972) yielded m_ν <55eV at 90% CL. First data (1976) from the ITEP group (Tretyakov et al[3]) were interpreted as m_ν <35eV (Aachen Neutrino '76). Subsequently measurements have continued and at this conference, the total data was broken down into 16 separate runs (Lyubimov et al[4]). For each run, a most probable value of m_ν is determined. The Kurie plot for a combination of 8 out of the 16 runs, as well as the distribution of m_ν values from different runs, is indicated together with the expected distributions for m_ν = 0 and m_ν = 34eV. The conclusion of Lyubimov et al is that at 99% CL, 14 <m_ν <46eV if no assumptions are made about the excited states of the daughter He^3 atom, and that 24<m_ν <46eV assuming a two-level structure.

This is a tremendously important result, and of course it will need to be confirmed in independent experiments. However, it is clear that the present result will remain in the data tables for a long time to come.

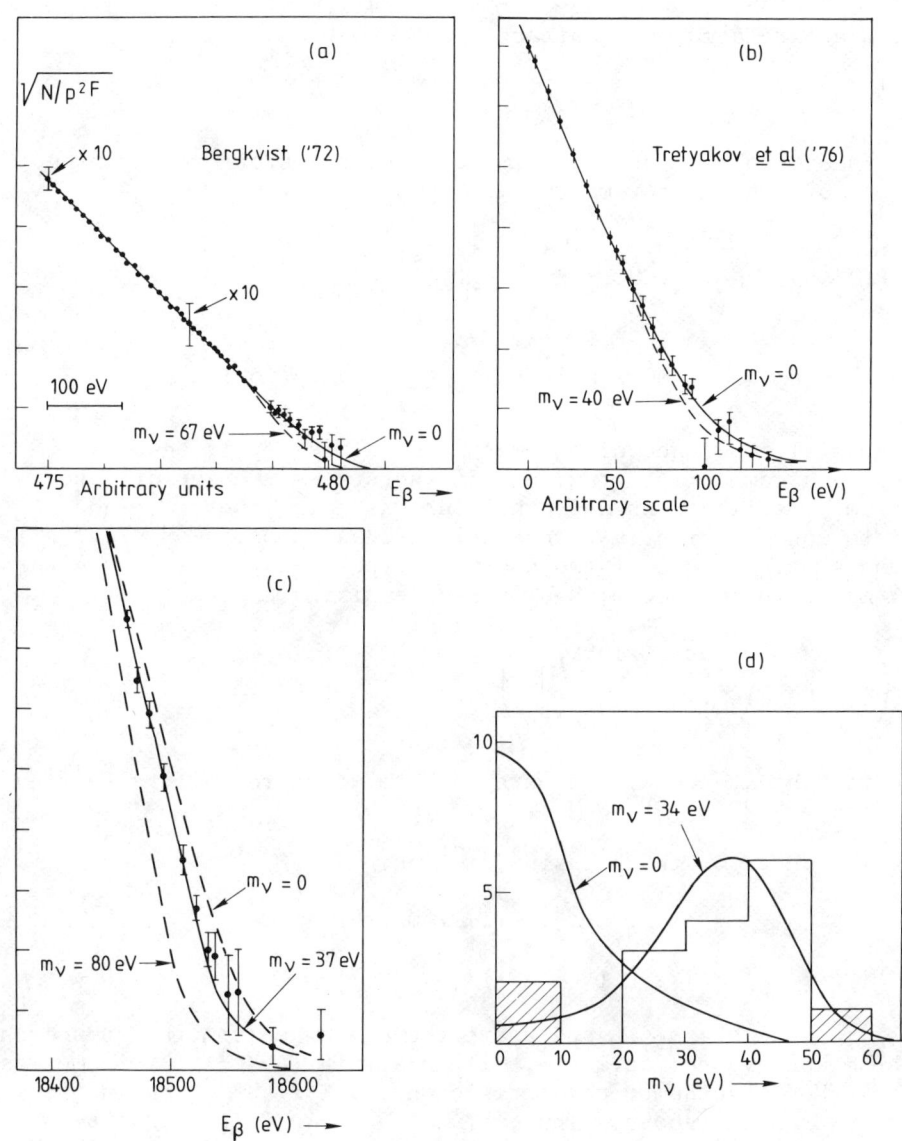

Fig. 1. Kurie plots and fits to the neutrino mass in tritium β-decay. Plots (c) and (d) are of the recent data of Lyubimov et al (ITEP).

Another hot topic of the conference was that of neutrino oscillations, proposed many years ago by Pontecorvo and others[5]. The weak-interaction eigenstates of neutrino are denoted by ν_α (where $\alpha = e, \mu, \tau$) and the mass eigenstates by ν_i (i=1,2,3). The two are related by a unitary transformation

$$|\nu_\alpha\rangle = U_{\alpha i}|\nu_i\rangle$$

A neutrino of type α at time t=0 can transform, via mixing, into one of type β at time t=t with probability

$$P(\nu_\alpha \to \nu_\beta) = \left| \sum_1^n U_{\alpha i} U_{\beta i}^\dagger \exp(-iE_i t) \right|^2$$

where $\qquad E_i = p + m_i^2/2p \qquad (p \gg m_i)$

Spatial coherence of the mass eigenstates ν_i is ensured if they have a common momentum, p. If the states ν_i have different masses m_i, oscillations $\alpha \to \beta$ can occur (assuming there are no absolute conservation laws on the individual numbers of ν_e, ν_μ, ν_τ). For the simplest case of two neutrino types - say ν_e and ν_μ - the 2 x 2 matrix U is specified by a single mixing angle θ:-

$$\begin{pmatrix} \nu_\mu \\ \nu_e \end{pmatrix} = \begin{pmatrix} \cos\theta & \sin\theta \\ -\sin\theta & \cos\theta \end{pmatrix} \begin{pmatrix} \nu_1 \\ \nu_2 \end{pmatrix}$$

Starting at t=0 with a pure ν_μ beam, the probability to find ν_e or ν_μ at time t is

$$P(\nu_\mu \to \nu_e) = \sin^2 2\theta \cdot \sin^2(\Delta m^2 L/4E)$$
$$= \sin^2 2\theta \cdot \sin^2(1.27 \Delta m^2 L/E)$$

$$P(\nu_\mu \to \nu_\mu) = 1 - P(\nu_\mu \to \nu_e)$$

where $\Delta m^2 = m_2^2 - m_1^2$, L and E are the distance from the source and the beam energy, respectively. If Δm^2 is in $(eV/c^2)^2$, L in metres and E in MeV, the numerical coefficient is 1.27 as given in the second expression above. Thus, $P(\nu_\mu \to \nu_e)/\sin^2 2\theta$ oscillates as shown in Fig.2, maxima occurring when $1.27 \Delta m^2 L/E = (2n+1)\pi/2$. No substantial oscillations will be observed if either the mixing angle θ is small or $\Delta m^2 \ll E/L$ (or both). Similar expressions apply for the case of three neutrino types, but more mixing angles are involved.

CONCLUDING REMARKS

Table 1 shows a summary of results using conventional neutrino beams from accelerators. These are formed from decay in flight of $\pi^{\pm}$ or $K^{\pm}$ mesons and are dominantly $\nu_\mu(\bar{\nu}_\mu)$ with a small (0.5-1%) component of $\nu_e(\bar{\nu}_e)$ from Ke3 decay. On the basis of the numbers of charged - current neutrino interactions with electrons or muons

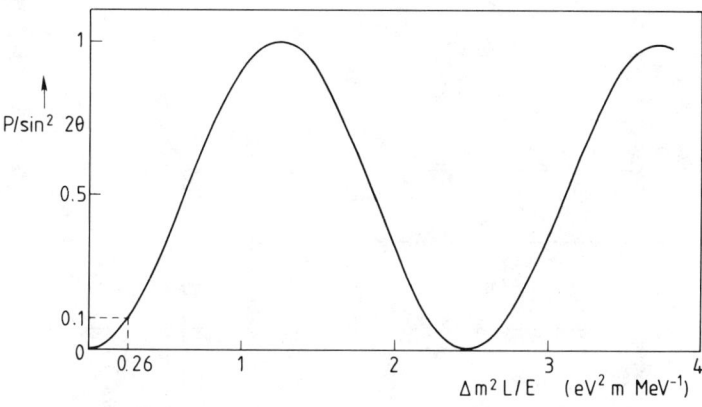

Fig. 2

in the final state, these experiments are therefore very sensitive to $\nu_\mu \to \nu_e$ mixing (simply by comparing the proportion of events with electrons with the number expected from the Ke3 'background'). The LAMPF and Gargamelle PS results are for low-energy neutrinos and give $P(\nu_\mu \to \nu_e) << 1\%$. They are not sensitive to $\nu_\mu \to \nu_\tau$ mixing, since any ν_τ would be below threshold for $\tau^{\pm}$ production and could therefore only lead to at most a 1% change in the rate of neutral current events. The FNAL experiments in the 15' chamber filled with Ne/H2 place limits on $\nu_\mu \to \nu_\tau$ if we assume (from Gargamelle) that $\nu_\mu \to \nu_e$ can be neglected, and make use of the known leptonic branching ratio for $\tau^+ \to e^+ + \nu_e + \bar{\nu}_\tau$. The BNL-Columbia group set conservative limits by observing that the number of electron events was within 30% of that predicted from Ke3 background. The Fermilab-Michigan-ITEP-Serpukhov antineutrino experiment exploited the fact that, in an inelastic $\bar{\nu}_e$ reaction on a nucleon, the e^+ receives the bulk of the energy, with a distribution $\sim(E_{e^+}/E_\nu)^2$; while in the reaction $\bar{\nu}_\tau + N \to \tau^+ + X$, followed by the decay

Table 1. Conventional Accelerator Results on Neutrino Oscillations

Experiment, Beam Energy	Transition	Limit(90%CL)	L/E(m/MeV)	Δm(min)
LAMPF(50MeV) [6]	$\bar{\nu}_\mu \to \bar{\nu}_e$	<0.065	0.3	~1eV
GGM PS (2GeV) [7]	$(\nu_\mu \to \nu_e)/(\nu_\mu \to \nu_\mu)$	<0.0013	0.04	~7eV
FNAL 15'(30GeV) νNe(BNL-Col [8]) νNe(FIMS [9])	$(\nu_\mu \to \nu_\tau)/(\nu_\mu \to \nu_\mu)$ $(\bar{\nu}_\mu \to \bar{\nu}_\tau)/(\bar{\nu}_\mu \to \bar{\nu}_\mu)$	<0.025 <.0075	0.04	~7eV
BEBC SPS νNe NB [10] (90GeV)	$(\nu_e \to \nu_e)/(\nu_\mu \to \nu_\mu)$	1.02±0.15	0.02	~10eV
νNe WB [11] (30GeV)	"	1.2 ± 0.25	0.04	~7eV
LAMPF	$(\nu_e \to \nu_e)$	1.1 ± 0.4	0.3	~1eV
GGM PS	$(\nu_e \to \nu_e)/(\nu_\mu \to \nu_\mu)$	0.92±0.21	0.04	~7eV

$\tau^+ \to e^+ + \nu_e + \bar{\nu}_\tau$, the e^+ receives only a small fraction of the incident neutrino energy. Thus by investigating the y distribution of electron events, they were able to set an upper limit $P(\nu_\mu \to \nu_\tau) < 1\%$.

The second half of Table I compares the observed number of events with electron secondaries, with that computed from detailed calculations of the ν_e flux from Ke3 (and $\mu^+ \to \bar{\nu}_\mu + e^+ + \nu_e$) decays in the parent beam. The most precise result is that from the narrowband experiment in BEBC, in which, in absence of oscillations the expected number of electron events (70) is equal within statistical and systematic errors to that observed (72).

In summary, the results of Table I are consistent with both $P(\nu_\mu \to \nu_\tau)$ and $P(\nu_\mu \to \nu_e)$ equal to zero and $P(\nu_e \to \nu_e) = 1$, that is, no oscillations. The experiments are however sensitive only to mass differences $\Delta m > 5\text{-}10 eV$, because of the values of L/E employed. Note that typical L/E values are the same at PS and SPS energies, because the length of the muon shield will be proportional to the (proton) beam energy employed. In view however of the quoted value for $m_{\nu_e} \sim 40 eV$, mass differences of order 10eV might not be so unreasonable. If so, the absence of oscillatory effects indicates that all the angles for mixing $\nu_\mu \to \nu_e, \nu_\mu \to \nu_\tau$ and $\nu_e \to \nu_\tau$ are small (<0.2 radians).

The main evidence for neutrino oscillations presented at this conference comes from a reactor experiment. Typical values of L/E are 1-10 m MeV^{-1}, thus a sensitivity to a mass-squared difference $\Delta m^2 > 1 eV^2$. The Irvine group[12] have made observations at the Savannah River Reactor, using He^3 gas-filled neutron counters to detect simultaneously the reactions

$$\bar{\nu}_e + d \to n + n + e^+$$

$$\bar{\nu}_e + d \to n + p + \bar{\nu}_e$$

They measure the ratio of (spectrum averaged) cross-sections for these charged and neutral current reactions, obtaining a value (for L=11m):-

$$r_{expt} = \frac{\bar{\sigma}(\bar{\nu}_e d \to nne^+)}{\bar{\sigma}(\bar{\nu}_e d \to np\,\bar{\nu}_e)} = 0.167 \pm 0.093$$

While this ratio is independent of absolute neutrino flux, it is not independent of the reactor spectrum shape, since the thresholds for the charged current and neutral current reactions are 4MeV and 2.2MeV respectively. The value of r is compared with the values expected theoretically, when averaged over the spectrum. Using the Avignone spectrum, $r_{th}=0.42$, while for the Davis-Vogel spectrum $r_{th}=0.44$. Thus the ratio of ratios

$$R = \frac{r_{expt}}{r_{th}} = 0.38 \pm 0.21 \text{ (Davis-Vogel)}$$
$$= 0.40 \pm 0.22 \text{ (Avignone)}$$

compared with the expected value R=1 in the absence of oscillations. Oscillations ($\nu_e \to \nu_\tau$, $\nu_e \to \nu_\mu$) will clearly reduce the charged current rate but leave that for neutral currents unaffected. An independent estimate of r_{th} was obtained using the spectrum determined in a separate experiment measuring the reaction $\bar{\nu}_e p \to ne^+$ - giving R = 0.47 ± 0.26.

In a second reactor experiment at ILL[13], the absolute rate for the reaction $\bar{\nu}_e p \to ne^+$ was compared with the expected rate using the Davis-Vogel spectrum. Recent measurements on β-ray spectrum from fission products suggest that the Davis-Vogel spectrum is fairly reliable, at least below 6MeV or so. For L=8.7m they obtained

$$\frac{\sigma_{exp}}{\sigma_{th}} = 0.87 \pm 0.14$$

which is consistent with no oscillations. The dependence of this ratio on energy is shown in Fig.3.

The Irvine and ILL results are clearly inconsistent with each other and this fact suggests systematic errors associated with one or both experiments. It may be noted that the Irvine group have also measured the cross-section for the charged current reaction $\bar{\nu}_e p \to ne^+$ at L=11m in a previous experiment. The ratio of cross-sections for $E_\nu > 4$MeV was found to be:-

$$\frac{\sigma(\bar{\nu}_e d \to nne^+)}{\sigma(\bar{\nu}_e p \to ne^+)} = 0.56 \pm 0.27$$

This ratio should of course be unity (since the two processes correspond to the same reaction, with and without a neutron spectator). The experimental method depends in the one case on detecting the two neutrons, and in the other on a delayed coincidence between gammas from positronium annihilation and those from neutron capture in gadolinium.

In summary, the reactor experiments on neutrino cross-sections do not, at present, provide precise information for or against

CONCLUDING REMARKS

oscillations. Some of the uncertainties arise from inadequate knowledge of the reactor spectrum, particularly at high energy. Spectrum-independent tests for oscillatory behaviour can be better made by varying the core-detector distance, and this the Irvine group are planning to do. However, such measurements will introduce new problems: the signal/noise ratio (determined by the reactor-off background) will be a strong function of distance.

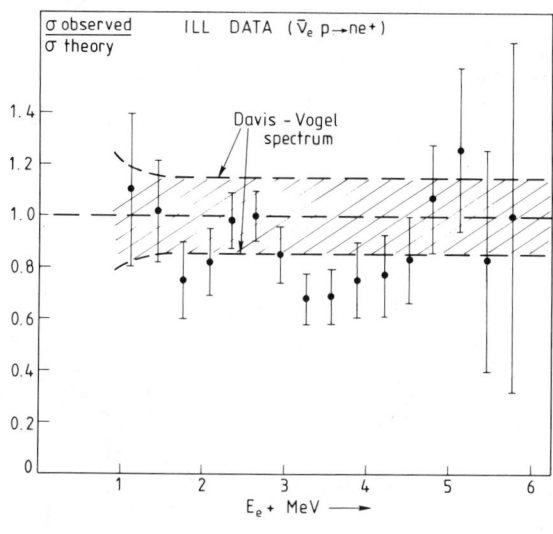

Fig. 3

Other experiments have been quoted as possible evidence for neutrino oscillations. The CERN beam dump[14] experiment on prompt neutrino production in hadron-hadron collisions gives for the ratio of events with electrons to those with (prompt) muons

$$\frac{N_e^{\pm}}{N_\mu^{\pm}(\text{prompt})} = 0.5 \pm 0.2$$

The values of L/E are very similar to those in the narrow-band experiment in Table 1, indicating no oscillations above the 10% level. Thus, the departure of the above ratio from unity (expected if the prompt neutrinos originate in D, D* semi-leptonic decay) must have some other explanation.

The deep underground experiments on interactions of neutrinos produced by cosmic rays in the atmosphere have also been mentioned. These experiments are also sensitive to $\Delta m^2 \sim 1 eV^2$. Both Krishnaswamy et al[15] and Crouch et al[16] find deep underground neutrino-induced muon rates somewhat smaller than expected. In the case of Crouch et al, the discrepancy amounts to a factor 1.6 ± 0.4. In view of the substantial ($\sim 30\%$?) uncertainties in the expected rates, this can hardly be claimed as significant.

In summary therefore, the most accurate measurements are of ν_e and ν_τ rates in ν_μ beams at accelerators, and no significant effects are observed. For the time being at least, the evidence for neutrino oscillations has melted away.

On the subject of neutral currents, there is little I need say. The world average of $\sin^2\theta_W = 0.23 \pm 0.01$ is the same as last year: it is still dominated by two experiments - the CDHS inclusive neutral and charged current neutrino cross-sections, and the SLAC measurements of asymmetry in the scattering of longitudinally polarized electrons by protons and deuterons. Significant parity-violating effects in atomic transitions now appear to be seen by practically all groups. In particular, the optical rotation measurements in bismuth by the Seattle, Oxford, Novosibirsk and Moscow groups all have the same sign, differ in magnitude by factors of only two or so, and seem to be drawing together. Professor Barkov showed a dramatic plot of the four sets of data against time, predicting unification of these numbers by about 1984!

Another hot topic nowadays is that of baryon stability. The progress of the various experimental projects looking for nucleon decay was summarized by Goldhaber. Just to remind you of the state of play, the relevant timescales are indicated in Fig.4. Radiochemical and geochemical analyses of abundances of rare isotopes which could originate from nucleon decay in more common nuclei give τ(nucleon) $>10^{26}$ years. Recent experiments on the stopping muon flux deep underground[17] give τ(nucleon) $>10^{30}$ years (under the assumption that π^+ - and hence μ^+ - are frequent decay products of the nucleon). The main news at this conference was the report by Miyake and co-workers, who have operated a small two-ton (lead plate plus flash tube) calorimeter in the Kolar Gold Mine (S. India) for 5 years. They have observed two events which look like $p \rightarrow e^+ + \pi^0$ and quote a lifetime $\sim 2.10^{30}$ years (assuming a 30% branching ratio). Cosmic neutrino background ($E_{vis} > 1 GeV$) would provide events at exactly this rate (0.12 events $ton^{-1} yr^{-1}$, one third of them induced by ν_e) although the number expected to simulate $p \rightarrow e^+ \pi^0$ would of course be smaller. This is at least encouraging news for the major calorimeter and water Cerenkov detectors which will shortly be operating. It needs to be emphasized that these projected experiments will be sensitive to proton lifetimes not exceeding 10^{32} years: beyond that, interactions of atmospheric neutrinos are dominant for the energy resolution (30%) and rather limited pattern recognition abilities which seem to be

CONCLUDING REMARKS

attainable. Even if it should turn out that the proton lifetime is too long for its decay to be detected, these experiments could

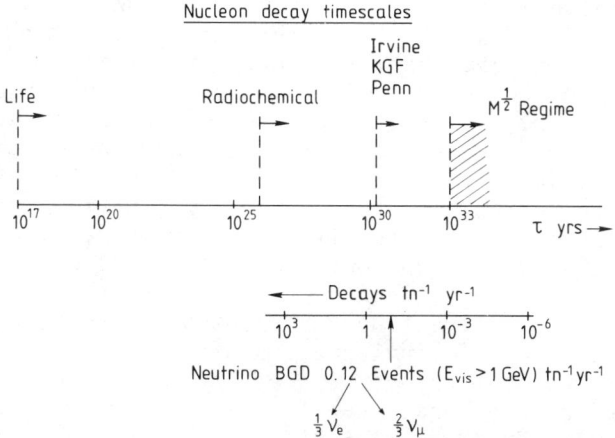

Fig. 4. The existence of life sets a limit $\tau > 10^{17}$ yr, the abundance of rare isotopes, $\tau > 10^{26}$ yr. The best limit, from underground experiments (Irvine, Pennsylvania[17] and Kolar Gold Field) is $\tau > 2.10^{30}$ yr. For $\tau > 10^{33}$ yr, neutrino background dominates and the limit attainable rises only as the square root of the detector mass.

nevertheless be very interesting. For the first time, detectors of 100's and 1000's of tons will be operated in very low background conditions deep underground, and totally new and unexpected phenomena might occur.

So much for the drama and excitement at this conference. I have to say that I was also impressed by the steadily improving precision, quality and depth of the bread and butter accelerator neutrino physics programme. Fig.5 shows, as an example, the most recent data on the structure function $F_2^{\nu N}(x,q^2)$ from the CDHS experiment. The detail and range of this data is at least as good as, if it is not superior to, that of the high energy muon experiments. Two years ago, I thought that the big new muon scattering experiments, like EMC at CERN, would wipe out the neutrino efforts. I was wrong. The different muon experiments get discordant results

and seem to have troubles with systematic effects (e.g. acceptance
uncertainties) as well as data analysis, and look to be in some
disarray. It is clear that neutrino experiments will provide
unique and worthwhile results - perhaps even the best - on
scaling deviations, for some time to come. Initially, some two
years ago, the BEBC (ABCLOS) experiment gave the first
exciting indications of agreement of the q^2 dependence of non-
singlet moments (of $xF_3^{\nu N}$) with perturbative QCD. Probably this
was accidental; non-perturbative effects are likely to be
important in the region $q^2 < 10 GeV^2$ and may be mimicking the perturbative behaviour. So, the comparison of the structure function
behaviour with theory is not as straightforward as we thought,
but it is not excluded that further and more detailed studies
may help to resolve the perturbative from non-perturbative
effects - or, it may need a really high q^2 machine, like the ep
collider HERA that we heard about, to sort it all out.

There were two interesting contributions from bubble
chambers on the study of final-state hadron distributions in
inelastic neutrino scattering. One of the most dramatic developments over the last year has been the observation of planar
"3-jet" events in $e^+e^- \to$ hadrons at PETRA. They have been interpreted in terms of hard gluon bremsstrahlung by one of the
quarks:- $e^+e^- \to Q\bar{Q}$, $Q \to Q+G$, where the gluon is emitted with a substantial part of the CMS energy, at large angle. Since such
observations are supposed to provide strong evidence in support
of the fundamental field theory of strong interactions (QCD) it
is important to obtain confirmation in independent experiments.

In the e^+e^- experiments, the direction of the jet axis
(along which the momenta of the Q and $\bar{Q}$ lie) is unknown and has
to be estimated by a minimisation procedure. For this purpose,
the eigenvalues Λ_1, Λ_2, Λ_3 of the momentum ellipsoid are found
by diagonalizing the matrix

$$M_{\alpha\beta} = \sum_{j=1}^{N} P_{j\alpha} P_{j\beta} \qquad \alpha,\beta = x,y,z$$

where the sum extends over the N secondary (charged) particles,
and to find Λ_1, Λ_2, Λ_3 one solves the cubic equation $(M-\Lambda I)=0$,
where I is a unit matrix. The normalized eigenvalues are

$$Q_{1,2,3} = \frac{\Lambda_{1,2,3}}{\sum P_j^2}$$

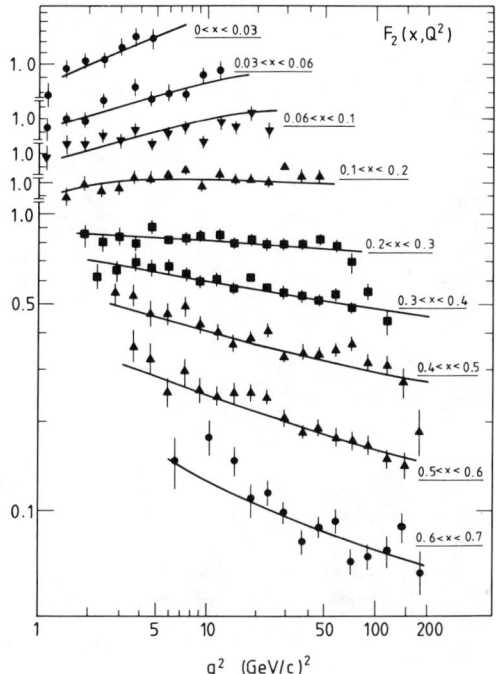

Fig. 5 Recent results on $F_2^{\nu N}(x,q^2)$ from the CDHS collaboration using the CERN narrowband neutrino beam.

where $Q_{1,2,3}$ are put in order $Q_3>Q_2>Q_1$, and $Q_1+Q_2+Q_3=1$.
Thus Q_3 is the axis along which (momentum component) is the largest, Q_2 that along which it is next largest, and Q_1 that along which it is smallest. The sphericity and aplanarity are defined as

$$S = \frac{3}{2}(Q_1+Q_2) = \frac{3}{2}(1-Q_3) = \frac{3}{2}\frac{\min \Sigma p_{Tj}^2}{\Sigma p_j^2}$$

$$A = \frac{3}{2} Q_1$$

Events with small S correspond to almost "collinear 2-jet events", those of large S are broad jets, and those with small A and large S correspond to broad jets in which the hadrons tend to lie in a plane. The findings of the TASSO[18] group at PETRA, for example, at CMS energy W ~30GeV, are shown in Table 2. The number of high sphericity (S> 0.25) planar events (A< 0.04) is larger than that expected from a 2-jet ($Q\bar{Q}$) Monte Carlo, but consistent with that expected from occasional radiation of a hard gluon at wide angle with high energy ($Q\bar{Q}G$). Such planar events constitute about 10% of all events.

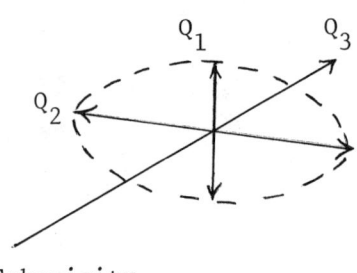

Sphericity axis

In e^+e^- experiments the mean value of p_T^2 of a hadron relative to the jet (sphericity) axis increases with W^2 (W=CMS energy). A

Table 2. Planar events in TASSO[18]

Selection	Observed	$Q\bar{Q}$ MC	$Q\bar{Q}G$ MC
S> 0.25 A< 0.04	18	4.5	17
S> 0.25 A> 0.04	38	38	35

similar effect is observed in neutrino-nucleon scattering[19,20,21]. As an example, Fig.6 shows ABCMO data[20], presented at this conference by Saitta, on the dependence of $<p_T^2>$ on $z = 2E^*/W$, the fraction of CMS energy carried by a hadron, for hadrons going forward or backward in the CMS. The dip near $z \sim 0$ simply reflects momentum-energy conservation ("seagull effect"). The distinctive

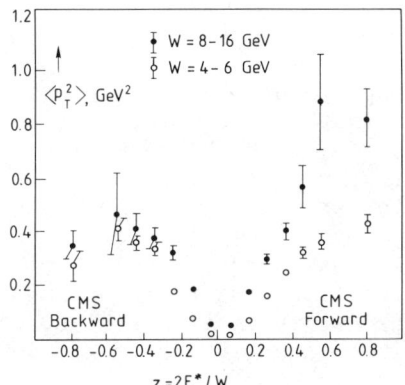

Fig. 6

feature is that $<p_T^2>$ at large z increases rapidly with W in the forward direction, while in the backward direction it hardly changes. This might be interpreted as a "jet-broadening" effect due to hard gluon radiation by the struck quark (current jet). In order to investigate further, values of A and S as defined above have been computed as a function of W. Table 3 gives the results from the ABCMO experiment, for events of charged multiplicity $\geqslant 6$. About 10% of events fall into the range S >0.25, A <0.04 used to specify planar events in the TASSO experiment. At high W, more planar events are observed than predicted according to the Monte Carlo program based on a 2-jet structure (1 target jet, 1 current jet).

Table 3. νH_2 events (ABCMO[20]) $n_{ch} \geq 6$

<W> GeV	4-6	6-8	8-16
All S, A	405	416	439
S >0.25, A <.04	58	35	26
Rotation about $\vec{q}$	61	31	25
MC prediction	83	31	15

The leptoproduction experiments have one great advantage over the e^+e^- experiments, in that the current vector ($\vec{q}$) direction is fairly well known. So, if each hadron is rotated by a random azimuthal angle about $\vec{q}$, any real excess of planar events (above those occurring by statistical fluctuation) should be destroyed. Upon recomputing the momentum eigenvectors, the numbers of events of high S and low A is not changed; the majority must therefore be due to random correlations, so that any randomising procedure creates as many planar events as it destroys. An upper limit on the fraction of genuine planar events at W=8-16GeV is ~1%.

The BFHSW[21] group obtained a large number (~10^5 events) of neutrino interactions in the 15' chamber filled with Ne/H$_2$ exposed to a quad-triplet beam. They selected a sample of ~1500 events with p_T^μ >4GeV/c and thus of high q^2, W^2 and mean energy (~100GeV). These authors take the $\vec{q}$ vector in an event as the resultant momentum vector of all the visible hadrons, and then find the principle momentum axes normal to $\vec{q}$ (the z-direction), defining the planarity as

$$P = \text{Max} \frac{\Sigma p_x^2 - \Sigma p_y^2}{\Sigma p_x^2 + \Sigma p_y^2} = \frac{Q_2 - Q_1}{Q_2 + Q_1}$$

They also define the transverse momentum dispersion

$$\Pi_N = \frac{A}{\sqrt{N_F}} \underset{F}{\Sigma} (p_T - <p_T>)$$

summed over the forward hadrons (N_F). $<p_T>$=0.32GeV is the average over all tracks and all events. Events of high Π_N include

CONCLUDING REMARKS

secondaries with large transverse momenta. As shown in Table 4, such events are associated with high planarity P, although phase space alone favours small P values. The values of W are similar to those in the highest bin (W=8-16) of the ABCMO experiment.

Both bubble chamber experiments therefore find planar events, (A <.04, P >0.8) associated with high W values (up to 15GeV or thereabouts) at the level of ≲1% of the total. On the basis of Monte Carlo calculations of gluon bremsstrahlung, it is expected that the proportion of planar events will increase rapidly with W, and this result is not inconsistent with the proportion of 10% found in e^+e^- collisions at W∼30GeV.

Table 4. BFHSW analysis (νNe) $q^2>2$, $W^2>4$, $E_\nu>10$, $n_{ch} \geq 3$

Planarity P	Π_N <2.4	Π_N >2.4
0 - 0.2	96	2
0.2 - 0.4	304	1
0.4 - 0.6	378	7
0.6 - 0.8	401	15
0.8 - 1.0	233	15

Whether any of the experiments (νN, e^+e^-) have proved that the planar events are definitely due to production of hard gluon jets - rather than something else, like baryon resonance production - is still open to question. The e^+e^- experiments and the νN experiment of BFHSW depend on comparison with Monte Carlo calculations, generating quarks and gluons according to QCD but relying on empirical parameterisations of "hadronisation" to predict hadron distributions. The ABCLOS νH experiment suggests that all the planar events could be due to random azimuthal correlations, but the sample is too small to detect genuine effects at the 1% level. There are clearly many subtle problems to be solved in the analysis, but the main point I wish to make is that accelerator neutrino experiments are making significant contributions in this field of research.

In summary then, neutrino physics in 1980 looks to be in a very healthy and exciting state. We have clearly a great deal to learn from low energy β-decay (Kurie plots and neutrino mass), from reactor experiments, and from the sun. There is also a great deal to be done with neutrino beams from the fixed target accelerators for many years yet. Ultimately, these will be supplanted by ep→νN experiments at colliders such as HERA towards the end of this decade. These colliders will cover a range in q^2 two orders of magnitude greater than we can attain with present machines, and they will undoubtedly uncover a whole new range of phenomena, the nature of which we cannot even begin to guess.

References

1. C. Rolfs, Bethe Symposium '40 years of Stellar Energy', Stony Brook (1979).
2. K.E. Bergkvist, NP B39, 317 (1972)
3. E.F. Tretyakov et al, Proc. Neutrino Conf., Aachen (1976)
4. V.A. Lyubimov et al., ITEP-62 preprint (1980). See also proceedings of this conference.
5. B. Pontecorvo, Sov.Phys. JETP 26, 984 (1968).
    Z. Maki et al., Prog. Th. Phys. 28, 870 (1962)
6. S.E. Willis et al., PRL 44, 522 (1980)
7. J. Blietschau et al., NP B133 205 (1978)
8. A.M. Cnops et al., PRL 40, 144 (1978)
9. B.P. Roe (private communication from Fermilab-ITEP-Michigan-Serpukhov collaboration)
10. Aachen-Bonn-CERN-Demokritos-London-Oxford-Saclay collaboration (CERN/EP/NPC-N80)
11. W. Venus (private communication)
12. F. Reines, H.W. Sobel and E. Pasierb, U. of Cal. (Irvine) preprint UCI-10P19-144 (1980).
13. F. Boehm, review talk at this conference on "Lepton Conservation"
14. H. Wachsmuth, Proc. 1979 Symp. on Lepton and Photon Interactions at High Energies, Fermilab (Ed. T.B.W. Kirk and H.D.L. Abarb
15. M.R. Krishnaswamy et al., Proc. Roy. Soc. A323, 489 (1971)
16. M.F. Crouch et al., PR D18, 2239 (1978)
17. K. Lande, report at this conference.
    J. Learned et al., PRL 43, 907 (1979)
18. R. Brandelik et al., PL 86B, 243 (1979)
19. H. Deden et al., (ABCMO collaboration) CERN EP report (1980)
20. B. Saitta, report at this conference
21. H. Bingham,   "   "   "   "

INDEX

Accelerators as source of neutrinos, 25, 36, 53 (see also future accelerators)
Atmospheric neutrinos, 53

Baryon conservation, 282-315, 408, 409
  experiments, 305-315
Beam dump experiments, 39, 55, 341-360, 407
  at BNL, 342-344
  at CERN, 346-358
    and the prompt Electron-Neutrino flux, 352-356
    and the prompt Muon-Neutrino flux, 349-352
  at FNAL, 344-346
Beauty production
  in hadron interactions, 99-102
  in neutrino interactions, 102-104
Beck and Sitte, 5
Beta decay, 5, 30, 31, 45, 46, 260, 261
  recoil, 6, 31
  spectrum 2-4, 7, 31, 45, 46, 260, 261, 400, 401
Big-bang theory, 42, 43
Bohr, 2, 12
Bromine experiment, 71

Cabibbo theory, 35, 49
Charge asymmetry, 35
Charm, 36
Charged Baryons, 81-90
  lifetime, 82-89
Charmed mesons, 91-97
  lifetime, 94-97
Charmed particles, 38, 48, 81-106
  lifetime, 40, 54
  production in hadron interactions 97-99 (see also Beam dump experiments)
Clorine-37 experiment, 62-64, 65, 66
Collapsing star neutrinos, 74-76
Compton wavelength, 2

Concluding remarks, 399-416
Cross sections for neutrino interactions, 12, 17, 22, 37, 147, 148, 208
CP, 33, 34, 35, 48
CPT, 34, 35
CVC, 34, 35

Davis experiemnt, 62-64
Dirac theory of radiation, 48

Elastic scattering of neutrinos, 12, 191-218
Electron spin, 3
Electro-weak effects at Petra, 324-332
Electro-Weak physics in e^+e^- interactions, 321-338, 410-412
Events in neutrino physics, 30-40

Fermi, 12, 14, 33, 46, 47
  interaction, 33, 47, 48
  statistics, 2
Future accelerators, 361-380
  Hera, 373-378
  Lep, 365-373
  $p\bar{p}$, 362-365

Gallium experiment, 72
GIM mechanism, 9

Hanford experiment, 16
Heisenberg, 4
Helium 3 - Helium 4 reaction, 66, 67

Indium experiment, 73, 74
Intermediate boson 38, 53
Inverse beta decay, 6, 11, 31, 244, 245, 246 (<u>see also</u> Neutrino
      oscillations)
Iwanenko, 4

J/ψ particles, 38 (<u>see also</u> New particles)

Lepton
  charge, 32, 43, 48
  conservation, 259-274
  pairs, 38, 42, 43
  polarization, 48 (<u>see also</u> Muon polarization)
Low energy neutrino interactions, 241-258

Majorana neutrino, 31 (<u>see also</u> Neutrino oscillations)
Meson discovery, 8, 31-34, 39
Michel parameter, 33
Multiplicative law, 40 (<u>see also</u> Lepton conservation)

INDEX

Muon
  capture, 35
  decay, 9, 261, 262
  discovery, 8, 32
  polarization, 39

Name of the neutrino, 5, 46
Neutrino
  in astrophysics and cosmology, 42-44, 61-80
  beams, 40 (see also Accelerators as source of neutrino
      beams)
  detectors, 32, 35-37, 41, 42, 44, 381-398
    background rejection, 393-395
    construction details, 395-397
    water Cerenkov technique, 388-393
  deuteron
    experiment, 246-255
    scattering, 39,40
  electron scattering, 38 (see also Weak neutral currents
      semileptonic interactions)
  emission from hot stars, 42 (see also Neutrino in astrophysics
      and cosmology)
  event rates, 384-386
  excitation of nuclei, 35
  experimental discovery, 21-25, 32 (<u>see also</u> Neutrino history)
  history, 1-60
  invention, 31 (<u>see also</u> Neutrino history)
  magnetic moment, 5
  mass, 5, 6, 31, 32, 38, 39, 55, 56, 317-320, 400, 401
  oscillations, 34, 37, 49, 52, 67-69, 244-255, 262-269, 386-388,
      402-408
    Grenoble experiment, 266-269
    Irvine experiment, 266 (<u>see also</u> Reactor neutrino experiments)
  polarization, 48
  stars, 42-44
  theory of light, 6
Neutron
  -antineutron oscillations, 282-297, 307
  and gamma ray contamination, 14
  radioactivity, 33
Nucleon decay, 38 (<u>see also</u> Baryon conservation)
New particles, 81-106 (<u>see also</u> Charmed baryons, Charmed mesons,
      and Charmed particles)

Parity violation, 9, 24, 34, 48, 183-185, 219-240
  in atoms, 36, 39
  in nuclei, 37, 219-240
    F-80 decay, 229, 230
    Li-6 decay, 225, 228
    Li plus alpha, 228-229

Parity violation (continued)
  in nuclei (continued)
    neutron capture, 225
    the two nucleon system, 219-224
    with T=1
Parton model, 37 (see also Weak charged currents)
Pauli, 2, 3, 24, 44
Perrin, 4
Pion factory experiments, 241, 242
Polarization
  of beta particles, 34
  of neutrino, 34
Polarized electron scattering, 39
Preons, 275-299
Proton decay (see Baryon conservation)

Quarks, 36 (see also Parton model)
  color, 37

Radiochemical method, 42
Reactor
  neutrino experiments, 11-29, 51, 243-258
  neutrinos, 11, 13, 39, 50

Savannah River
  experiment, 17   (see also Reactor neutrino experiments)
  plant, 11, 17
Scintillation detectors, 13-15, 32, 44 (see also Neutrino detectors)
Scintillation telescope, 43
Solar model, 64, 65
Solar neutrino experiments (future), 69-78
Solar neutrinos, 42-44, 50, 52, 61-74, 399, 400 (see also Neutrino oscillations)
Strange particles, 33, 48
Supersimmetry particles, 332-336

Underground experiments, 42-44 (see also Baryon conservation)
Unification of forces, 34-37, 49, 55, 196-207, 275-300, 332-336
Upsilon meson, 39

Tau lepton, 38, 48, 54, 332-324
Two neutrino problem, 53, 54

V-A coupling, 7, 34, 35, 39, 48, 49
Violation of B, L and F, 277-282

Weak charged currents, 48, 49, 192-195, 409, 411, 413-415
  hadron final states, 107-142
    experimental results, 121-128
    fragmentation functions, 117-128

Weak charged currents (continued)
  hadron final states (continued)
    gluon searches, 134-138
    higher twist, 120, 121
    multiplicity and charge distribution, 108-115
    QCD modifications, 118-120
    transverse momentum, 128-134
    vector meson production, 115-117
  structure functions, 148-161
    antiquark distributions, 152, 153
    higher twist terms, 159-161
    QCD, 158, 159
    scaling violation, 157-161
    strange sea, 156, 157
  total cross sections
    on isoscalar targets, 147, 148
    on protons and neutrons, 144-146
Weak neutral currents, 38, 49, 50, 54, 194, 195, 408
  leptonic interactions, 191-218
    experimental results, 200-210
    the future outlook, 210-215
    the Savannah River experiment, 208
  semileptonic interactions, 165-190
    coupling constants, 185, 186
    distributions in the inelastic variables, 183
    elastic scattering, 168-171
    inclusive interactions on isoscalar targets, 177-183
    inclusive interactions on protons and neutrons, 174, 175
    Lorentz structure, 165-168
    parity violation, 183-185
    seminclusive reactions, 171-174
Weak nuclear forces, 36